MAKING ONE'S WAY IN THE WORLD

The footprints and trackways of prehistoric people

by

Martin Bell

OXBOW | books
Oxford & Philadelphia

Published in the United Kingdom in 2020 by
OXBOW BOOKS
The Old Music Hall, 106-108 Cowley Road, Oxford, OX4 1JE

and in the United States by
OXBOW BOOKS
1950 Lawrence Road, Havertown, PA 19083

Hardback Edition: ISBN 978-1-78925-402-0
Digital Edition: ISBN 978-1-78925-403-7 (ePub)

A CIP record for this book is available from the British Library

Library of Congress Control Number: 2019954268

Printed in Malta by Melita Press

Typeset in India for Casemate Publishing Services. www.casematepublishingservices.com

For a complete list of Oxbow titles, please contact:

UNITED KINGDOM
Oxbow Books
Telephone (01865) 241249
Email: oxbow@oxbowbooks.com
www.oxbowbooks.com

UNITED STATES OF AMERICA
Oxbow Books
Telephone (800) 791-9354, Fax (610) 853-9146
Email: queries@casemateacademic.com
www.casemateacademic.com/oxbow

Oxbow Books is part of the Casemate Group

Front cover: A hollow way at Chideoak, Dorset (photo: M. Bell).

Contents

List of figures and tables ix

Acknowledgements xiii

1 Steps towards understanding: routeways in practice, theory and life 1
- Background 1
- Introduction 1
- False paths 2
- Talking stock and steps forward 4
- Environmental and geoarchaeology 5
- Landscape archaeology 7
- Theoretical perspectives 9
- Anthropology: the contribution of Tim Ingold 11
- Landscape change and clues to movement 14
- Agency and niche construction: human and non-human 16
- Cognition: thinking through things 18
- How literature and art help us think about movement 19
- Timescale, dating and spatial scale 22
- Terminology 23
- Conclusions 24
- Chapter organisation 26

2 Walks in the temperate rainforest: developing concepts of niche construction and linear environmental manipulation 27
- Introduction: why the American North-west coast? 27
- The Douglas Map 28
- The area and its archaeology 29
- Trails and prairies 33
- Plant utilisation 40
- Elsewhere in North America 42
- Palaeoenvironmental perspectives 44
- The social significance of routes 45
- Conclusions 46

3 Niche construction and place making: hunter-gatherer routeways in north west Europe ... 49
Introduction ... 49
Anthropological perspectives ... 51
Topographic factors and 'natural' routeways ... 53
The wildwood, disturbance factors and routeways ... 54
Woodland manipulation and management ... 57
The broad spectrum revolution and niche construction ... 59
Hunter-gatherer plant use ... 60
Hunter-gatherer vegetation disturbance in Britain ... 60
Star Carr ... 63
Case Study: Kennet Valley ... 63
Case Study: A Welsh model of river valley based mobility ... 66
Continental Europe ... 67
Mobility and sedentism ... 68
Artefact areas and 'monuments' ... 69
Isotopes and mobility ... 71
Material culture and movement ... 71
Conclusions ... 72

4 Footprints of people and animals as evidence of mobility ... 75
Introduction ... 75
Trace fossils ... 77
Formation processes and terminology ... 77
Recording methodology ... 81
Dating and timing ... 81
Identification and interpretation ... 82
Associated animals ... 85
Palaeolithic footprint-tracks on open sites ... 85
Holocene hunter-gatherer-fishers ... 90
Case Study: Mesolithic paths in the Severn Estuary ... 90
Footprint-tracks in later prehistoric contexts ... 98
Case Study: seasonal pastoralists in the Severn Estuary ... 99
Other later prehistoric examples ... 101
Footprint-tracks in the Americas ... 102
Footprints: perceptual and symbolic aspects ... 103
Conclusions ... 104

5 Early farmers: mobility, site location and antecedent activities ... 111
Introduction ... 111
Case Study: the Ice Man ... 111
Skeletal, isotopic and DNA evidence for Neolithic mobility ... 113

Neolithic landscapes in Britain ... 114
Neolithic monuments in Britain ... 116
Case Study: Avebury henge, Wiltshire ... 123
Case Study: Stonehenge, Wiltshire ... 125
Geological evidence for Neolithic mobility ... 131
Conclusions ... 132

6 Wetland trackways and communication ... 133
Introduction ... 133
Wheeled vehicles ... 136
Trackways dates ... 138
Mesolithic trackways? ... 138
Neolithic trackways in mainland Europe ... 139
Neolithic trackways in the British Isles ... 140
Bronze Age and Iron Age trackways in Northern Europe ... 142
Bronze Age and Iron Age trackways in the British Isles ... 148
Case Study: Somerset Levels ... 148
Case Study: Severn Estuary ... 149
Later prehistoric trackways in Ireland ... 151
Bridges, post alignments and associated ritual deposits ... 154
Conclusions ... 160

7 Barrow alignments as clues to Bronze Age routes ... 163
Introduction ... 163
Denmark ... 163
Case Study: Kilen, a Bronze Age cross roads in Jutland ... 165
Germany ... 167
Netherlands ... 168
Case Study: Veluwe barrow roads ... 168
North European connections ... 170
England and Wales ... 170
Conclusions ... 172

8 Trackways in later prehistoric agricultural landscapes ... 175
Introduction ... 175
Recognising tracks in agricultural landscapes ... 177
Dating tracks in agricultural landscapes ... 178
Agents of transformation: horses, carts and chariots ... 183
Hollow ways ... 184
Coaxial fields and tracks in moorland ... 186
Yorkshire Wolds ... 188
Coaxial fields and droveways in lowland Britain ... 188

Survival of coaxial field systems ... 190
Ridgeways ... 192
Case Study: the Wiltshire and Oxfordshire Ridgeway ... 193
The Icknield Way ... 195
The origins of Roman roads in Britain ... 197
Conclusions ... 200

9 Maritime and riverine connectivity and the allure of the exotic ... 201
Introduction ... 201
Riverine transport ... 202
Log boats ... 203
Hide boats ... 205
Sewn plank boats in the British Isles ... 206
Possible wrecks round Britain ... 208
Landing places in Britain ... 209
Artefact distributions in Scandinavia ... 210
Transported things in Britain and Europe ... 210
Scandinavia: ships and rock art ... 214
Conclusions: Maritime connections and cultures ... 216

10 A case study of the Wealden District in south-east England ... 219
Introduction ... 219
The South Downs ... 221
Case studies: Bishopstone and Bullock Down, 'ghost routes' ... 223
Other Downland routes ... 225
The Rother valley ... 230
Land allotment, tracks and fields in the Low Weald ... 232
The North Downs ... 232
Case study: multi-method dating at Lyminge, Kent ... 237
Riverine and maritime connections ... 239
Conclusions ... 239

11 Conclusions: why paths matter ... 241
Bodily engagement, perception, anthropology and literature ... 241
Steps forward ... 242
Multi-scalar and multi-disciplinary approaches ... 242
Landscape structures and retrogressive analysis ... 244
'Natural routes' and ridgeways ... 245
Droveways ... 246
Ethnohistory of Lesser Transhumance ... 247
Excavation ... 247
Linear environmental archaeology ... 248
Geoarchaeological approaches to human and landscape connectivity ... 248

Movement as niche construction 249
Critical thresholds 251
Routes to sustainable heritage and nature conservation 252

Bibliography 255

Index 292

Online resource available at http://books.casematepublishing.com/Making_Ones_Way_in_the_World.pdf

Supplementary Appendix 3.1 Mesolithic sites in Britain with evidence of vegetation disturbance: date, types of evidence, OD height, and sources. The evidence behind Figures 3.1 and 3.8.

Supplementary Appendix 4.1 Catalogue of human and animal footprint-tracks worldwide with evidence for location, date, sedimentary context and associations. The evidence behind Figures 4.2 and 4.15.

Supplementary Appendix 6.1 Catalogue of wetland trackways in Britain and North-west Europe: date, construction type, finds and sources. The evidence behind Figure 6.2.

Note on dating and additional references in the above appendices. As noted on p. 22, Holocene radiocarbon dates in this book are given in the form cal BC without the full radiocarbon date and laboratory numbers where these are given in the reference cited. The exception to this is online Appendices 3.1 and 6.1 which include full details of the radiocarbon dates, laboratory numbers, calibrated ranges and other information on the sites listed. There is an additional bibliography at the end of each Appendix.

List of figures and tables

Fig. 1.1 Footprint-track of a Mesolithic child aged 8–9, *c* 5500 cal BC, Goldcliff, Wales

Fig. 1.2 Microscopic evidence for human and animal activity and movement

Fig. 1.3 The cover of W.G. Hoskins' book *The Making of the English Landscape* (1974)

Fig. 1.4 A model of time-space geography, with time on the vertical axis, space and place on the horizontal axis

Fig. 1.5 Images of the lifeworld

Fig. 1.6 Thinking through one's feet: path at the Hunebedder Centre, Borger, Netherlands

Fig. 1.7 A deeply incised hollow way in North Chideock, Dorset

Fig. 1.8 Hollow ways on Jazira Plain, Iraq

Fig. 1.9 *Chalk Paths* by Eric Ravilious (1935)

Fig. 1.10 A beaver path at Nature Park, Lelystad, The Netherlands

Fig. 2.1 A man of Nootka Sound, Vancouver Island

Fig. 2.2 James Douglas (Hudson Bay Company) map of 1840

Fig. 2.3 The North-west Coast of North America

Fig. 2.4 A Haida settlement reconstruction at the Museum of Anthropology, University of British Columbia, Vancouver

Fig. 2.5 A trail marked by a curvilinear depression in the rock and a small, probably later, cairn, Sooke Hills, Vancouver island

Fig. 2.6 Small stone cairns along trail, Padern Lake, Vancouver Island

Fig. 2.7 'Sea wolves' at the Petroglyph Park, Nanaimo, Vancouver Island

Fig. 2.8 Fishtrap dated *c* AD 1450 crossing a channel at Mud Bay, Washington

Fig. 2.9 Bear faeces on a trail, Sooke Hills, Vancouver Island

Fig. 2.10 First Nations trail blazing from the Americas

Fig. 2.11 Willamette valley: a painting by Henry Ware 1845

Fig. 2.12 Beacon Hill Park, Victoria, Vancouver Island; park savanna with stone burial cairns

Fig. 2.13 Camas plants, a favoured First Nations food item at Cattle Point, Vancouver Island

Fig. 3.1 Mesolithic sites in Britain and evidence of vegetation disturbance

Fig. 3.2 Ethnographic evidence for patterns of movement by non-agricultural communities

Fig. 3.3 Creswell Crags, UK, a limestone gorge with many caves occupied in the Middle and Upper Palaeolithic on a 'natural' routeway
Fig. 3.4 Submerged forest tree, Goldcliff, Wales which was associated with charcoal
Fig. 3.5 Multiple trees felled by a storm, Brighton, Sussex
Fig. 3.6 Neolithic beaver-gnawed wood from Shapwick Heath, Somerset, UK
Fig. 3.7 The Avebury area, Wiltshire, UK, showing modelled pre-Neolithic disturbance and paths
Fig. 3.8 Mesolithic vegetation disturbance in Britain
Fig. 3.9 Kennet Valley showing the location of Late glacial and Mesolithic sites
Fig. 3.10 Hypothetical model of Mesolithic annual movement based on evidence from western Britain
Fig. 3.11 Early Mesolithic sites in Britain and North-west Europe with evidence for fire and vegetation disturbance
Fig. 4.1 Goldcliff East, Wales: the Mesolithic footprint-track of a young person aged 10–12
Fig. 4.2 The geographical distribution of Pliocene, Pleistocene and Holocene human and animal footprint-tracks in Britain, Europe, Africa, Asia, Australasia and the Americas
Fig. 4.3 Terminology used in the description of human and animal footprint-tracks
Fig. 4.4 Photogrammetric recording of Mesolithic human footprints at Goldcliff East, Wales
Fig. 4.5 Graph of the relationship between age and foot size in modern human populations
Fig. 4.6 Laetoli, Tanzania: footprint-tracks of three hominid individuals crossed by the trail of a three-toed horse *Hipparion* and its foal
Fig. 4.7 Happisburgh, Norfolk, footprint-tracks
Fig. 4.8 Willandra Lakes, Australia: multiple human footprint-tracks dating to 23–19 ka BP
Fig. 4.9 Uskmouth, Wales: Mesolithic red deer footprint-tracks
Fig. 4.10 Footprint-tracks at Goldcliff East, Wales: landscape and stratigraphic context
Fig. 4.11 Footprint-tracks revealed within laminated sediments, Goldcliff East, Wales
Fig. 4.12 Goldcliff East showing sites with Mesolithic footprints in relation to Quaternary sediments and direction of movement
Fig. 4.13 Goldcliff Site N showing Mesolithic footpaths
Fig. 4.14 Middle Bronze Age seasonal settlement at Redwick, Wales
Fig. 4.15 The emerging picture of human and animal footprint track evidence
Fig. 5.1 Monte Bego, Italy: prehistoric rock art with two cattle drawing a plough
Fig. 5.2 Neolithic chambered tomb at Gwernvale, Wales showing successive phases of activity
Fig. 5.3 Rudston, Yorkshire: five cursus monuments in relation to topography and the river Gypsey Race
Fig. 5.4 Thornborough Henges, Yorkshire
Fig. 5.5 Avebury, Wiltshire: the henge in relation to the avenues and other key prehistoric sites
Fig. 5.6 Stonehenge landscape, Wiltshire

Fig. 5.7 Dorchester area, Dorset, a linear pattern of Neolithic and Bronze Age monuments

Fig. 5.8 Monument alignments at Fengate and Catholme

Fig. 6.1 Examples of the diverse types of prehistoric wood trackways

Fig. 6.2 The dates of trackways in areas of North-west Europe

Fig. 6.3 The Sweet Track, Somerset, UK

Fig. 6.4 Borremose, Denmark, paved road across wetland leading to enclosed settlement

Fig. 6.5 Broskov, Denmark: stone roadway of several phases between AD 300 and 1000 with a cup marked stone at one end

Fig. 6.6 The Emmen arca of the Netherlands showing wetland and dryland activity in later prehistory

Fig. 6.7 The Somerset Levels, UK, showing the Bronze Age trackways and other finds

Fig. 6.8 Goldcliff, Wales: Iron Age trackways

Fig. 6.9 Corlea, Ireland, trackway 1

Fig. 6.10 The Fengate, Flag Fen and Must Farm landscapes, UK

Fig. 6.11 Must Farm alignment formed of two rows of large posts crossing a palaeochannel and underlying the excavated Bronze Age palisade and settlement

Fig. 6.12 Harter's Hill alignment, Somerset, hypothetical reconstruction of the alignment

Fig. 6.13 The Ebbsfleet Valley, Kent, a hypothetical Bronze Age routeway along the saltmarsh edge represented by a possible bridge, short lengths of brushwood track, pit clusters and barrows

Fig. 6.14 Yarnton, Thames Valley, UK Iron Age stone causeway

Fig. 7.1 Denmark, early routeways

Fig. 7.2 Barrow alignments in the Fousing/Kilen area in west Jutland, Denmark

Fig. 7.3 Barrow alignments in Veluwe, Netherlands

Fig. 7.4 Penmaenmawr, north Wales, UK

Fig. 8.1 Map showing the distribution of sites in England discussed in Chapters 7 and 8

Fig. 8.2 Sketches of types of evidence for trackways in agricultural landscapes

Fig. 8.3 Isometric diagram showing some typical relationships between colluvial deposits, lynchets, trackways and hollow ways

Fig. 8.4 Retrogressive landscape analysis as a guide to relative dating

Fig. 8.5 Multiple hollow ways descend the escarpment from the Marlborough Downs to the Vale of Pewsey at All Cannings, Wiltshire

Fig. 8.6 Bronze Age landscapes on Dartmoor, Devon

Fig. 8.7 Yorkshire Wolds: trackways, linear ditches and Iron Age cemetery

Fig. 8.8 Bronze Age landscape of fields and trackways revealed by excavations at Heathrow Terminal 5

Fig. 8.9 Uffington hillfort, the Ridgeway, hollow ways and the White Horse

Fig. 8.10 Holme seahenges in relation to the line of the Icknield Way

Fig. 8.11 Sharpstone, Shropshire: Roman road overlying roads of Iron Age date

Fig. 9.1 Dug out boat under excavation at the Mesolithic site of Hardinxveld De Bruin, Netherlands

Fig. 9.2 Reconstruction of a Danish Mesolithic dug out canoe of the type found at Tybrind Vig, Denmark; the locations of dug out canoe finds and paddles in Denmark; and a decorated paddle from Tybrind Vig

Fig. 9.3 Sewn plank boats: reconstructions of the Ferriby and Dover Boats; distribution and dates of prehistoric sewn plank boats

Fig. 9.4. Denmark showing the distribution of 'wheatsheaf' motif on Mesolithic bone objects

Fig. 9.5 Maritime connections indicated by metalwork in North-west Europe

Fig. 9.6 Boats in rock art: Soletorp, Tanum, Sweden

Fig. 10.1 The Weald in south-east England

Fig. 10.2. Fields and trackways at Plumpton Plain middle Bronze Age settlement, Itford Hill and Itford Bottom

Fig. 10.3 Bishopstone, Sussex, UK: the relationship between barrow alignments, Neolithic pits, an Iron Age enclosure and later features indicating a 'ghost' routeway

Fig. 10.4 Fields and trackways at Bullock Down, East Sussex

Fig. 10.5 Peacehaven, East Sussex: extensive excavated area with Neolithic and early Bronze Age linear alignments, Bronze Age and Iron Age trackways, fields and settlements

Fig. 10.6 The South Downs, East Sussex in the Saddlescombe/Ports Way area

Fig. 10.7 Trackways at Saddlescombe, East Sussex

Fig. 10.8 Thundersbarrow Hill, East Sussex showing the relationship between enclosures, routeways and fields

Fig. 10.9 East Dean Woods: Lidar image and section across double lynchet trackway

Fig. 10.10 Sunken lanes in the valley of the River Rother, West Sussex, UK

Fig. 10.11 The North Downs east of the River Medway, Kent

Fig. 10.12 Saltwood Tunnel, Kent: barrow alignments, prehistoric to Roman ditches and trackways showing long term continuity into recent tracks and boundaries

Fig. 10.13 Lyminge, Kent: map showing area; section of lynchet; land molluscs and dates from sediments in the lynchet

Fig. 11.1 Human footprints in estuarine silts at Goldcliff, Wales

Fig. 11.2 Bronze Age rock art Tanum, Bohuslan, Sweden: iconic images associated with movement

Table 3.1 Mesolithic sites in the Kennet valley outlining date, environmental evidence and sources

Table 8.1 Hypothetical types of later prehistoric trackways in Fig. 8.2 linked to illustrated examples in Chapters 8 and 10 and other publications

Acknowledgements

Several colleagues have been kind enough to provide comments on earlier versions of chapters: Prof Duncan Garrow kindly commented on the whole book; Prof. Richard Bradley read Chapters 5–11; and other colleagues who have provided comments on individual chapters are Prof. J.R.L. Allen, Prof. John Boardman, Dr Richard Brunning, Dr Dale Croes, Dr Petra Dark, Prof. Nancy Turner, Dr Darcy Mathews and Prof. Nicki Whitehouse. I am especially grateful for advice from anonymous reviewers and Prof. Stephen Rippon. I have benefited from discussions on mobility and collaborative fieldwork in the Vale of Pewsey with my colleague Dr Jim Leary who shares my fascination with these topics. All have been most helpful in suggesting improvements and things I had missed. Time did not allow me to follow up every one of the valuable leads they provided and I remain responsible for the limitations of what is here.

My research on this topic started to take shape for a conference at Olympia, USA in 2003 and it came towards a conclusion with the presentation of some of the results at the Society of American Archaeologists conference at Vancouver, Canada in 2017; those two visits were especially stimulating, providing introductions to new research areas. Between them lectures on this topic were given at several conferences in the UK and Europe. I am most grateful for the opportunities these provided for discussions with colleagues. My research has benefited in particular from discussions with Prof. Nick Barton, Dr Damian Goodburn, Dr Alasdair Barclay, Dr Mark Knight, Dr Matija Cresnar and Dr Jette Bang. Prof. John Boardman has been especially helpful and stimulating in discussing our mutual interest in hollow ways, soil erosion and the Wealden evidence. Several of my PhD students have shared with me the exploration of aspects of this theme, including Dr Rachel Scales and Dr Kirsten Barr (footprints); Dr Simon Maslin (Lyminge case study); Elspeth St John Brookes (geochemistry); Dr Lionello Morandi (non-pollen palynomorphs); Dr Alex Brown (wetland-dryland relations); Dr Scott Timpany (wetland botany); Dr Chris Speed (experimental aspects); Claire Nolan (wellbeing); Katie Whitaker (stone mobility); and Dr Tom Walker (Mollusca). Dr Stuart Black and Prof. Phil Toms have collaborated on dating aspects of hollow ways. The South Downs National Park, The Heritage Lottery Funded Secrets of the High Woods Project and Cotswold Archaeological Trust (especially Buz Busby) are thanked for collaboration at East Dean Woods. David Rudling, Dr John Manley, Peter Herring and Dr Malcolm Lillie have answered questions about aspects of their research. Richard Chappel has kindly allowed me to include material from his catalogue of Irish Radiocarbon dates in Supplementary Appendix 6.1.

Believing that first-hand experience of sites is important in understanding patterns of movement I have tried to visit nearly all the sites discussed in any detail, but there remain a few that I have yet to explore. I am grateful to the many people who have helped to make visits possible and have told me about aspects of their work: in Canada and North America: Professor Dale Croes, Jill and Wyn Taylor, Daryl Fedje, Joanne McSporran, Duncan McLearon, Al Mackie and Nicole Smith;

in Denmark: Prof. Søren Andersen; in Germany: Dr Helmut Schlichterle, Dr Bodo Diekmann and Dr Harald Lübke; in the Netherlands: the late Dr J.A. Bakker, Prof. Leendert Louwe Kooijmans, Prof.Annelou van Gijn and Dr Hans Peters. Professors John and Bryony Coles have done much to stimulate and encourage my researches on wetlands and many of the other topics which occur in this book.

My fascination with this topic began long ago in south-east England where my work was encouraged by the late Prof. G.W. Dimbleby, Dr Ken Thomas and the late Prof. Peter Drewett. Since those days I have benefitted from a shared interest in many of these topics with Dr Mike Allen.

Thirty-five years of research in the Severn Estuary has played a big part in developing my interest and ideas in this subject and I am grateful to Cadw, NERC and the British Academy for funding this research, and the late Dr Rick Turner for facilitating our research. Thanks are also due to the hundreds of people who have assisted with the excavations. Collaboration with the following was particularly important in developing the Severn Estuary cases: Prof. John Allen, Dr Alex Brown, Dr Richard Brunning, Dr Heike Neumann, Prof. Steve Rippon, Dr Rick Turner and Dr Tom Walker. The Severn Estuary Levels Research Committee has provided a collaborative framework for discussion and research. I am grateful to the University of Reading for allowing me research leave during which much of this book was written.

My wife Dr Jennifer Foster provided major assistance throughout; she accompanied me on many of the site visits that went into preparing this book, helped clarify the developing ideas, provided most helpful comments on the text and prepared nearly all the graphics and the index – as ever I am most grateful.

Martin Bell

1. Steps towards understanding: routeways in practice, theory and life

Background

I arrived at this study along an experiential path. As a child I explored the sunken and mysterious hollow ways of the chalk downland where I grew up and which have exercised my curiosity ever since. Near the beginning of my archaeological activities my first excavation revealed the ghost of a trackway which survived only as an alignment of entrances and features. As an environmental archaeologist and geoarchaeologist I studied traces of people and animals in the landscape and went on to excavate wooden trackways. More recently my interests in traces of human movement have been crystallised by the magical experience of discovering 8000 year old human footprints in intertidal silts (Fig. 1.1). Eventually these seemingly unconnected topics coalesced unexpectedly during walks in the temperate rainforest of the American North-west Coast. That led to a conviction that we need to develop an approach to environmental and landscape archaeology which has a greater emphasis on connectivity than the current preoccupation with sites. This approach would illuminate the most ubiquitous, and probably the earliest, way in which people structure and comprehend landscape, through the movement of their bodies.

Introduction

The objective is to demonstrate that the study of prehistoric patterns of movement is important, achievable and relevant in all areas and periods. For some sections of the book, particularly regarding hunter-gatherers, vegetation changes and footprints, examples are drawn globally. For other sections concerned with burial monuments, fields and trackways, where the evidence base is vast but is seldom considered in detail, the focus is mainly on examples drawn from the area the writer knows best: Britain and North-west Europe, although many of the approaches adopted are much more widely applicable. The book aims to demonstrate that, notwithstanding the pessimism of many previous scholars, there is an ever-increasing range of ways in which we can investigate patterns of movement in the past. The intention is to encourage archaeologists, and others who share a fascination with landscape, to raise their sights from the individual dots on the map that we call sites to the ways in which those foci were networked together by patterns of habitual movement constituting living landscapes. The approach is necessarily multi-scalar, from the information contained in the individual footstep or walk to issues of longer distance communication.

One reason prehistoric routeways have often seemed intractable is that they have been approached from the perspective of one source of evidence, be it landscape analysis, historical sources, phenomenological inference and so on. Archaeologists have long learnt to adopt a multi-disciplinary approach to the study of settlements, burials, ritual sites etc. Routeways are of such significance they are also deserving of a full battery of techniques applied in an integrated way, using archaeological features, artefact distributions, environmental archaeology, footprint studies, geochemistry, sediments, social theory and the application of a range of dating techniques. The approach also needs to be multi-site so that we can think beyond the boundaries of our site, focusing not so much about how landscapes were, but about how they worked and interacted with others through axes of movement.

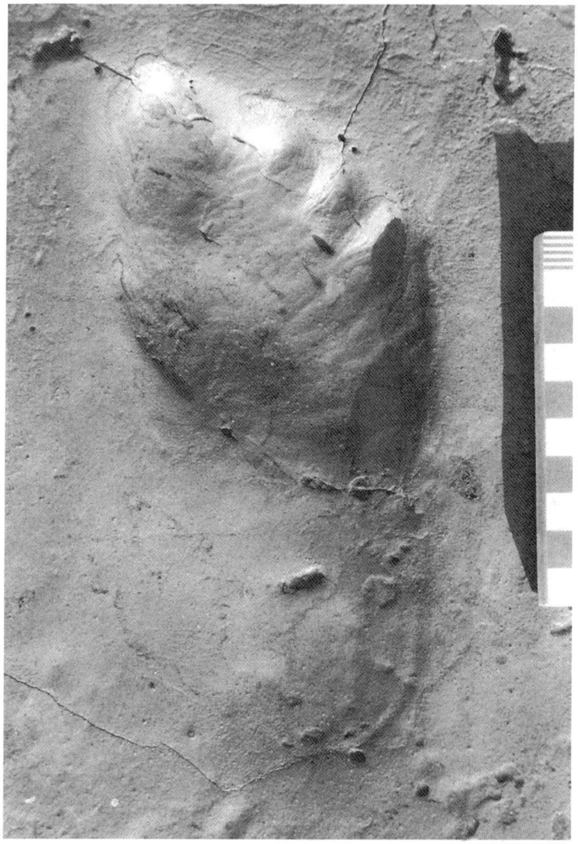

Figure 1.1 Footprint-track of a Mesolithic child aged 8–9, c 5500 cal BC, Goldcliff, Wales (photo: E. Sacre).

A multi-period approach is necessary since some tracks once established are perpetuated over millennia and thus significantly influence the encounters which successive generations have with that landscape and the perspective and angle from which their perception occurs. Here the focus is mainly on prehistory. The justification is that until recently there has been much less emphasis on routeways in prehistory as compared to those of historic periods. Consequently the case that prehistoric routes are capable of investigation needs to be articulated and appropriate techniques identified. When it comes to detailed investigation of routeways in specific areas, however, our research can seldom be confined to prehistory. One of the key approaches, certainly in areas where there has been a long history of settlement and activity, is retrogressive analysis, beginning with the present landscape, using air photographs and maps and diverse other sources to work progressively back in time from historic to prehistoric periods. Thus we can effectively peel away successive layers of landscape revealing the underlying structure of earlier prehistoric landscapes. For this reason we will necessarily stray in places onto the routeways of historic periods, when considering, for instance, the ethnohistorical records of people's movement in the American North West in Chapter 2 or British droveways in Chapter 10.

False paths

One factor which helps to explain a puzzling neglect of trackways for some 80 years, at least in Britain, is a book published by Alfred Watkins (1925) *The Old Straight Track.* He was an amateur archaeologist and his book was grounded in the Herefordshire countryside in which he grew up and reflects his empathy for that landscape. It is illustrated by some fine photography, a medium in which Watkins was a notable pioneer. He nonetheless drew totally erroneous conclusions from the landscape, observing that some historic places could be joined by dead straight lines. His argument was fatally undermined by the very varied character and date of the sites involved and the special and unsubstantiated pleading which permeates the work. The monuments include Neolithic and Bronze Age barrows, prehistoric settlements, boundary and waymark stones, Christian churches, medieval moated sites, avenues of trees, even isolated pines, and many others. The dead straight lines joining these places he called ley lines, which he regarded as ancient communication or trading routes. They went up hill and down dale with no reference to topographic barriers. He rationalised the inclusion of sites of wildly different dates by arguing that, for instance, Christian churches were put on previously significant sites. However, why this should apply to moated sites and many others was not explained. Suffice to say there is no convincing evidence for the ancient ley routes which Watkins claimed.

Robert Macfarlane, whose significant literary contribution to the study of paths will be noted below (p. 20), has recently provided an introduction to a new edition of Watkin's book. Macfarlane (2014) sees his work, naturally enough, from the perspective of the leading English scholar of landscape writing. He says 'Watkins re-enchanted the English landscape, investing it with fresh depth and detail and prompting new ways of looking and new reasons to walk'. Watkins was certainly a lyrical and persuasive writer on a landscape he clearly loved, and that must have contributed to the popularity of his writing. Macfarlane sees Watkins as somebody who opened up the countryside to the popular imagination, whilst he acknowledges the highly dubious nature of Watkins' interpretations. Watkins' message was that anybody could be a landscape historian; all they needed was a map. Macfarlane describes Watkins' ideas as going 'viral'; they provoked widespread interest and to this day are elaborated in a whole host of New Age theories. One can only speculate as to whether, if archaeologists in the 1920s had engaged more actively in critique of Watkins, his ideas would have proved so persistent. The pioneering field archaeologist O.G.S. Crawford, Archaeology Officer of the Ordnance Survey, dismissed Watkins' ideas (Hauser 2008) but refused to review *The Old Straight Track* in the journal *Antiquity* which he edited, or to debate with Watkins. A half sentence dismissal of Watkins' ideas was included in the Ordnance Survey (1973, 157) *Field Archaeology in Britain* which Crawford originally wrote. The first really substantive critique of Watkins' ideas was published 58 years after his first edition by Williamson and Bellamy (1983) and that provides a systematic demolition of ley lines and the subsequent new age paraphernalia which has been built upon them.

Whilst Watkins could be argued to have encouraged thought about long term structures in the landscape, his ideas about them were so significantly in error that they have proved a Upas tree which poisoned the ground for research on past communication routes for three generations. Two pieces of evidence demonstrate the extent to which this occurred. Before 1925 the study of prehistoric routeways had been quite an active field with very good empirically-based field surveys by pioneering archaeologists: Curwen and Curwen (1923); Williams-Freeman (1915); Crawford (1922); and Fox (1923). After publication of Watkins' book this promising area of research virtually died. Instead, archaeologists like the Curwens focused on settlements, fields and burials; they noted the existence of tracks but after 1925 seldom made very much of them in terms of wider patterns of communication. It is interesting to compare Watkins to the very measured approach of the great archaeologist E.C. Curwen (1929, 119) who, in considering prehistoric routes in Sussex, was at pains to separate out evidence which he regarded as conclusive from circumstantial inconclusive evidence, and thereby showed a pioneering appreciation of both the problems of dating early routes and how those problems may be addressed.

A second piece of evidence for the Upas tree effect of Watkins' book is provided by comparison with continental Europe. Here the pioneering studies of Sophius Müller (1904) were followed by a steady stream of archaeological writing on past routeways, of which the work of J.A. Bakker (1976) is especially notable, and draws on evidence for alignments of barrows and other monuments reviewed in Chapter 7. Perhaps the main lesson from Watkins is that a feel and empathy for the landscape, whilst something of great value, is insufficient for an adequate appreciation of its origins. It comes back to the need for detailed examination of individual features, critique of ideas and interpretations and the need to develop a robust chronology and interpretative framework.

Taking stock and steps forward

Evidence of past trackways is extensive but often fragmentary, dispersed and difficult to interpret and still in many ways neglected. We are still well short of a toolbox of methods for the study of prehistoric routes and this book attempts tentative first steps in that direction. As the Dutch scholar Bakker (1991, 518) observes, in Britain the phenomenon of roads marked by monuments is 'regarded with scepticism and its study seems somewhat neglected'. Illustrative of scepticism in the British literature is Coles' (1984, 1) observation that discussion of roads allows 'the prehistorian to indulge in conjecture unencumbered by the need to pay attention to observable evidence'. Fowler (1998, 25) describes tracks as 'the haunt of the romantic, the irrational and the obsessional'. Bradley (1997, 81) says: 'the recognition of ancient roads or trackways is notoriously subjective and all too often turns out to be based on circular argument'. As Fleming (2012) notes, archaeologists have never felt completely comfortable handling old roads and the subject has been left to amateurs. One of the most notable British studies is Christopher Taylor's (1979) *Roads and Tracks of Britain*. He begins, somewhat disarmingly, by saying that 'most popular books on ancient trackways are nonsense'. Taylor was the greatest British landscape archaeologist of his generation; the main strengths and most detailed treatment in his book concerns Roman and later roads and it repeatedly concludes that 'all but a few [prehistoric tracks] are quite impossible to date' (Taylor 1979, 1). Similar views are echoed by other British syntheses which have a Roman and later emphasis with just a few pages on prehistory, e.g. Hindle (1993; 2001) and Morriss (2005). A topic, which partly accounts for the scepticism of previous British writers, is the ridgeways, which have long been supposed to represent ancient routes along the ridge crests. Of these the best-known example is the Wiltshire Ridgeway, and its continuation the Icknield Way in eastern England. These will be more fully discussed in Chapter 8 but suffice it to say there is remarkably little supporting evidence for the early origins of some claimed ridgeways. These rather pessimistic views actually reflect the position 20 or 30 years ago.

Today archaeologists are starting to use a new and imaginative range of concepts and techniques to investigate past mobility as shown by several edited surveys of diverse geographical areas and periods including site and area specific case studies from British prehistory (Cummings and Johnston 2007; Preston and Schorle 2013; Leary 2014; Leary and Kador 2016). Alcock *et al.* (2012) provide diverse case studies from many parts of the world mainly from historic periods. Sellet *et al.* (2006)

and Snead *et al.* (2009) consider mobility from the combined perspectives of field archaeology and anthropology, the latter with a focus mainly on middle and north America. Jim Leary is preparing a book (*Roam*), an accessible account of the role of mobility in human affairs generally. Written from an archaeological viewpoint, it has multidisciplinary relevance and will help introduce a wider audience to the necessity for detailed examination of multiple sources of evidence and case studies attempted in the present book. Field archaeology is also providing a wealth of evidence particularly from extensive landscape scale excavations, and there is also the increasing deployment of a range of scientific and dating techniques. Consequently there is within our sights an understanding of prehistoric movement in the landscape that seemed out of reach a generation ago.

Environmental and geoarchaeology

Environmental archaeology has an ecological emphasis and draws on an ever-increasing range of biological evidence to address archaeological questions (Evans and O'Connor 1999; Dincauze 2000). Its emphasis has been on the role of people in transforming environments, for instance by clearance and agriculture, and evidence of their economy. Geoarchaeology uses earth science concepts and techniques to address archaeological questions (Allen, J.R.L. 2017). It has mainly been concerned with the sedimentary context of sites and the use of geological resources (French 2015). Both approaches have contributed to many aspects of archaeological investigation, but neither has been especially focused on questions of routeways and mobility, to which it is argued in the following chapters they can make a significant contribution. Some types of biological evidence, which are particularly abundant in waterlogged contexts, have already contributed to studies of routeways and figure in later chapters. Several cases are considered in which pollen and charred plant macrofossils (e.g. seeds or charcoal; Fig. 1.2a) contribute to the identification of more open corridors associated with possible routeways. Likewise, beetles may point to concentrations of dung on routes used by herbivores and mites may identify areas where animals congregated (Schelvis 1992). Plant macrofossils, beetles and molluscs originating in wetland environments, but found on dry ground sites, may also point to patterns of connectivity between these environments.

There are a number of other sources which, whilst not widely applied in the past, have potential in future environmental investigations of routeways. An introduction to several of these is provided by Nicosia and Stoops (2017) in the context of their presence in sediment thin sections (micromorphology). Non-pollen Palynomorphs (NPP; microscopic organic particles), such as fungal spores, include carbonicolas (carbon associated) fungi indicative of burning (Fig. 1.2b) and coprophilous (dung associated) spores indicative of the presence of herbivore dung (Fig. 1.2f and g). NPP and pollen are increasingly used in a complementary way to establish the role of natural disturbance and human activity in Mesolithic environmental change (Innes *et al.* 2010; 2013; Ryan and Blackford 2010). Dung of herbivores has been shown to provide clues as to where animals had been grazing (Caseldine *et al.* 2013) and also the seasonality of grazing (Aberet and Jacomet 1997) and could potentially provide valuable clues to the vegetation zones through which animals had recently moved on transhumant routes (van Asperen 2017). Faecal spherulites (spherical calcareous particles found in dung; Fig. 1.2e) may also be present where herbivores are abundant, depending on alkaline soil conditions (Canti and Brochier 2017). The eggs of human intestinal parasites (Fig. 1.2c and d) could also contribute to the identification

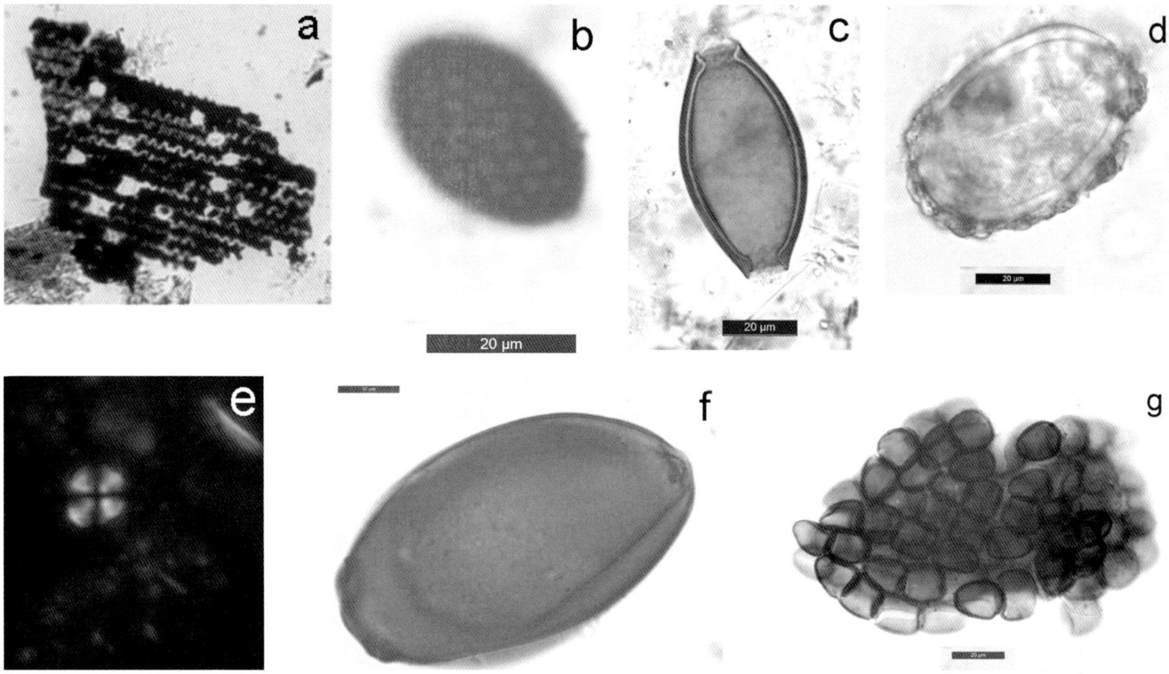

Figure 1.2 Microscopic evidence for human and animal activity and movement: (a) macro-charcoal particle resembling Phragmites *reed (image: courtesy Dr P. Dark); (b)* Gelasinspora *(carbonicolous fungus); (c) egg of the intestinal parasite* Trichuris*; (d) egg of the intestinal parasite* Ascaris*; (e) Faecal spherulite; (f)* Podospora*, dung spore; (g)* Sporormiella *(dung spore) cluster (images b–g: courtesy L. Morandi).*

of frequented routeways. Of these *Trichuris* occurred in pollen samples around activity areas at the Mesolithic site of Goldcliff, Wales (Dark 2004; 2007) and provided insights to defecation behaviour, an aspect of the human use of space seldom considered in prehistory outside the arid zone contexts where coprolite evidence is well preserved (Sobolik 1996). Micromorphological thin sections also provide evidence of layered and trampled dung and the effects of animal presence and traffic which can assist in the identification of routeways (Goldberg and Macphail 2006; Rentzel *et al.* 2017). A range of analytical techniques, including biomolecular studies, gas chromatography and mass spectroscopy, can be applied to the identification of faecal and bile acid biomarkers (Shillito 2017). Sedimentary DNA has been used to identify the presence of grazing animals in agricultural contexts in the French Alps (Giguet-Covex *et al.* 2014) and could be used more widely to identify areas of concentrated animal movement and transhumant routes. Where it is possible to employ a combination of the foregoing techniques as part of a multi-proxy investigation the interpretations will be most persuasive.

In foregrounding scientific approaches it is important to avoid an overly deterministic position which assumes that societies or sites can necessarily be explained by their environmental context. Although routeways are clearly functional, they are not solely so, and cannot simply be explained by concepts of least effort and the distribution of resources. Where we have the benefit of ethnohistorical sources, as in the case of Middle-American routes discussed in Snead *et al.* (2009), it is evident that many were far from utilitarian and had social, symbolic and ceremonial roles. The same is self-evidently the case with pilgrimage routes wherever they occur (Maddrell *et al.* 2015) and the Buddhist concept of pathways to enlightenment (Neelis 2012). Investigations need to take account of cognitive and social aspects including perception, cosmology, religion and ideology (Flannery and Marcus 1993) as well as issues such as habitual usage and taboo areas (Jordan 2003a; Seitsonen *et al.* 2014).

Landscape archaeology

Landscape can be defined as environments understood and modified by human agency, in terms of effects on vegetation, the creation of fields, paths, monuments, etc. The extent to which agency is implicated is highlighted by the root of the term landscape in Old Frisian, meaning coastal land drained and protected from the sea (Stilgoe 2015). That author describes landscape studies as interdisciplinary and strongly linked to the experience of walking and fieldwork. Landscape archaeology is one of the ways in which archaeologists have succeeded in moving beyond the straight-jacket of individual site-based studies. It draws on a map-based approach and historical documents to complement the results of air photography and field survey which reveal traces of settlements, fields, trackways, ancient woodland, old pasture and other living biological communities that tell us about past landscape (Aston and Rowley 1974; Aston 1985). A pioneering study was W.G. Hoskins (1955) *The Making of the English Landscape*, the cover of my much thumbed 1974 paperback edition (Fig. 1.3) exemplifies the retrogressive landscape approach, demonstrating how early topography and vegetation patterns affect the subsequent patterns of tracks, fields and woods, features which become in part fossilised in the subsequent urban plan. The approach is exemplified by a trilogy of books by Christopher Taylor: on fields (Taylor 1975), settlements (Taylor 1983) and tracks (Taylor 1979). In the latter Taylor argued that much of the pattern of roads in Britain is essentially the same as 900 years ago. In a comment that can be seen as three to four decades ahead of its time, foreshadowing a key theme of this book, Taylor (1979, 153) made the observation that 'roads are not just changed by the demands of external pressures and events, they change their own environment and the social organisation of their users.' Oliver Rackham has also made a major contribution to the study of landscape history, initially on botanical aspects, such as ancient and managed woodland (Rackham 1980), later exploring a diversity of landscape features including many perceptive observations on the early origins of some roads and tracks (Rackham 1986). Landscape archaeology can achieve particularly detailed reconstructions in medieval and later periods for which documentary sources are also available. In the following chapters we will review several cases where large landscape-scale archaeological projects have enabled dated and excavated features to be spatially related to surviving routeways and field banks demonstrating that existing boundaries and tracks are in some cases of prehistoric origin.

Andrew Fleming (1988) has pioneered prehistoric landscape archaeology on Dartmoor and elsewhere and has recently developed a particular research focus on the trackways of early historic periods which can also contribute to an understanding of prehistoric routes (Fleming 2012). He argues that we should focus not just at the very local scale but search for coherence

Figure 1.3 The cover of W.G. Hoskins' book The Making of the English Landscape *(1974) illustrating the concept of retrogressive landscape analysis.*

and integrity of neglected long-distance networks which have, for the last two millennia, been associated with what he calls horse elites (Fleming 2010). In a Devon case study he makes the important observation that these routes often seem to bypass villages and farms that lie on spurs off the route. Fleming also observes that many open fields, which were established around the tenth century AD, respect existing roads. He further observes that early inscribed stones of pre-Saxon date (i.e. in Devon pre- *c* 700 AD) are on long distance routes, including those he associates with elites, and also those associated with long distance transhumance routes which Fox (2012) has shown go back to before the Norman conquest.

Theoretical perspectives

Johnson (2007) has criticised what he describes as the atheoretical and romantic approach of the landscape archaeology of Hoskins and his successors and has advocated an emphasis on practice and agency, in which he explicitly mentions routes and hollow ways. One of the contentions of this book is that in order to overcome the problems of investigating past routeways we need to draw on the perspectives provided by social theory, anthropology and science. At times these approaches have seemed to be in opposition. The proponents of scientific method were dubbed processualists, and those who advocated the primacy of social factors, post-processualists (Johnson 2010). Processualists were mainly concerned with very detailed investigations of specific sites as a basis for generalisation, the post-processualists with philosophical approaches and particular classes of prehistoric monument which illuminated the role of social factors. Neither initially was particularly interested in mobility and the connections between sites. However, this later became an increasingly significant strand of thought from those with a social emphasis (e.g. Tilley 1994), less so within the scientific strand. Increasingly in the last 20 years, theory and science have come together, a key work being Hodder (1999) *The Archaeological Process* which outlined the complementary perspectives which they can provide, as illustrated by the philosophical framework for excavations at Çatal Höyük, Turkey. Increasingly, scientific techniques are being employed to answer, not just questions about the environment and dating, but social questions.

It is appropriate to begin by considering some of the theoretical discussions of recent decades which have brought the study of mobility to the forefront of archaeological thought. A key influence on much theoretical writing has been Heidegger's (1962) *Being and Time,* which focuses on dwelling in the world, the context-specific situatedness of the human condition in terms of the relationship between people, place and time. This challenged the tendency of processual archaeologists to look for universally applicable patterns and for environmental archaeologists, consciously or unconsciously, to slip into deterministic reasoning. However, as Ingold (2011, 12) observes, dwelling should be less about a situated place and more to do with movement along paths constituting a way of life. Gosden's (1994) *Social Being and Time,* which built closely on Heidegger's work, highlighted the role which monuments and artefacts play in creating patterns of recursiveness by which the past is used to create present and future action. Gosden (1994, 17) argues that 'thought arises from within the flow of life helping us to cope with problems encountered by habitual action', a point which clearly highlights the significant role of walking and thus the marks made on the landscape by habitual patterns of movement. Also strongly influenced by Heidegger was Thomas (1996, 89) who argued: 'in creating paths in the forest they also created links between both places and people'. Myths were then told and the place became historical.

Perceptions of place relate not just to observation with the eyes but to the bodily experience derived from all five senses; these include the haptic sense derived from touch and walking. Spatial skills are developed through the conduct of various tasks, including habitual action. In navigating landscapes, people remember a sequence of patterns of bodily movement, changes of direction and topography which provide prompts. Tuan Yi-Fu (1977) described this by reference to diverse ethnographic examples ranging from understandings of architectural space to the spatial skills of Eskimos and the navigational skills of Pacific islanders. Just as architecture structures space and choreographs patterns of movement (Barrett 1994), so, on a much larger spatial scale, do the structures, including

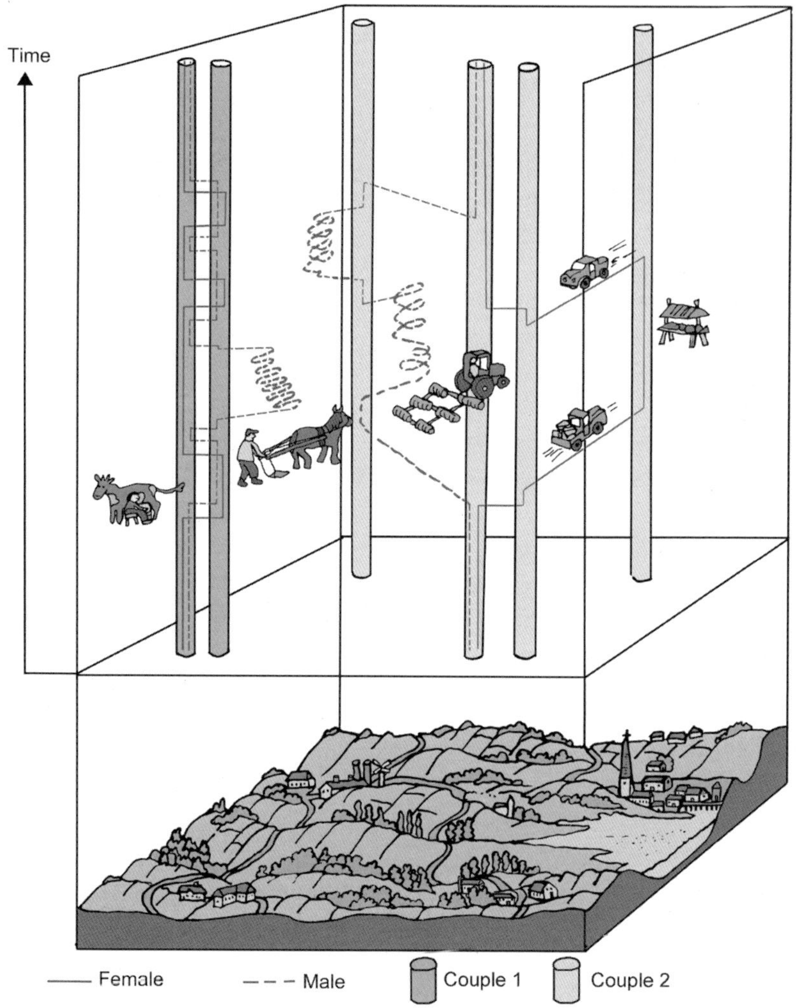

Figure 1.4 A model of time-space geography, with time on the vertical axis, space and place on the horizontal axis. It depicts a day in the life of four rural individuals, two males (dashed lines) and two females (solid lines); places are vertical pillars (graphic: J. Foster after Carlstein 1982, fig. 2.7).

routes, which exist in landscapes, many of them created by past human agency. The spatial information acquired by the senses is augmented by stories, myth and legend as exemplified for instance by the Dreamtime of aboriginal Australians (Chatwin 1989). Together these all form part of the Lifeworld which Simmons (1993) defined as the sum total of a person's involvement with the cosmos, landscape being as much the work of mind and memory as the topography which is created by layers of rock.

Landscapes have both spatial and temporal dimensions, since all movement in space is movement in time which is the essence of time-space geography. Hägerstrand (1975) developed a social-environmental web model in which individual paths flow through time and space. An example is shown in Figure 1.4 with space on the horizontal plane, and time on the vertical plane. Places are marked by vertical columns and the model depicts a day in the life of four individuals in a rural landscape. At times paths converge, they meet and move around together in what has been called a space – time aquarium (Carlstein 1982). In this chronogeographic approach places are but pauses in movement (Parkes and Thrift 1980).

A focus on an individual's situatedness within the landscape led to an emphasis on phenomenology, which is concerned with the world as perceived by an observer and their embodied experiences of the landscape, its rocks, mountains, rivers and coasts. It also looks at the influence of such features on the placing of monuments and the structuring of people's lives and patterns of movement (Tilley 1994). Phenomenology facilitates thought, not just about how landscapes were in abstract, but about how landscapes appeared to people in the past and how that experience related to patterns of movement through the landscape. As Aldred (2014) says, 'It is only by using our own bodies that we can begin to enter a dialogue about actual past movements'. The approach has led to many prehistoric case studies and phenomenological approaches have been increasingly adopted (Cummings and Whittle 2004; Tilley 2010). Inevitably, however, consideration of prehistoric environmental perception can be quite subjective; how do we know which particular route, or direction, was taken through a landscape and which particular features were important to people at the time? Tilley (2010) contends that he does not claim to be able to see landscape phenomena in the way in which people did in prehistory. However, unless that is the objective it surely limits the value of the approach. To counter these difficulties Evans (C. 2009) has argued that, rather than adopting a single phenomenological perspective, multiple perspectives need to be evaluated. Developing a phenomenological approach requires investigation of the precise relationship between prehistoric monuments and natural features and identification of actual evidence for routes of movement. Fleming (1999; 2006) has demonstrated that this demands a significantly more critical and rigorous methodology than is evident from some published phenomenological studies. Since perception is clearly influenced by directions and patterns of movement (Ingold 2011), it is important that these patterns are identified in a precise and defined way using a methodology which is explicit and open to critique by others.

Phenomenological studies have also seldom taken into account the ways in which the landscape has changed since the period in question, for instance by using the available palaeoenvironmental evidence to evaluate past vegetation patterns, visibility and the influence which these patterns would have exerted on routes taken through landscape (Chapman and Gearey 2000). These are increasingly attainable objectives as the density of palaeoenvironmental sites in some landscapes increases and they become more precisely dated, such that Geographical Information Systems (GIS) and modelling can be used to upscale site specific data creating landscape scale reconstructions. The environments around Neolithic complexes at Avebury and Stonehenge, Wiltshire have been reconstructed in this way and are discussed in Chapter 5 (Allen, M. 1997; Gillings *et al.* 2008). In the Netherlands Doorenbosch (2013) has reconstructed the environment associated with heathland barrows and their relationship to routeways. Models of past landcover are now being produced using a multiple scenario approach to reconstruction which generates multiple computer models based on the various parameters and estimates which best fits the pollen data (Bunting and Farrell 2017).

Anthropology: the contribution of Tim Ingold

Of the other disciplines which interface with archaeology in the study of past mobility perhaps the most pertinent is anthropology, which shares with archaeology the central mission of exploring the rich diversity of human existence. Particularly influential, from this writer's perspective has been the work of Tim Ingold who, in a series of papers over the last 20 years, has written from the perspective of anthropology in a way which resonates with, and informs, the work of the landscape archaeologist. His much quoted paper 'The temporality of landscape' (Ingold 1993) animated the landscape which he saw as 'the congealed form of the taskscape' in which 'paths and tracks impose a habitual pattern of movement on people' by the patterns of recursiveness they establish (Ingold 1993, 162, 167). How people perceive the landscape is to a large extent seen as a result of the tasks undertaken within it and thus the associated patterns of movement. In *The Perception of the Environment* Ingold (2000) brought together a series of papers on the perceptual theme, including one which contrasted the life-world as seen from an experiential centre with the externally viewed world perspective of the present-day environmentalist (Fig. 1.5a and b). Thus the experiential world of Yupik Inuit cosmology is illustrated diagrammatically in Figure 1.5c in which the home is surrounded by concentric environmental zones to which it is linked by paths. Two other graphical representations which convey the importance of paths come from Walbiri aboriginal Australian iconography. Figure 1.5d represents their world as a network of concentric camps with waterholes at the centre linked by paths which were originally laid down by the movements of ancestral beings (Munn 1973). Figure 1.5e is a representation of kangaroo footprints and tail impressions moving towards a camp. The ancestral and mythical origins of these routes demonstrate one of the ways routes become entrenched over extended timescales. Such routes structure the encounters of successive generations, and thus to a degree their environmental perception, since each generation sees the landscape from the particular perspective provided by that route.

Ingold (2007) has increasingly focused on the importance of lines which human beings generate wherever they go; as he says, life is lived along paths and it is along paths that people grow in the knowledge of the world around them, 'inhabitants journey from place to place along a way of life' (Ingold 2000, 193). Ingold (2011) acknowledges that his own intellectual journey has taken him in recent years from a focus on the experiential centre of place or dwelling, to recognise the importance of lines of movement in perception. He argues that 'the path and not the place is the primary condition of being or becoming' (Ingold 2011, 12). Paths he sees as ways of animating or bringing the statics of anthropology to life. The same can be said of archaeology's need to escape from a rather static preoccupation with sites investigated in depth. Ingold contends that those who walk with bare feet are naturally more in touch and engaged with their environment and that for them the feet are as important as the heads and hands in their perception of the environment. A similar concept is translated to an educational experience at the Hunybedder Centre at Borger in the Netherlands, an open air museum focused on prehistory which includes a footpath along which children are encouraged to walk barefoot, experiencing the contrasting tactile properties of wooden trackways, mud, woodchips, charcoal, pebbles, peat and shell etc (Fig. 1.6). Following an established route recalls previous patterns of past movement and tactile experiences (Ingold and Vergunst 2008, 17). That activity is seen, not purely as a set of individual acts, but a form of social activity whereby people's feet and patterns of movement are responsive to the presence and activity of others, whether they are actually present, or have created memories of their former presence in the landscape; these could be monuments, or traces of former presence in vegetation. Two studies should be noted as putting some of these ideas into practice by adopting a joint anthropological and archaeological approach. One concerned evidence for mass migration of the Tamu-mai people of the Annapurna Highlands in Central Nepal; the project looked at both the archaeological traces and the social memory of that

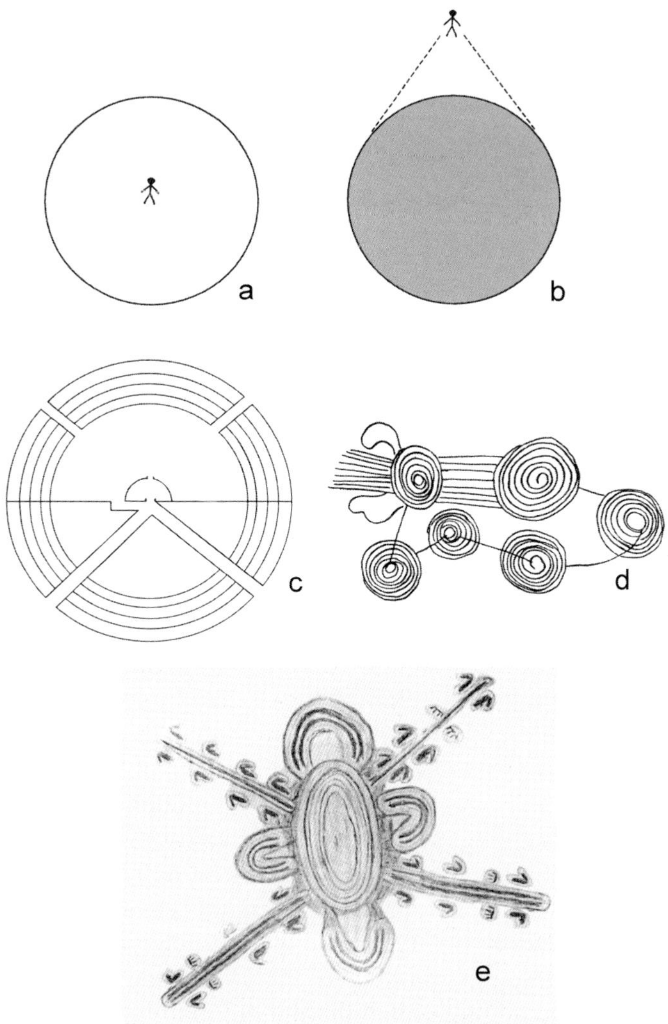

Figure 1.5 Images of the lifeworld: (a and b) the lifeworld as seen from an experiential centre compared to the global world that the environmentalist envisages today (after Ingold 2000, fig. 12.1); (c) Yup'ik cosmology in cross section (after Ingold 2000, fig. 12.3); (d) the cosmology of Walbiri aboriginal communities in Australia representing the relationship between camp sites (circles) and paths (lines) (after Munn 1973, fig. 1); (e) Walbiri iconography showing kangaroo footprints and tail trails moving towards a camp (after Munn 1973, fig. 2).

movement (Evans C. 2009). The other is Jordan's (2003a) anthropological study of the Siberian Khanty which challenges the frequent assumption that hunter-gatherer-fishers operated in a pristine 'natural' world, by documenting the rich social symbolism of their encultured landscape and the ways in which physical transformation occurs through the making of places, shrines, paths and artefact deposition. Ethnographic evidence for the mobility of recent hunter-gatherers is increasingly being used by archaeologists to conceptualise the potential scale, nature and frequency of movement and its effects on the landscape (Sellet *et al.* 2006).

Figure 1.6 Thinking through one's feet: path at the Hunebedder Centre, Borger, Netherlands where children can experience walking barefoot on a range of surfaces (photo: M. Bell).

Landscape change and clues to movement

Environments are not just a given product of particular factors such as climate, soils, geology, flora and fauna. Progressively through the Holocene they have been subject to significant transformation by human agency (Dincauze 2000; Bell and Walker 2005). Deforestation, drainage and embankment against the sea are some of the most obvious, but it goes right down in scale to the marks on the landscape made by paths or modification of individual trees. Environmental archaeological studies draw on a wide range of proxy sources to reconstruct landscapes and their changing nature. Where multiple sites have been investigated in a study area then it becomes possible to create a high resolution picture of the spatial variation of environment types. Where sources of evidence are fewer, spatial variation is less identifiable and there can be an unfortunate tendency to assume landscape homogeneity. However, certain locations would have provided particular attractions, such as the edges of lakes and rivers. The effects will also be concentrated in frequented routeways, and in places these can be readily predicted, for example where natural topography funnels movement, particularly in the passes between valleys. Such are often referred to as 'natural routes' and ridgeways along uplands are frequently cited examples. We need, however, to be cautious in identifying 'natural routes' given the risks of circular reasoning; we should not accept a route simply because it seems suitable. The hypothesis of its existence needs to be tested using other evidence, be it the spatial relationship to the distribution of monuments, such as barrows, or rock art, or dating evidence for particular sections of track.

Frequent travel along paths on a slope gives rise to erosion, forming a linear depression which during times of high rainfall acts as a channel for runoff and thus become more deeply incised, creating a pronounced hollow way (Fig.1.7). Such features are widespread on slopes in North-west Europe, particularly on more erodible geologies such as chalk, limestone and sandstone. Boardman (2013) has recently shown that hollow ways can act as alternative lines of drainage and thus represent humanly-created geomorphic features, an example of geomorphic connectivity from slopes to valleys brought about by human agency (Wohl *et al.* 2014). In North-west Europe hollow ways can be difficult to date, but not always impossible, a problem explored in Chapters 8 and 10. In arid areas of the American South-west major erosion gullies created by runoff were, in some cases, initiated by earlier routeways such as waggon trails. Similarly in semi-arid Near Eastern contexts, a landscape survey by Wilkinson (2003) has revealed extensive hollow ways. Some represent local routes radiating from tell sites of the third millennium BC, at Tell al Hawa, Iraq (Fig. 1.8a), where they extend up to 5 km from the tell, connecting it to its fields, a relationship evidenced by sherd scatters. Other hollow ways represent longer distance routes connecting major centres and providing clues to hierarchies of connectivity between settlements, some being connected, whilst minor sites lie away from these routes (Fig. 1.8b–d; Wilkinson 1992).

Figure 1.7 A deeply incised hollow way in North Chideock, Dorset. This was the landscape in which Household (1939) set Rogue Male *(photo: M. Bell).*

On a wider spatial scale, clues to patterns of movement can be obtained from materials which have been transported, whether these are geological materials, plants, shells or animals. A decade ago we would have been more inclined to interpret the distributions of artefacts, such as stone or metal axes, mainly in terms of down the line exchange whereby the object was passed from one community to the next without much long distance movement of people. However, studies of isotope geochemistry have provided a contrasting perspective. In the Neolithic and Bronze Age of Britain, for instance, there is growing evidence that some people and animals came from areas of very different isotope geochemistry to that in which they were buried (Evans *et al.* 2006) implying much greater mobility than previously supposed.

Brown (2014) shows how a combination of isotope and DNA analysis of human skeletal remains can contribute to an understanding of female or male mobility and exogamy (marriage out of one's community). Schulting (2003) showed that in Mesolithic Brittany some women had moved from inland locations, where they grew up, to coastal environments, again reflecting exogamy. Transported materials, or contrasting isotope geochemistry, provide clues to patterns of connectedness but will rarely inform on the precise routes followed. However, in some cases routes can be predicted from topography and may also be attested from later historical sources. The Silk Route between China and south-west Asia is well attested in later historical times but was clearly significant much earlier, as evidenced by the movement of artefacts (Cunliffe 2015; Frankopan 2015) and the introduction of millet from China to south-west Asia and wheat from south-west Asia to China (Jones 2016). Likewise, the astonishing diversity of artefacts from east and west deposited in Scythian graves in Central Asia (Simpson and Pankova 2017).

Since the 1960s archaeologists have tended to interpret changes in material culture in terms of either development or acculturation, i.e. adoption from neighbouring influences, having rejected the waves of population movement favoured by earlier generations. However, the increasing capacity and sophistication of ancient DNA analyses presents a major challenge to recent assumptions. Re-evaluation has been especially marked in relation to the arrival of the Beaker complex in North-west Europe (Olalde 2018) which demonstrates a significant ingression of steppe ancestry from eastern Europe. The change is particularly marked in Britain where in *c* 2400 cal BC the

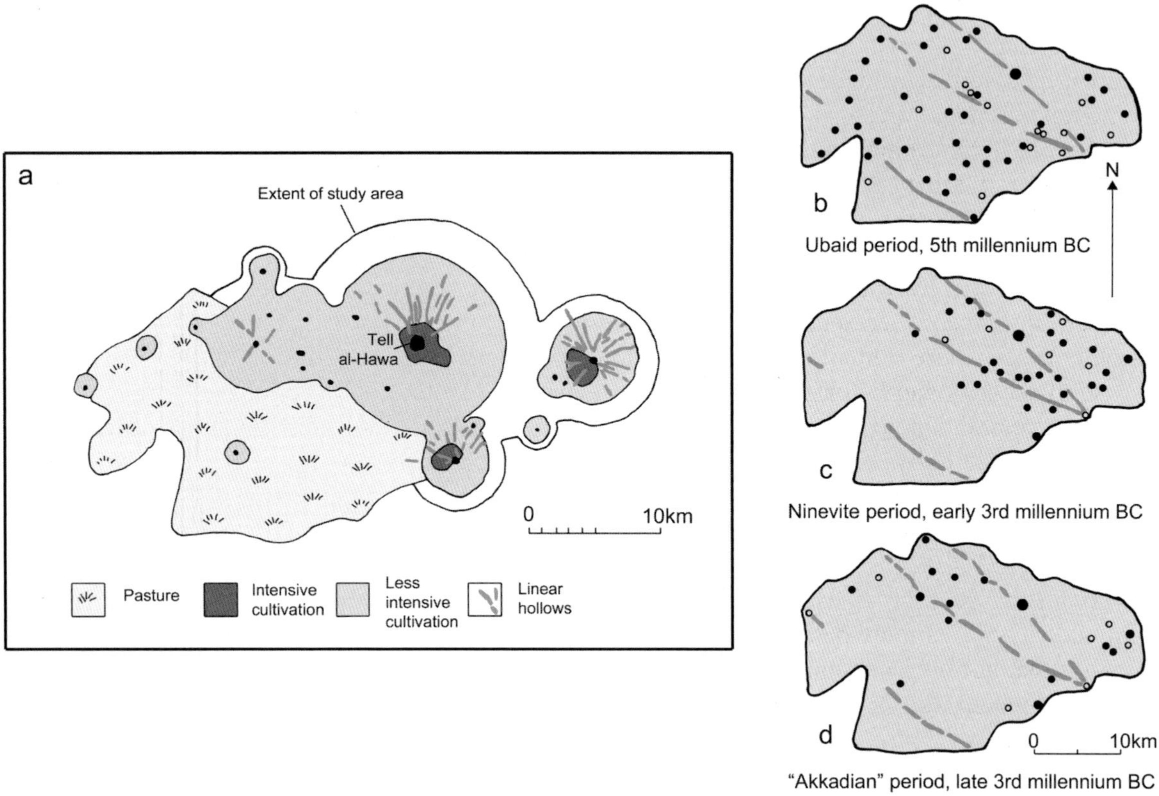

Figure 1.8 Hollow ways on Jazira Plain, Iraq: (a) Radiating hollow ways at Tell al-Hawa which provided local access to cultivated areas, identified from sherd scatters, around tells (after Wilkinson 2003, fig. 6.17), (b-d) Longer distance routes on the plain related to settlements between the fifth and third millennia BC; large circles towns, smaller circles definitely occupied (solid) or probably occupied (open) sites (after Wilkinson 1992, fig. 3) (graphics: J. Foster).

genetic evidence indicates a 90% population replacement within a few centuries. The same study has indicated that the earlier arrival of Neolithic farmers in Britain *c* 4000 cal BC involved significant genetic input from Iberia rather than just the Danubian farmers of central Europe. If further studies confirm these results it will bring about a fundamental rethink of the extent of population mobility and the causes of cultural change.

Agency and niche construction: human and non-human

Pierre Bourdieu (1977) introduced the concept of 'habitus', underlying structures created by practice, as people interact with environments and with other people; the ways they do so influence future interactions. A development of this is Giddens' (1984) structuration theory whereby actions are influenced by pre-existing structures, but with the opportunity for active agents to act reflexively in making independent choices to change the social and environmental structures which they inhabit.

Van de Noort and O'Sullivan (2006, 19) define agency as people's ability to make their own way, to choose and experience different aspects of identity. Thus, people make their own history but under particular circumstances and conditions inherited from previous generations (Simmons 1993, 154). There are two types of influence acting in opposition: one the creation of structures trending towards long term continuity and stability and the other a capacity for radical change. As Kador (2013) has emphasised, the study of mobility must embrace both, the everyday routine actions as well as periods of change and their social consequences. This is essentially the distinction which the geographer David Harvey (1989) has made between routine and transformative action. As subsequent chapters will show both continuity and radical change are evident in the history of trackways. There are instances where people seem to have followed the same route for many millennia, and yet other cases where routes are totally abandoned and very different patterns emerge. People gain a sense of security and belonging from the familiarity of habitual daily actions and the reassuring marks on the landscape which these create. However, there is another trait of human nature, perhaps stronger in some individuals than the majority, to always want to look over the horizon, to beat a new path and to explore the new and unfamiliar. We will look at examples of this in Chapter 9 on maritime communication where at great, often to our minds, perhaps irrational, personal risk, people travelled distantly to obtain exotic goods and esoteric knowledge.

In preceding sections we have seen how human agency alters environments through its effects on plant and animal communities and the creation of routeways. Increasingly researchers are coming to realise that not only people, but also other entities, modify their own environments in ways that influence the activities of other organisms. The term ecosystem engineering has been employed to convey the way in which various organisms contribute to the development of ecosystem structure (Smith 2011). A wide range of organisms can be implicated as 'engineers' e.g. earthworms, ants, termites, beavers and many of these contribute to the ways in which the archaeological record itself is structured (Bell 2015).

The reciprocal and coevolutionary relationships between environment and culture have tended to be obscured by false dichotomies set up between people and nature in identifying the causes of environmental change (McGlade 1995). An integrated solution to this problem is provided by Niche Construction Theory (NCT): 'the process whereby organisms, through their metabolism, their actions and their choices modify their own and/or each others niches' (Olding-Smee *et al.* 2003, 419; Laland and O'Brien 2010). Feedback loops within a niche alter the dynamics of the system creating change. Niche construction envisages that organisms have a triple inheritance, from genetics, culture and ecology. These influence future generations, including evolutionary processes, which can be seen as driven, not just by external factors such as climate, but for instance by the effects of human cultural practices which favour particular genes (Laland and O'Brien 2010). An example would be the transmission of environmental knowledge resulting in a modification of behaviour and selective advantage. Human niche construction involving the manipulation of wild plant and animal resources often mimics natural processes, *e.g.* the opening of woodland as a result of storms, floods etc. In these ways the effects of people and non-human agents reset succession to create more productive ecosystems (Smith 2011). This theory has been especially important in the study of coevolutionary relationships evident in the domestication of plants and animals (Smith 2011; Rowley-Conwy and Layton 2011; Zeder 2012). This is seen, not just as the one-way street of people domesticating organisms, but a mutual process whereby organisms are in some senses domesticating people.

Through recognition of the mutual relationships between people and organisms comes an appreciation that both can be seen as having agency in the sense that they can instigate change. In these ways Niche Construction Theory provides a significant bridge between science and culture which is attracting increasing cross-disciplinary interest. Ingold (2011, 6) expresses a similar point with his usual clarity and insight: 'producers, both human and non-human do not so much transform the world, imposing their preconceived designs upon the material substraight of nature, as play their part from within in the world's transformation of itself'. Ingold and Vergunst (2008) highlight the case of pastoralists accompanying animals: Which is leading which? And where does the agency reside? Or is it in some way mutual? The close associations between people and animals, whether it be herders accompanying reindeer (Lorimer 2006), or naturalists engaged in the close and intimate observation of animal behaviour (ethology; Lorimer 2010), cause people and animals to develop entwined lives, each experiencing the agency of the other. In considering pastoralism, Chadwick (2007, 134–5) has similarly argued that 'agency is the outcome of relationships between humans and the living and material world'. Herding, he suggests, should not be seen in purely functional terms because of the role it plays in the construction of identity and memory (Chadwick 2016). Nor can this active role be restricted to people and animals. Jones and Cloke (2002) critique the tendency to separate people and nature and argue for the agency of non-human life forms such as trees, thus making a case for what has been called hybrid geographies which are inclusive of non-human agents. A similar argument for the active role of woodland in Neolithic landscapes has been made by Noble (2017). Comparable cases have been made for the agency of inanimate material culture by Robb (2004) and by Van de Noort (2011) in terms of the influence exerted by the creation of powerful and iconic items such as boats.

Cognition: thinking through things

Recent developments in cognitive studies are especially pertinent to the routeway theme of this book because it is increasingly evident that thinking is not just an internal cerebral process but a process which is acted out through the engagement of our bodies with the material world. In short we think through things; our interactions with things and with the world provide clues and prompts influencing how we remember, think or act in a given context. These are the concepts of the extended mind and material engagement theory which so far have been mainly worked through in terms of the active role of material culture in the ways in which people act, perceive and think (DeMarrais *et al.* 2004; Malafouris and Renfrew 2010). In the case of some routine actions this influence may be almost unconscious. We may speculate that habitual actions contribute to a sense of security and well being, that in a sense an individual is at one with the world through patterns of familiar and repeated engagement. The hippocampus is the part of the brain which is particularly concerned with long term memory and spatial navigation whereby cognitive maps of the life world are created. This part of the brain has been shown to be especially enlarged in highly experienced London taxi drivers who are trained in 'The Knowledge', a detailed understanding of the street map of the city (Maguire *et al.* 2000). Thus habitual actions develop neural pathways, whereby parts of the brain become enlarged (Gosden 2010). Clearly these concepts are relevant, not just to the clues and prompts provided by material culture, but potentially to a greater extent at a landscape scale, where clues are provided by landscape features and the ways they are modified by human agency, the paths, monuments, and altered vegetation etc which influence the ways subsequent actors encounter that landscape. A pioneering application of these ideas has given new significance to barrow alignments in Himmerland, Denmark and suggested how movement through, and perception of, that landscape could have been prompted by the disposition of burial monuments (Lovschal 2013).

How literature and art help us to think about movement

Archaeologists cannot expect to fully understand past routeways by looking at site plans, maps and air photographs. A fuller appreciation comes from the three-dimensional experience of walking routes on the ground (Stilgoe 2015). Our studies can be further nourished and given fresh stimulus by writers and artists who have thought deeply about, and recorded, their experiences of landscape from different points of view. Literature on movement, walking and observation has a long pedigree, much of it of great value to the landscape historian; two Welsh examples are Giraldus Cambrensis' (1191) *Itinerary Through Wales* and much later George Borrow's (1862) *Wild Wales.* Examples where a journey provides the vehicle for social and religious commentary include Geoffrey Chaucer's (*c* 1400) *The Canterbury Tales,* recounting a pilgrimage from London to Canterbury, and John Bunyan's (1678) *The Pilgrims Progress,* which was partly anchored in the Bedfordshire landscape Bunyan knew. The ancient hollow ways of the Weald around Selborne fascinated the pioneer of Natural History writing, Gilbert White (1788; Mabey 2006), and later the rural reminiscences of Flora Thompson (Mabey 2014). The inspiration of both writers clearly derives from walking, meticulous observation and a deep empathy with landscape. Edward Thomas (1909) was similarly inspired by walking the paths of southern Britain and wrote on the Icknield Way (Thomas 1916), just as his fellow poet Hilaire Belloc (1910) wrote on the Pilgrims Way in Hampshire, Surrey and Kent. The supposed ancient origins of these long distance ridgeway routes has influenced subsequent generations but will be critically examined in Chapter 8.

What we might call the pedestrian literature of the early twentieth century reflects a world of increasing leisure, education and access to the countryside by railway and road as well as the wider availability of tourist maps. It was an interest fuelled by nostalgia for a rural pre-urban and pre-industrial idyll, exemplified by the writing of H.J. Massingham (Mabey 2014). The significance of a sense of place and the history of place in environmental writing and ecological understanding is particularly demonstrated in a North American context by the writings of Cronon (1983) and Worster (1990). Lang (1999) argues from an American North-west coastal perspective that an ethical environmental history has to be informed by a developed sense of place.

Recently, the somewhat nostalgic walking-based literature of the twentieth century has been eclipsed by a more experiential and biographical literature, in which movement, encounters with landscape, individual memoir and a deep time perspective are frequent elements. The appeal of this literature must be, in part, because it maps onto what people actually do: thinking on their feet and through movement. This current literature defies disciplinary classification. It is not primarily archaeological, yet it so often exhibits the fascination which archaeologists have for landscape, frequently drawing on history and archaeology for evidence and inspiration. It encourages archaeologists to think in new ways about mobility and is introducing ideas and concepts which are helpful in doing so. An influential early example is Schama's (1995) *Landscape and Memory,* a wide ranging discourse on art, history, literature and nature, how they are perceived, altered and 'recreated' by successive generations. The symbolic associations of landscape are seen to be imbued with a wealth of myth and recollection, a bulging backpack of memories which we carry around. Many begin as highly personal stories, indeed Schama starts the book with the landscapes of his Essex childhood, then progresses backwards to those of his ancestors in eastern Europe. Writing from a more modernist and political North American perspective, Rebecca Solnit's (2014) *Wanderlust* promotes walking as a way of thinking and remembering, her argument punctuated by accounts of her own very varied walks through landscapes and life. Her *Book of Migrations* (Solnit 2011) concerns reflective self-discovery through travel, partly on foot, and a historical meditation on the Irish transatlantic diaspora. Focusing down on a local scale, Julia Blackburn's (2012) *Thin Paths* recounts walks around an Italian mountain village; through this she tells the story of its people and the last vestiges of seasonal transhumance, involving flocks moving for the summer to high mountain pastures, and how that way of life engendered the anti-fascism of an independent-minded mountain people. Her latest work recounts a personal quest to understand the vanished landscape of Doggerland, art and poetry inspired by the quest and its relationship to points and relationships in the writer's own life (Blackburn 2019). Juxtaposing the local and global is Gooley's (2012) *The Natural Explorer,* which begins each chapter with a segment of one day's walk near his home; this leads to wide-ranging reflections on diverse landscape themes around the notion that a traveller becomes a natural explorer by turning observation into art. Gooley's (2010) *Natural Navigator* concerns the ways in which people use evidence within the landscape to situate themselves and find their way around. He proposes that environmental clues create a sort of 'sixth sense' whereby we partly unconsciously navigate our way in the world (Gooley 2018). Stroud (2017) *The Wild Other* resonates with a significant theme of this book, the transformative effects of the horse on human connectivity, through an account of the role of horses in a young woman's self-discovery set against the backdrop of the Oxfordshire Ridgeway and its Uffington White Horse (Chapter 8).

Recent writing is particularly concerned with the relationship between nature and culture and has often been stimulated by a desire to connect with nature and discover a greener, more ecologically balanced way of life. A recent example is George Monbiot's (2013) *Feral,* which is a passionate case for rewilding with native trees and the reintroduction of regionally extinct animals, which challenges many of the tenets of agricultural and nature conservation policy. Such ideas gain increasing traction from the ecological success of current species reintroductions, especially of the beaver, and also pioneering rewilding projects (Tree 2018). Ecological emphasis is also strongly reflected in *Wildwood; a journey through trees* by Roger Deakin (2007), which is not so much a book about natural trees as trees and woods altered and perceived by people, just as other aspects of his empathy for landscape and its history is exhibited in his two other books (Deakin 2000; 2008). Deakin was very much an activist whose main writing was at the very end of a varied life. He was one of the founders of the organisation Common Ground, which aimed to encourage communities to empathise with their landscape context.

Robert Macfarlane is from a more traditionally academic background, a Fellow in English at Emmanuel College, Cambridge, who has published what he describes as a loose trilogy about people and landscapes: on mountains (Macfarlane 2003); wild places (Macfarlane 2007); and, especially pertinent to the subject in hand, *The Old Ways* (Macfarlane 2012). The latter he describes as about 'walking as a way of knowing'; it concerns 'the subtle ways in which we are shaped by the landscape through which we move' (Macfarlane 2012, xi). As an English scholar his perambulations are in part about understanding and situating other writers, such as the poet and essayist Edward Thomas, whose book on the Icknield Way was noted above (Thomas 1916). A recent short work on a title very pertinent to our theme is *Holloway* (Macfarlane *et al.* 2013), an account stimulated by Geoffrey Household's (1939) novel *Rogue Male,* an adventure story of one man's battle against totalitarianism in which the ultimate hiding place was an overgrown and forgotten Holloway, a device which highlights the lost and mysterious nature of these enigmatic features (Fig. 1.7). In *Landmarks,* Macfarlane (2015) considers the vernacular names given to landscape features and how those specific local names contribute to our sense of place. He lists 38 regional vernacular British names for tracks and paths (Macfarlane 2015, 202). Many of these are slipping out of use with the loss of the knowledge which those names convey concerning local perception and ways of life.

Macfarlane's writing, particularly on mountains and wild places, and most recently his journeys underground and in the Arctic (Macfarlane 2019), are characterised by remarkable feats of endurance and hardship through which he sought to escape from the paraphernalia and comforts of modern life to connect with nature and roots. Perhaps the most extreme of this genre is Nicholas Crane's (1997) gruelling walk along the mountain chains of Europe between Cape Finisterre in western Spain and Istanbul, Turkey. During this 17 month continuous walk he encountered surviving remnants of the way of life and ethnographic artefacts of seasonal herders who used the mountain pastures and paths. It is a book nostalgic for the old Europe and its folk life, and of historical value for insights to the rapid changes and economic challenges faced by many East European remote rural communities following the collapse of communism. Later in his career Crane (2010) went on to present the BBC television series *Coasts*, recounting journeys round the coast of the British Isles which brought many aspects of coastal ways of life and archaeology, including the footprints mentioned at the start of this chapter and Chapter 4, to a wide audience. Also on a coastal theme but local in its spatial ambition, is Jean Sprackland's (2012) *Strands,* which records the thoughts, observations and discoveries made by walking during 1 year along the same stretch of beach from Southport to Formby in north-west England, a distance of 15 km. That shore is especially significant in the present context because, as Sprackland observes, it is where some of the best examples of prehistoric human footprints have been found in Britain (Chapter 4). The highly individual and local accounts of writers such as Sprackland and Blackburn are more aligned to the experiences of the average walker for leisure. The bold adventurers in the writings of Macfarlane and Crane are more akin to Homer's *Odyssey* or medieval pilgrimage whereby endurance and hardship contribute to status building, demonstration of faith, or writers' credibility. Thus landscape writing contributes in myriad ways to our capacity to think about mobility past and present.

Art, just as powerfully as literature, provides a way of thinking about the traces we leave in the landscape including the vestiges, both physical and mental, of patterns of movement. Paths are a prominent aspect of the paintings of both Paul Nash (1889–1946) and Eric Ravilious (1903–1942), whose work reflects a preoccupation with the paths of chalkland landscapes (Russell 2009), where routes stand out as white markings against the downland turf, as shown in his *Chalk Paths* painted in 1935 (Fig. 1.9). The hollow ways, terrace ways and ridgeways of the downlands will be a significant theme in Chapters 8 and 10. Land artists such as Andy Goldsworthy and Richard Long also work in the landscape, creating instillations of rocks, patterns in sand and objects which mark the land

(Malpas 2007), and a comparison is often drawn between their work and the monuments and symbols of prehistory (Renfrew 2003). A work particularly pertinent to the theme here is Richard Long's photograph *A Line Made by Walking* taken in 1967, a path in grassland evidenced only by the effects of light catching moderate trample on the grass (Malpas 2007). The connectedness between these varied art forms is exemplified by Nicholas Crane's (1997, 157) rather startling revelation that his epic mountain journey was in part inspired by the linear land art of Long. Especially redolent of the coastal footprint theme developed in Chapter 4 is some of Anthony Gormley's work, especially *Another Place* created 1997, which comprises 100 actual-size cast iron images of the artist's body in the intertidal zone at Crosby, all gazing out to sea and covered and uncovered by the ebb and flow of tides, just as the prehistoric footprints on the same foreshore at Formby 9 km to the north are uncovered and covered (Roberts *et al.* 1996). As far as the writer is aware there is no link between the location selected by Gormley and the footprints, though both excite similar fascination. A work by Gormley more directly revealing of his interest in the past is a life-sized statue buried upside down with only the soles of the feet exposed in the pavement outside the Archaeology Department, Cambridge University where Gormley was a student. A photograph of those feet was used by Jim Leary to illustrate the introductory chapter of his *Past Mobilities* (Leary 2014, fig 1.2).

Figure 1.9. Chalk Paths *by Eric Ravilious 1935.*

So what more specifically do literature and art contribute to studies of past mobility? First, in a multivocal sense there are the different perspectives provided by those whose primary focus is much broader than archaeology. Secondly the imperative of applying the particular skills of the archaeologist, or landscape historian, in ways which are relevant to a wider audience and contribute to broader debates concerning people, nature, sustainability, quality of life, connectivity and migration. The biographical nature of much current landscape writing demonstrates how reflections on, and encounters with, heritage and nature contribute to self-discovery and well-being.

Timescale, dating and spatial scale

Fortunately today, we have many advantages over the pioneering archaeologists, including the availability of scientific dating techniques which can be applied to trackways, footprints etc. A guide to each of the main dating techniques is provided by Walker (2005). Most widely applicable of these is radiocarbon dating. The dates given in this book are all calibrated to dates BC/AD. The original uncalibrated radiocarbon dates and laboratory numbers are not given when they are cited in the original published source referenced; to include them here would make the text cumbersome. Exceptions to dates BC are those of the pre-Holocene (i.e. before 9500 cal BC) where BP (Before Present 1950) is used, as are abbreviations ka, thousands of years, and ma, millions of years. Some sections, notably on wooden trackways, make use of dendrochronological (tree ring) dates which are given in the BC form. Also increasingly relevant to the dating of sediments associated with trackways is Optically Stimulated Luminescence dating (Duller 2008). Some sediments associated with footprints and trackways are also dated by the Uranium Series method. Dates derived from these various methods are frequently quoted in the text as a way of helping the reader to relate sites to a sequence and one site to another. This is necessary because the large number of site examples, from many areas and periods, means that it is generally not possible to introduce the local archaeological cultural sequences.

A particular concern in this book is the relationship between movement in space/time continua and issues of perception as related to movement. We therefore need to keep in mind that scientific dating is only one way of measuring time. There is a distinction between linear chronometric time, as measured by scientific dating techniques, and time as perceived by people, which may for instance be circular or cyclical (Dietler and Herbich 1993; Gosden 1994) and marked not by dates or calendars but by the traces that people leave on the landscape. As regards spatial scale any form of movement is clearly relevant, ranging from the information we can gain from the individual footprint, through the activity space associated with specific tasks, to the scale of movement associated with the daily life of a settlement, or the more distant seasonal rounds and transhumant patterns of its occupants. There is also the more distant communication between peoples and places by walking, using animal traction, or travel on water. All are relevant to our understanding of the past, although in the space available only a selection of examples can be given.

Terminology

Mobility refers to the study of movement by people (and animals). *Route* and *routeway* are used here for any frequented axis of movement. *Paths* refer to movement on foot, and *Trails* movement on foot and by horse. *Lane* is a local route from settlement to its fields. *Trackways* generally imply constructed routes in wetland. *Footprint-tracks* are the traces left by the feet of people and animals. *Roads* are prepared surfaces generally allowing year-round passage by wheeled vehicles. *Hollow ways* are routes incised by erosion. Use of the word *droveway* requires special clarification. Here it is used for routes along which domestic animals were moved on any spatial scale: some went locally from farm to field; some were long-distance seasonal movement of tens of kilometres; and some went much further. A potential distinction needs to be recognised between these practices and medieval and post-medieval practices of droving involving the movement of animals from remote rural areas to urban and port markets. Animals were, for instance, moved from Wales to London along routes which acquired names such as 'Welsh Way'. Although growing evidence from isotope studies for the long-distance movement of animals in prehistory is noted in subsequent chapters, we should make a distinction between this and the market-oriented droving of recent centuries. Prehistoric long-distance animal movement may have had very different motivation, including ritual, symbolism, cosmology and tribute. *Transhumance* is used to refer to the seasonal movement of animals and people generally over long distances, tens to hundreds of kilometres, usually from lowland to mountain uplands, a practice which was particularly prevalent in the Mediterranean. *Lesser transhumance* implies smaller scale seasonal movement of a few kilometers to *c* 30 km between contrasting ecological zones, such as between lowland and upland mountains and moors. It also includes movement from lowlands and hills to seasonally available coastal saltmarsh pastures which in prehistory, before seawalls, reclamation and erosion, were far more extensive and more socially significant than they are today. Stilgoe's (2015) approach to landscape puts particular emphasis on the names and origins of landscape features and what that can tell us about the perception of landscape. This work and Macfarlane's (2015) *Landmarks* highlights the importance of preserving and recording local vernacular terminology as part of local identity and a source for understanding its origins.

Conclusions

It has been argued that, by comparison with many other aspects of archaeology, trackways and mobility have often been neglected. New strands of evidence from many disciplinary sources, augmented by perspectives from the arts and literature help to chart a way forward. It is timely, therefore, to review the evidence and consider the challenge represented by translating field traces into an understanding of prehistoric people's encounters with landscape. The conscious act of translation which requires careful and critical analysis of the field evidence, the testing of ideas and the integration of evidence from diverse disciplines to establish whether, what may be an interesting and stimulating theory, is borne out by the evidence on the ground.

The disciplines, or strands of knowledge, which can contribute to the study of how people moved around the landscape include: ethnohistory; tracking skills; ichnology (traces of animal behaviour, e.g. footprints); field archaeology; botanical and other biological traces; historical analogues; geoarchaeological and geochemical traces, etc. The solution to the problems of identifying prehistoric patterns of movement advocated here is the adoption of a multidisciplinary approach, or what the environmental archaeologist refers to as a multi-proxy approach. Above all, the challenge is to integrate perspectives which are both social and scientific, using one source to test and develop ideas derived from the other. In anthropology Wylie (2002) has described such interpretations as being based on cables, comprising multiple strands of evidence, and as tacking, a dialectical process between the contrasting perspectives of the anthropologist and the subject (Wylie 2002, chapter 11). This is analogous to the process which Wylie (2002) has called 'triangulation', whereby an interpretation, which may not be entirely convincingly demonstrated from any one source, becomes more precisely grounded by triangulation with other sources of varying independence. The writer has recently reviewed these concepts in terms of their application in experimental archaeology (Bell 2014). The academic resilience of such a multi-stranded approach is significantly enhanced by two key concepts introduced by Hodder (1999). The first of these is multivocality, which values the contributions of a diversity of voices and approaches, both scientific and social. The second is self reflexivity, a critical awareness of the effects of scientific and archaeological assumptions on the data obtained and the effects of these on the communities involved. Together these two concepts challenge tendencies to scientific dogmatism and the arrogance of the 'expert view' which has made some people resistant to a scientific approach. Later, in Chapter 4, when reviewing Robbins' work on the forensic use of human footprint evidence, we will see rather dramatically the dangers of uncritical acceptance of the expert view.

Many of the points made in foregoing sections, concerning archaeologists' previous neglect of trackways, are made with specific reference to dryland situations and certainly do not apply to wetlands where there is a long history of successful studies, reviewed in Chapter 6. In wetlands trackways are readily identified and dated and there is a wealth of associated environmental evidence. In these wetland contexts the challenges are rather different, they particularly concern the relationship between wetland tracks and the wider picture of dry ground activity so that the wetland evidence can become more fully integrated into archaeological discourse (Van de Noort and O'Sullivan 2006). To a significant extent this is about bridging the gap between wetland archaeology, which is rich in evidence, but where social theoretical concepts have not been so much employed, and dryland prehistoric research, where the theoretical concepts have often been to the fore, and the available field evidence has sometimes been less rigorously engaged with. Chapter 6 shows that there are areas where both dry ground and wetland evidence is beginning to be integrated and the development of this field potentially offers new ways of identifying and dating past trackways.

Figure 1.10 A beaver path at Nature Park, Lelystad, The Netherlands (photo: M. Bell).

A substantial part of the way forward is about patterns of connectivity between people, places and ideas. The proposition has been put forward that not just people but plants, animals, material culture and landscape features such as tracks influence behaviour and have agency. Each component, both people and the many facets of the world in which they live, contribute to Niche Construction. Interconnected relationships are increasingly foregrounded by recent writers who use related, but somewhat different, labels, what Ingold (2011) called meshwork, and Hodder (2012) refers to as entanglement. Establishing networks involves integration of evidence from multiple sources, linking wetland and dry ground perspectives, using one strand of evidence to test ideas derived from another, and in particular connecting the scientific and social evidence. In a landscape context it is often not possible, nor academically at all desirable, to separate the natural and the cultural; they are, in Ingold's (2011) terms, an interwoven meshwork. The path a hunter-gatherer follows will often be influenced as much by the activities of animals as any human decision-making. People may follow paths originated by animals and vice-versa (Fig. 1.10). Macfarlane (2019, 273) visited a long-abandoned settlement in remote north Norway at Lofoten where thin paths linking the doorways of ruined houses survived because they are still used by otters and other animals. Cleared areas which attract grazing by animals, foraging and settlement by people, come about through the combined effects of animals, people and other environmental processes such as storms and floods. The hollow ways which form in oft-frequented routes on certain geologies channel storm water and become geomorphical features. Routes which people take through a landscape are the result of interacting factors including issues of human environmental relationships, such as the food quest, patterns of contemporary social interaction and recursiveness to the past. Routes are often perpetuated over extended timescales creating a legacy effect for subsequent generations (Wohl *et al.* 2014). Bradley (1993) has shown how sites of one period are frequently referenced by much later monuments, their status being legitimated by association. Thus, routes are created and maintained by an interaction of natural and cultural factors and may be said to have an agency of their own in influencing the subsequent structure of the cultural landscape and human activity.

Chapter organisation

Succeeding chapters look first at the ways in which non-agricultural communities use and mark paths and landscapes in ways which influence movement. Chapter 2 focuses on the North-west Coast of North America where archaeological evidence is augmented by a rich ethnohistorical record enabling the development of two key themes: niche construction and a more mobility-based or linear environmental archaeology. Chapter 3 considers the more ephemeral traces of hunter-gatherer movement in North-west Europe. Then in Chapter 4 there is a distinct scalar shift down to the footprints of humans and animals, a relatively new topic for archaeologists and one which is considered drawing on evidence worldwide and from all periods. Traces of movement by early farmers are reviewed in Chapter 5 drawing on evidence from Britain and North-west Europe and mainly focusing on the spatial relationships between monuments. Chapter 6 looks at the abundant evidence of wooden trackways in North-west Europe and attempts to draw some lessons from these about wider patterns of movement. Bronze Age alignments of monuments in North-west Europe are considered as evidence of routeways in Chapter 7. Chapter 8 looks at the very abundant evidence for trackways in agricultural landscapes; so abundant is the latter that it is mainly focused on the British Isles. In Chapter 9 we then move from land to water to consider communication by river and sea which, it is argued, was a more significant aspect of long distance communication than often appreciated, its routes logically complementing and integrating with those on land. Discussion of communication by sea is comparatively brief because it has been well served by many recent publications. Chapter 10 presents a case study from the Weald District of south-east England utilising many of the lessons derived in preceding chapters and relating these to patterns of prehistoric and later historic landscapes. The concluding Chapter 11 pulls together evidence from diverse contexts and periods to identify general concepts and criteria which will help to advance the study of early routeways. In most of the chapters after this, certain examples are designated as Case Studies which indicates that they are especially central examples of the key themes of this book, and they are outlined in a little more detail. The student who recently asked me which page of my last monograph that I was recommending that he should read will get some idea of the theme of this book by reading Chapters 1 and 11 and the designated case studies between, and my hope is this may whet the appetite to read the rest!

2. Walks in the temperate rainforest: developing concepts of niche construction and linear environmental manipulation

Introduction: why the American North-west Coast?

This chapter in part derives from thoughts stimulated by walks in the remnants of the temperate rainforest and in other areas of remote woodland on Vancouver Island, Canada and the Olympic Peninsula, Washington State, USA. Observations provide a framework within which to review evidence from a range of sources concerned with the archaeology, ethnography, botany and environmental history of the area. Though remote from North-west Europe, which is the main geographical focus of this book, it is an ideal area in which to explore some of the themes identified in Chapter 1, notably the ways in which trails and places are created through the interaction of human agency and the activities of animals, the role of non-agricultural communities in ecosystem change and how these factors relate to the paths along which First Nations people walked (Fig. 2.1). The presence of a strong ethnohistoric record means that there is direct oral and written evidence of intentionality and the social significance of routeways, which in prehistoric European contexts we can only infer from very fragmentary clues. The North-west Coast of North America offers these advantages because of the relatively late eighteenth century arrival of Europeans. Furthermore, some explorers, naturalists and early colonists had good scientific educations, so that they could compile detailed accounts of the way of life of First Nations people. Today many of the traditions of plant use and manipulation still survive and are well documented (Turner 2014). It is also an extremely culturally rich area which has been the subject of extensive archaeological and ethnographic work and there are areas of surviving and historically documented semi-natural vegetation which themselves provide clues to the activities, patterns of movement and ecosystem management of past communities.

This chapter illustrates the role of people in modifying vegetation communities and how this might lead to a more linear and mobility focused approach to environmental archaeology. This is certainly not to imply that evidence from the American North-west Coast can be directly extrapolated to pre-agricultural Europe. Clearly the context and biological resources are very different and there are evident philosophical and practical dangers in using the ethnographic recent to populate the prehistoric past (Finlayson and Warren 2010). The case is not for direct analogy but that the richness of the north American evidence provides the thinking space within which the writer, and hopefully readers, can come to appreciate the diversity of ways in which routes can be formed and marked. Many of these have been little considered in a European context.

Figure 2.1 A man of Nootka Sound, Vancouver Island; drawing by John Webber on the Cook expedition of 1778 (Gift of the Estate of Belle J. Bushnell, 1941; courtesy Peabody Museum of Archaeology and Ethnology, Harvard University, PM 41-72-10/496).

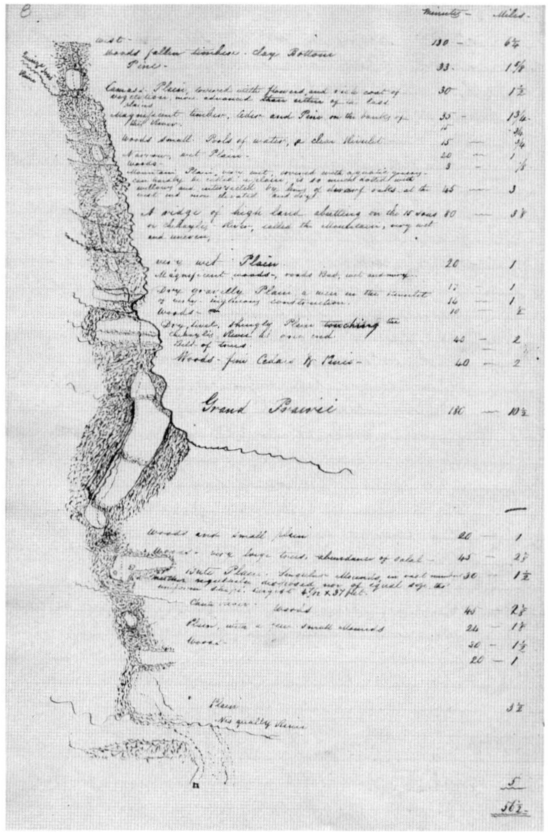

Figure 2.2 James Douglas (Hudson Bay Company) map of 1840 showing the route from the Cowlitz Plain to the Nisqually River, as a chain of small prairies within forest (courtesy: Information Management Services BA Archives, Victoria; Call no AB40 D75. 2a).

The Douglas Map

The potential is demonstrated by a trail documented in a diary of 1840 by James Douglas, then of the Hudson's Bay Company, who went on to found Victoria and was later the first Governor of British Columbia. He recorded a route north of the Willamette Valley between the Nesquilly River and the Cowlitz Plain (Leopold and Boyd 1999). His map (Fig. 2.2) is especially revealing in depicting a chain of small prairie areas in a wooded landscape, resembling the beads on a necklace. It is recorded that the prairie areas were burnt by First Nations groups to encourage the growth of berries and other plants. The burning took place as they moved to their winter camps in anticipation of rejuvenated plant growth when they returned the following year. These burnt prairie patches would also have attracted deer.

The area and its archaeology

The coastal belt of the north-western part of the United States west of the Cascade mountain range and western Canada west of the Coast Mountains, referred to here as the North-west Coast (Fig. 2.3), is dominated by maritime influences with annual rainfall up to 2.5 m in parts of the Olympic Peninsula and the west coast of Vancouver Island. The Canadian coast is indented with numerous estuaries extending far inland and there are many islands, the largest being Vancouver Island and the group of islands known as the Haida Gwaii (formerly Queen Charlotte Islands). The maritime climate gave rise to a coastal temperate rainforest of which small areas have survived the effects of rapacious logging in the nineteenth and especially twentieth centuries. These remnants of Old Growth Forest are mainly characterised by western hemlock, Sitka spruce, Douglas fir and western red cedar. Some red cedars and Douglas firs may be up to 1000 years old (Nuszdorfer 2000). Trees modified by people, and known as Culturally Managed Trees (CMT), bear tangible witness to past ways of life.

The earliest reliably attested settlement of this area, and indeed in North America, is around 12–15 ka BP (Fagan 1991). Over the last few thousand years before present there had developed a remarkably complex and sophisticated fisher-gatherer-hunter society not employing agriculture, in which the only domesticate was the dog. For this reason it is an especially instructive, and well documented, case of the environmental relationships of communities where the economy was not based on domesticates but nonetheless had extremely sophisticated and complex relationships with plant resources involving landscape niche construction on a scale which is only now becoming clear. The richness of coastal resources meant that this area achieved the highest population densities of any area in pre-contact North America. The people of the North-west Coast had substantial winter villages; some virtually sedentary communities were established as early as 5000 cal BP. The houses were made of giant western red cedar posts and boards, which in the northern part of this area were fronted by totem poles often ranged along the shore, both houses and poles being richly decorated with painted carvings (Fig. 2.4; Matson and Coupland 1995; Ames and Maschner 1999). The cedar houses and organic material culture of the area is most effectively demonstrated by the site of Ozette in Washington State, buried by a mudslide *c* AD 1750 (Kirk 2015). Here 90% of the material culture was wood and fibre. Anthropologists recorded that this was a winter village with people moving elsewhere to summer villages; however, faunal evidence indicates a more complex picture showing that part of the Ozette population remained year round (Samuels 1994). The evidence seems to point to sedentism on some coastal sites rich in resources, and greater mobility among inland groups (Deur 1999). Some groups moved five or six times in a year, dispersing in summer to smaller camps from which particular resources of fish, shellfish, game, root vegetables, berries and greens could be exploited (Ames and Maschner 1999). The patchiness of resources, spatially and seasonally, created a need for seasonal movement and the preservation and storage of food, such as salmon, halibut, shellfish, berries and plant foods.

Whilst in dietary terms plant resources contributed less than fishing and hunting, their use is especially significant in the present context because there is such a comprehensive record thanks to the activities of anthropologists such as Franz Boas (1921), and because plant resources have continued to be utilised up to recent times and have been carefully documented by ethnobotanists.

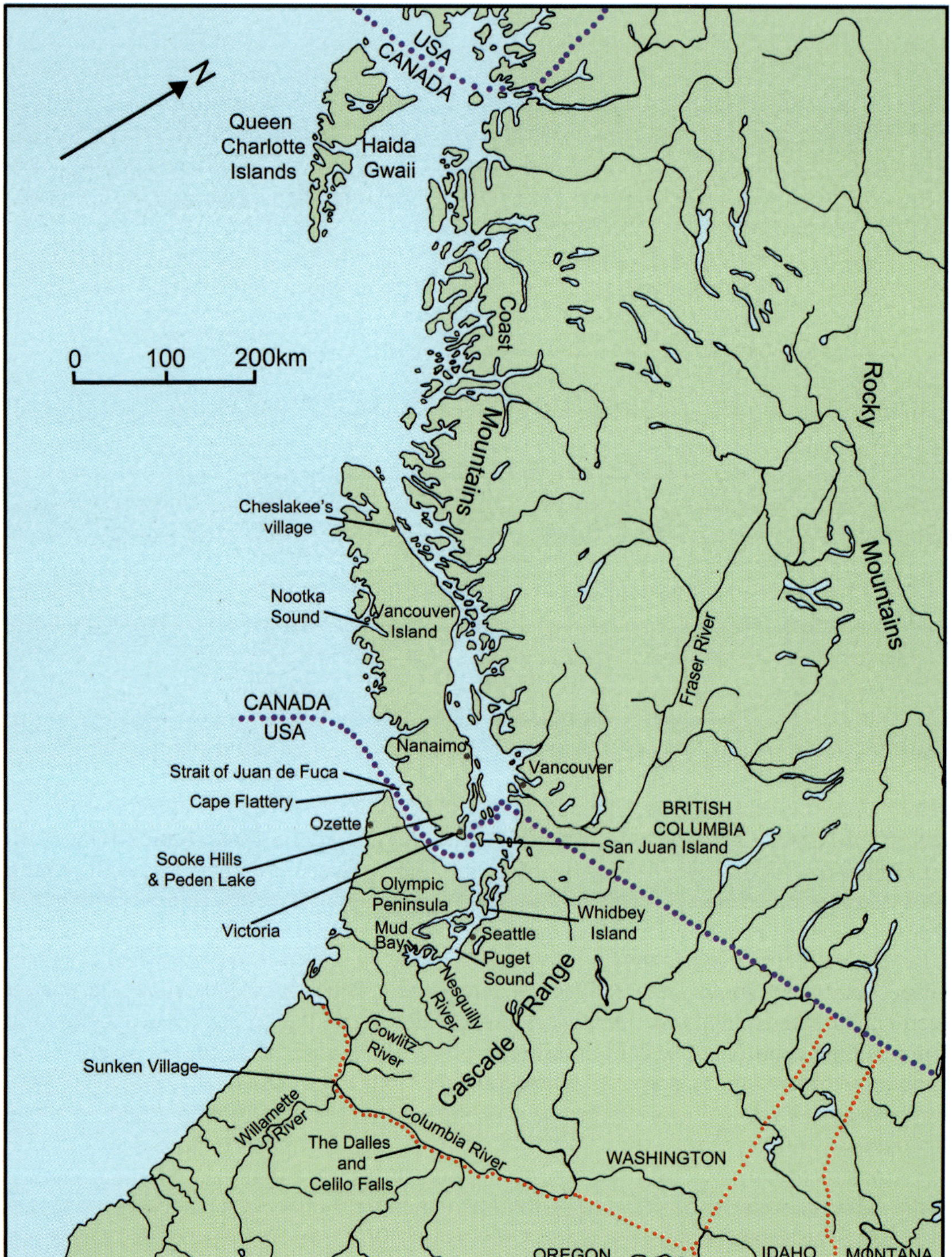

Figure 2.3 The North-west coast of North America: Oregon, Columbia and Washington, USA and British Columbia, Canada showing key sites mentioned in the text (graphic: J. Foster).

Figure 2.4 A Haida settlement reconstruction at the Museum of Anthropology, University of British Columbia, Vancouver, Canada, showing winter cedar plank houses, totem poles and mortuary post (photo: M. Bell).

A recent comprehensive survey by Nancy Turner (2014) provides a wealth of information about the traces in the landscape left by plant utilisation. First Nations communities were also characterised by a respect and empathy for nature. This indigenous understanding of biological interrelatedness has been called ethnoecology, or traditional ecological knowledge. This provides insights to the environmental perception of the communities involved and has informed recent debates about the sustainable use of resources, especially when compared to the ways in which Europeans have used the same areas in the last 200 years (Turner 2005; 2014; Goble and Hirt 1999).

This area has such a wealth of anthropological information due to the lateness of encounters with explorers from the Old World. It appears to have been visited by Juan de Fuca in 1592, after whom the straits between Canada and the USA were named, but only a limited second-hand account of that voyage exists. Much later Russian expeditions led by Vitus Bering in 1728 and 1741 began the fur trade in North-west America. The Spanish began to colonise California in 1769 and sailed north to the Straits of San Juan de Fuca in 1778. There followed the expeditions of Captain James Cook in 1778 (Cook 1784) and Captain George Vancouver in 1792–4 (Vancouver 1801). European contact resulted in very rapid cultural change, which initially magnified certain characteristics of the indigenous communities. The ready availability of iron tools led to a fluorescence of wood carving and the production of totem poles which seem to have been less frequent previously, their production now being partly fuelled by the wealth generated by the fur trade (Kramer 1999). This also magnified the scale of potlatch ceremonies during which goods were competitively distributed by chiefs, often accompanied by the erection of totem poles. The fur trade initially conducted by Russia, then Britain and fuelled by demand from China, illustrates the severe consequences of early globalisation for First Nations communities. Such exploitation was unsustainable and rapidly brought one of the main hunted animals, the sea otter, to the verge of extinction (Riedman 1997) with the result that trade with some areas, such as Haida Gwaii, virtually ceased. The damage had been done, however; European infections, particularly smallpox in the north, led to major epidemics and a catastrophic reduction in the native American population (Ames and Maschner 1999). Malaria had an equally devastating impact in Oregon. Many settlements were abandoned, and survivors moved to a smaller number of large settlements.

These dramatic cultural and settlement pattern changes inevitably make it difficult to separate pre-contact from post-contact practices, much of the evidence being drawn from anthropological and ethnobotanical accounts of the twentieth century. Particularly difficult to disentangle, given the settlement pattern changes of the last three centuries, are the patterns of seasonal movement before and after contact. However, Brody's (2002) anthropological investigations, related to land claim issues, have shown that traditional economies, and a seasonal hunting and gathering round, remain important to this day in some areas, notwithstanding the effects of European contact.

Although maritime and plant resources in certain areas, and at certain times of year, were very abundant, the environment was, and is, patchy both in time and space, varying with successional stages of forests, topography, moisture and habitat type. These communities adapted to this by developing sophisticated techniques for preserving, e.g. drying fish and plant foods and storing them in bentwood cedar boxes and in cedar packbaskets and storage baskets. Transport of edible roots, berries and other food stores, from harvesting sites to winter camps was a significant undertaking (Hunn 1999). Resource patchiness also gave rise to seasonal movements and to exchange with groups whose environments were differently favoured. Both Captains Cook (Cook 1784) and Vancouver (1801, 110) incorrectly inferred the wandering nature of these groups from the many seemingly deserted settlements and habitations observed on their voyages. Lewis and Clarke, in their expedition from the east down the Columbia River in 1804–6, made many observations concerning the seasonal movement of village groups to exploit particular resources during which temporary huts of matting are recorded as used by some communities (McMaster *et al.* 1814). These expeditions noted that communities had already been decimated by small pox.

Helm (1972), writing of the Dogrib, a subarctic boreal forest group further north, records that each group was defined in terms of the main routes or axes of summer–winter movement. In the maritime western parts of the North-west Coast, movement was often by canoe, whether locally to collect food for a day (Turner 1995), or as part of longer seasonal movements. Water transport would have been essential to move the cedar planks for housing, placed across canoes in catamaran fashion and stacked with supplies, children and dogs; these planks were often taken to summer encampments leaving only the massive cedar trunk frames of houses for the autumn return. Dugout canoes of cedar would also have been essential to move the bentwood boxes and woven baskets of preserved food. The flexibility which water transport confers means that the reconstruction of these patterns of movement is to a significant extent reliant on ethnographic records, supplemented by oral history and increasingly by materials analysis, such as the sourcing of lithic artefacts. Ethnohistorical sources show people travelled to the Cascade Range and Columbia River to obtain some lithic resources, particular types and colours of rocks being used for specific purposes (Stein 2000). Long distance trade networks are also known from Vancouver Island to the south carrying acorns, salt fish, shells, clothing, baskets and dogs. Obsidian came from eastern Oregon as established analytically by X-Ray Fluorescence. Groups from the south coast of Vancouver Island paddled 15 km across the Straits each summer to San Juan Island where they fished and dug camas bulbs (Stein 2000; Fig. 2.13). The Straits Salish groups exchanged camas bulbs with people of the wetter rainforest areas on the west coast of Vancouver Island for other goods (Beckwith 2004). To the north the Haida people traded with mainland communities exchanging dried seaweed, crab apples and native tobacco for a range of berries and other products. In the south, communities on both sides of the Cascades Range met at portage sites (where canoes were carried past obstacles) and very rich salmon fishing stations on the Columbia River at The Dalles and Celilo Falls to trade products of the different environmental zones, coastal resources from the west being exchanged with those derived from the bison prairies to the east (Ewert 1999). On the Columbia Plateau, winter was spent in the river valleys and groups went up into the mountains in spring and summer (Deur 1999). The extensive seasonal mobility which characterised the Columbia Plateau has been highlighted as a key factor ensuring that resources were not depleted, whilst at the same time supporting free access to resources facilitating socially significant large seasonal gatherings occasioned by the digging of camas bulbs (Hunn 1999).

Trails and prairies

An individual insight to patterns of movement is provided by a body preserved in the ice of a glacier in northern British Columbia 800 km north of the area shown in Figure 2.3. We now know that person was male aged about 20 and died AD 1400–1490 (Beattie 2000). He is called Kwaday Dan Ts'inchi, which in the local First Nations Tutchone language is 'long ago dead person found' (Hebda 2017). His clothing included a hat of woven plant fibres and a robe of squirrel fur. He was found 80 km inland from the nearest coast but isotopic studies showed that 90% of his diet in the preceding 5–10 years had been of marine derivation and, furthermore, his gut contents included sea fish and crab. The pollen of maritime plants on his clothing showed he had been on the coast shortly before his death (Dickson *et al.* 2004). He had also spent part of his life in inland locations and the ingested mineral particles in his gut identified the geology he passed through on his last journey from coast to high mountains. It is inferred from his movements that he was probably on a regular trading route because in the same area artefacts of more recent date have been recovered from the melting glacier. Ice, which formed much earlier in the Yukon some 7000 years ago, is today melting due to global warming; as it does so large amounts of equipment from the hunting of caribou are revealed; the animals have a habit of congregating on snow patches as their dung in the melting ice shows (Greer 2002). That illustrates one of the factors which would have taken people to the mountains seasonally.

Journeys would have been undertaken on foot to obtain resources and for hunting expeditions. Routes in the mountains are in places marked by sunken paths (Wells 1987) and here and there

Figure 2.5 A trail marked by a curvilinear depression in the rock and a small, probably later cairn, Sooke Hills, Vancouver island (photo: M. Bell).

Figure 2.6 Small stone cairns along trail, Padern Lake, Vancouver Island (photo: M. Bell).

routes or changes of direction are marked by small piles of stones (Fig. 2.5). Routes may also be marked by small cairns marking the position of burials (Fig. 2.6), or events. In the case of animal tracks the convergence of routes would have pointed the ways to watering places and river crossings. Thus, in addition to the physical traces of paths and cairns, plant communities may have provided subtle clues to a route, very obvious to the experienced tracker alert to the merest traces.

First Nations people marked places with petroglyphs which were spiritually charged and associated with shamanic practices. Figure 2.7 shows examples from a concentration of petroglyphs at Nanaimo, Vancouver Island, the social significance of which is emphasised by the Coastal Salish name which means 'meeting place'. Wood structures would also have marked activity areas and special places; of these, fish traps of many different kinds are recorded among First Nations communities (Stewart 1977). Figure 2.8 shows an example near the Mud Bay settlement in Washington State which is dated *c* AD 1450. The scale of niche construction is also demonstrated by landscape modifications including the placing of lines of large boulders on beaches to promote sedimentation and extend habitats suitable for clams, known as 'clam gardens' (Williams 2006; Moss 2013). There is also evidence for saltmarsh environments being modified to create favourable conditions for the growth of plants such as springbank clover, Pacific silverweed and northern riceroot (Deur 1999; Turner 2014).

Figure 2.7 'Sea wolves' at the Petroglyph Park, Nanaimo, Vancouver Island (photo: M. Bell).

When one walks along the rarely used paths in mountain uplands one is often aware that the path one is following is not only a human path but one which is far more regularly used by animals; footprints and faeces of Black tailed deer, bears, etc (Fig. 2.9) are regularly found on the path. Writing about present day First Nations hunting practices on the British Columbia/Alberta border, Brody (2002) records how the lines followed by trappers out from the winter camps follow the hunting trails of fur-bearing predators from one stream bed to the next. The tendency to use paths frequented by animals would be particularly evident in areas of very dense vegetation, such as the remaining areas of temperate rainforest where the forest floor is covered with decaying dead trees, and broken up by depressions, some water filled, resulting from tree throw etc. As Liebenberg (1990) writes, concerning the tracking skills of people in South Africa, they utilise a wide range of pieces of evidence including: footprints, slight disturbance, vegetation bent in the direction of travel, cracked twigs, kicked pebbles, scent, urine, faeces, saliva, pellets and feeding signs. Jaeger (1948) describes the many clues employed by Native American trackers and the way the strong nimble feet of these trackers literally feel their way along a trail. In North America generally, Jaeger also describes the technique of trail blazing whereby trees were marked to indicate routes (Fig. 2.10). Practices included: bending a sapling in the direction of a journey; removing patches of bark on trees or groups of stones to indicate right or left turns; pieces of bark slotted into trees; carvings indicating rights to beaver trapping; stakes indicating crossing places over water, or the direction to be taken at a river fork. Saplings were also bent in the direction of travel and these indications could be perpetuated in the form of the tree's growth for centuries (Blackstock 2001). Dale Croes (pers. comm.) reports that Squaxin Indians split conifers so that they grow in a fork, and split evergreen bushes three ways, braiding them so that the braided growth marks territories, events or burials.

One of the most striking and increasingly well documented ways in which communities in the temperate rainforest marked their landscape and provided clues to settlement and routeways is through the cultural modification of trees (CMT). At least 12 species have been recorded as showing evidence for various forms of cultural modification, with cedar the most important (Stewart 1984; Mobley and Eldridge 1992; British Columbia Ministry of Forests 2001). This cultural modification is particularly significant because of the extreme longevity of trees in the rainforest, some living trees providing evidence of tree management practices in pre-contact times. On the west coast of Vancouver Island some 15% of CMT go back to pre-contact (1778) times and one example is as old as AD 903 (Earnshaw 2017). Living trees had part of their bark stripped for fibre to make clothing, baskets etc. Planks were also split from living trees. Provided only part of the bark circumference was stripped, the tree continued living but was left with distinctive scars on the bark which are visible centuries later. The trees provide a fascinating example of the sustainable use of woodland resources. They are now recognised as important cultural artefacts and are protected under heritage conservation legislation. For archaeologists they provide evidence of trail locations and population movements and their

Figure 2.8 Fishtrap dated c *AD 1450 crossing a channel at Mud Bay, Washington (photo: M. Bell).*

Figure 2.9 Bear faeces on a trail, Sooke Hills, Vancouver Island (photo: M. Bell).

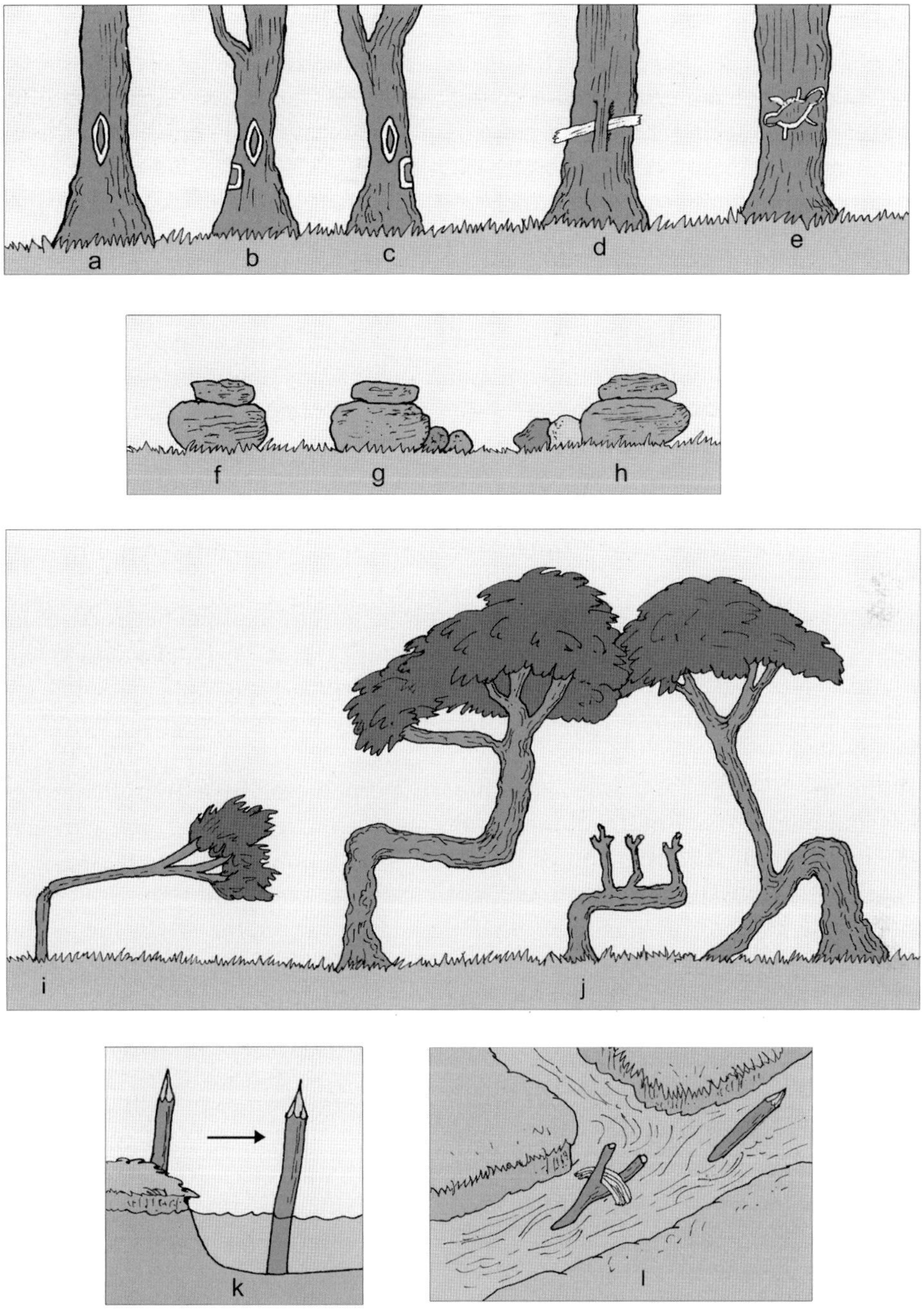

Figure 2.10 First Nations trail blazing from the Americas generally: (a–e) bark removal: (a) marks the trail, (b) left turn, (c) right turn, (d) bark blaze, (e) beaver trapping rights; (f–h) stone blazes: (f) marking a trail, (g) right turn, (h) left turn; (i) bent sapling indicates route; (j) three trees after two centuries; (k) lopped sticks mark river crossing; (l) cleft stick with grass marks direction and pointed stick marks branch to follow (graphic: J. Foster after Jaeger 1948, pl. 11).

presence is increasingly invoked as evidence of former possession in First Nations land claim cases (Turner *et al.* 2009).

We may also speculate that routes which were particularly frequented, either by people or animals, may over time have developed a distinctive flora. Grazing animals would have removed new shoots preventing regeneration and gradually a somewhat more open corridor is likely to have come about, even without any deliberate intention by people. Frequented routes on impoverished soils would have nutrient enhancement from urine and faeces of animal or human origin, favouring certain plants e.g. of nitrogen-rich conditions. Seeds introduced from elsewhere on clothing, or the coats of animals, are likely to have sprouted along routeways. Seeds which remained viable after they had passed through the gut may have propagated. One may also speculate that some degree of transplantation is likely to have occurred, perhaps a stick from a favoured nut tree thrust into suitable ground beside a trail, or a stem from a berry shrub pushed in the ground at a sunny woodland edge. In these ways it may be supposed that some frequented paths could have developed a distinctive trail-side plant community which in time would provide valued resources for passers-by. Although these are to varying degrees suppositions, they are borne out by the observation in many geographic areas of useful plants along frequented routes. They are also in line with what we know of First Nations plant use in general (Turner 1995; 1998; 2014) and the specific association with trails can be suggested in some cases.

Three examples in particular provide evidence of an association between manipulation of plant resources and trails. The first is the Douglas Map (Fig. 2.2). The second is the Willamette Valley, Oregon where the first European travellers described a pleasant park-like oak savannah landscape (Fig. 2.11). A land survey of the 1850s shows prairies covering much of the valley. Leopold and Boyd (1999) show that the oak savannah had existed from 4000 BC originating in a drier period until 2000 BC, but was subsequently perpetuated by a combination of grazing by wild animals and burning by First Nations groups. Both grazing and burning will inevitably have been concentrated along favoured routes of animal and human movement (Boyd 1999). When burning ceased in the 1850s, after European colonisation, woodland regenerated in some areas, showing that the park savannah was fire climax vegetation. From the 1840s the Willamette valley was a key destination of the Oregon trail by which 350,000 Europeans colonised the west coast, the ruts of their waggon trains today being designated as cultural treasures. Colonists described the valley as a sort of untouched Eden, the fertile soil of which was brought to civilisation by the establishment of European farming. In reality it had been managed by First Nations communities for millennia and the fertile prairie soil was of their creation, although not necessarily deliberately so (Robbins 1999).

A third example concerns some well-documented 'grease trails' which were especially associated with the transportation of fish oil from coastal to inland sites. In Tsimshian territory the routes were owned by a chief who commissioned his likeness to be carved on trees along the way, which is also marked by culturally modified hemlocks (D. Mathews, pers. comm.). In the Fraser valley First Nations trade networks are well documented and have been mapped (Schaepe 2001; Oliver 2010).

Brody (2002, 193) describes former First Nations practices of burning in northern British Columbia, in this case during spring, of very carefully chosen areas, to encourage new growth of browse for moose and deer, and to renew pasture for horses. He also recounts the burning of dead fallen trees on paths and trails. Prairies inland of the famous Ozette settlement have been shown to have been maintained by First Nations burning, although with the abandonment of this practice they are subject to incursion by trees. Here early historical sources make it possible to map trails connecting prairies and other resources (Anderson 2009).

In voyaging north along the coast from what is now northern California, Oregon and Washington State, George Vancouver (1801) recorded a predominantly forested landscape but with discrete cleared areas and settlements. South of Cape Flattery he noted cleared spaces in the forest 'which gave the resemblance of a country in an advanced state of cultivation' (Vancouver 1801, 27). The islands in the Strait of Juan de Fuca and Puget Sound, such as Widbey Island, were especially noted for their open grassy aspect with just scattered trees within the forest which again gave the appearance of having been cleared (Vancouver 1801, 66). Vancouver says of the park-like landscapes in parts of the North-west Coast, 'a picture so pleasing could not fail to call to our remembrance certain delightful and beloved situations in old England' (Vancouver 1801, 66). Only later does he seem to have come to understand that these people were not farmers and the cleared areas were a result of Native American burning practices (Boyd 1999). Alexander

Figure 2.11 Willamette Valley: *a painting by Henry Ware 1845 showing oak savanna (courtesy: the Amon Carter Museum, Fort Worth no 1996.4.10).*

Henry described in 1814 'a most delightful situation for a fort, on a prairie 2 miles long and broad … with pines at the rear' and deer in abundance; a decade later that became Fort Vancouver, Washington (Leopold and Boyd 1999, 146). Similarly, in 1842 the previously mentioned James Douglas was looking for a site for a Hudson's Bay Fort and selected Victoria at the southern end of Vancouver Island. It was, he described, 'a perfect Eden in the midst of the dreary wilderness of the North-West coast' (Turner 2005, 147). Grass and herbs covered the area and there were stands of Garry Oak trees. In some small areas, such as Beacon Hill Park and Cattle Point in Victoria, remnants of this park-like savannah landscape remain today (Fig. 2.12; Huck 2006). By 1858 it was evident that this parkland was not entirely natural; 'miles of ground burnt and smokey' are described. The Hudson Bay Company records at Victoria document the firing of prairies every year before the first autumn rains. Thus, both the modern city of Victoria and Fort Vancouver, Washington were originally selected because the sites exhibited discrete patches of park-like landscape of a type familiar and attractive to British colonists, although in fact they were the human artefact of a very different First Nations way of life. This is a significant point demonstrating that modification of landscape by one group, including the creation of routeways, may create conditions favourable for another human group, whether or not there is any direct continuity between them.

Plant utilisation

George Vancouver (1801, 123) describes all the inhabitants of a village of 80–100 people 'engaged like swine rooting up this beautiful verdant meadow in the quest of a species of wild onion and two other roots'. The account of the Lewis and Clark expedition of 1806, down the Columbia River, describes small alluvial prairies within woodland, these being characterised by many of the plants used by First Nations communities, including areas of almost pure wild onion or camas lily (McMaster *et al.* 1814, 24 and 59). Such pure stands are little found today and are considered to have been the product of First Nations practices. The blue camas bulb was an especially favoured food plant (Fig. 2.13). Good plots were owned and inherited, areas being marked by modified trees/plants; they were looked after in a way which can be described as cultivation (Beckwith 2004; Turner 2014). Plants were encouraged by burning and the removal of rocks and scrub (Turner 1999). The poisonous Death camas was weeded out at flowering time. The larger blue camas bulbs were dug using digging sticks and small bulbs and seed heads returned to the ground; disturbance encouraged propagation. There is also evidence that camas was transplanted to areas where it had not grown naturally (Lepofsky *et al.* 2005). Harvest was followed by burning which helped to fertilise the ground as did deer attracted to the later lush growth of young plants. The bulbs were cooked by steaming with hot rocks in pits and could be stored for the winter. Charred camas are found on sites as early as 5000 BC and earth ovens used in camas processing occur from 3000 BC in the Willamette Valley (Beckwith 2004). Today the park-like landscape characterised by camas fields and Garry Oaks is reduced by some 98% but areas of this distinctive landscape survive in Beacon Hill Park where there are also burial cairns made by removing rocks from the camas fields (Fig. 2.12; Huck 2006).

Camas provides one of the clearest cases of cultivation and deliberate environmental modification but many other plants were associated with similar practices. Berry patches were burnt and pruned. Turner (2005, 220) documented the fertilisation of berry patches using waste from cleaning salmon and animals and the scattering of ashes and broken clam shells to improve the

Figure 2.12 Beacon Hill Park, Victoria, Vancouver Island; park savanna with stone burial cairns (photo: M. Bell).

Figure 2.13 Camas plants, a favoured First Nations food item at Cattle Point, Vancouver Island (photo: Edwin Taylor).

soil. Berry bushes were also transplanted to favoured sites. Nettles were extensively used for medicine, dye, fibre, string and nets and were effectively cultivated in areas of nutrient-rich conditions created by long occupation. To ensure a good fibre supply the best plants were transplanted to these areas. Captain Cook (1784, 310) records the consumption of bracken fern rhizomes by Nuu-chah-nulth people at Nootka Sound. Later sources record how bracken was particularly found in disturbed and burnt areas and the rhizomes were used to make flour for bread. Both grinding stones and rhizomes were found on the wet site at Hoko River, *c* 1000 cal BC (Croes 1995). Estuarine root gardens were established to expand areas where valued plants grew and these were owned and inherited. Fruit and fibre plants were transplanted to settlements (Turner *et al.* 2013). A range of plants, cedar bark and roots, grasses, etc were carefully husbanded to provide materials for baskets used for transporting and storing foods. Many of those which survive in museums are of superb quality. The designs and techniques of basket making were distinctive of a group's social identity for millennia (Croes 2015). Acorns were also very important resources in some areas, as demonstrated by the excavation of 114 acorn leaching pits in the Sunken Village site, Oregon (Croes *et al.* 2009) and the discovery of many acorns during excavations at the Qwu?gwes site (D. Croes, pers. comm.)

Many First Nations groups made use of well over 150 plant types (Turner 2014). Where there was a concentration of plant resources, or fish, different communities came together and routeways converge. Turner (2014, vol. 2, 23) called these 'Cultural Keystone Places' because of their role in sharing social knowledge. The range of documented management and enhancement practices includes burning, trimming or pruning, coppicing, thinning, bark ringing, selective harvesting, water diversion to encourage growth, sparing (setting aside for later use), replanting of propagules, fertilising, mulching, weeding and transplanting (Turner 2014). There was a sort of empathetic belief among First Nations communities that plants need to feel human presence as they do when people walk on them, prune, harvest, burn etc. These practices are best described as cultivation (*i.e.* the encouragement of growth) rather than domestication (control of reproduction) (Harris 1969). The only apparent domesticated plant may be tobacco, the growing of which from seed is documented by the Haida and Tlingit just 13 years after European contact in 1776 (Deur 1999).

Elsewhere in North America

In considering the relationship between people and plant resources in North America as a whole, Smith (2011) has drawn on ethnographic evidence to identify six main categories of ways in which people act as Niche Constructors: (1) general modification of vegetation; (2) broadcast sowing of wild annuals; (3) transplanting of perennial fruit-bearing species e.g. near settlements and along major trails; (4) in-place encouragement of economically important perennials; (5) transplanting and in- place encouragement of root crops; (6) landscape modifications to increase abundance in specific locations. In California, as in the North-west Coast, where there were no pre-contact domesticated crops except tobacco, there is extensive evidence for the use of fire and the manipulation of plant resources (Blackburn and Anderson 1993; Fowler 1996). Acorn crops in this area were particularly important, management of this resource by some groups amounting to tree agriculture. Willows were coppiced near springs or seeps to provide long straight stems for basketry. Pinon pine nut trees were whipped, which stimulated bud and thus nut production, and the areas of Pinon pines were kept clear of dead wood to facilitate nut collection. Some groups sowed wild seeds, irrigated favoured food plants and tended root crops. In the semi-arid areas of the American South-west routes can sometimes be followed for hundreds of kilometres and there are ethnographic records of some being associated with the acquisition of salt and others served primarily spiritual and ceremonial purposes (Snead *et al.* 2009; Snead 2011). In the Mojave Desert Native American routes are ethnographically recorded and networks of them have been mapped (Fowler 2009).

On the east coast of North America, Europeans arrived very much earlier than on the west coast, beginning with John Cabot in AD 1497, so the early contact period historical sources are often less detailed. Early explorers, such as Drake in the 1570s and Hudson *c* 1600, observed clouds of smoke arising from Native American burning. The cultivation of domestic crops such as maize, beans and squash occurred in New England and some burning was to prepare ground for crops. However, this burning was documented as occurring much more widely in connection with hunting and gathering, including in areas to the north beyond the limits of crop growing. William Wood (1635), writing of New England, described a landscape which was open and park-like with a greater abundance of fruits and wild animals. The accounts of early writers highlight the role which fire played in hunting animals, improving pasture for wild grazers, driving animals for the hunt and increasing the ease with which people could move through the landscape. Drawing both on the accounts of early travellers, and recent evidence of management for nature conservation purposes, Mellars (1976) showed that burning can increase the carrying capacity for herbivores between 300 and 700 per cent. In areas where crops had been grown, continued burning created conditions favourable to strawberries, blackberries, raspberries and other plant foods.

By the mid-nineteenth century Native American management practices had been prevented or abandoned through exclusion of people from their traditional lands, and many areas of earlier fire climax vegetation became colonised by secondary woodland. Introduced plants provided clues to patterns of movement. The broad-leaved Plantain (*Plantago major*) thrives on trampled paths, hence Native Americans calling the plant 'English Man's Foot'; wherever the white men went it was found (Mabey 1972). Following the introduction of the horse, weeds such as fat hen proliferated along trails dispersed in horse dung (Turner 2014, vol. 1, 209). In the Great Lakes region of Canada, First Nations people often made use of animal paths and dried stream beds to move through woodland (Coles 1984). Intriguing evidence for the influence of wild animals on patterns of movement, settlement and land use is shown by beavers, as their dams provided crossing places over rivers and wetlands. Following European colonisation the sites of some beaver dams were perpetuated as crossing places for roads (Cronon 1983, 106). Where bodies of water were retained behind beaver dams Europeans harnessed the water to power mills. Thus in New England many of the first European settlements were in the more open areas of landscape created by Native American burning, departed beavers, or in some cases by the effects of seasonal floods (Cronon 1983). These rivers, once considered by geomorphologists as models of natural meandering steams, have now been shown to be the result of patterns of sedimentation resulting from successional niche construction by beavers and later people (Edgeworth 2011). Goldfarb (2018) assembles evidence showing that in many parts of North America beavers are a keystone species, i.e. they often determine the character of entire ecological communities to the advantage of their biodiversity. In Pennsylvania it has been possible to map networks of Native American paths using the journals of early travellers, maps and land records (Wallace 1965); some routes served as documented war paths (Snead 2011). Levy

(2007) has considered how these trails provided a context for the interaction of Native Americans who initially guided Europeans but were progressively supplanted by them.

Palaeoenvironmental perspectives

The foregoing discussion has presented evidence for the role of human agency in environmental change in pre-contact North America which is a significant part of this book's argument that the effect of human activity will, more often than generally realised, have been concentrated along trails. Evidence for the manipulation of vegetation, particularly by burning, comes both from groups on the east coast which combined small-scale crop husbandry with hunting and foraging and also from non-agricultural groups on the west coast which have been the main focus of this chapter. The evidence comes from the accounts of the early European explorers who describe discrete prairie patches and burning and anthropologists who describe plant manipulation by First Nations communities (Turner 1999). That these changes occurred on some scale is evident from the widely reported secondary successions from prairie to woodland that followed the cessation of burning in some areas, one such case is the Nisqually prairies between 1850 and today (Leopold and Boyd 1999). Although these practices go back well before European contact, their origins and time depth are more difficult to identify. Pollen studies provide only limited evidence for the effects of First Nations peoples. Kruckeberg (1999, 58) states that, prior to the arrival of Europeans, people made hardly any detectable sign on the pollen record of the Puget Sound Basin, whilst also acknowledging that some bogs may show an increase in fire. This seems oddly at variance with the ethnohistorical evidence from the Puget Sound area. A partial explanation may lie in the contrasting spatial scales and locations of ethnobotanical knowledge and pollen analysis and the fact that much North-west Coast pollen analysis has been concerned with the broad scale, very much off-site, patterns of Holocene vegetation colonisation and change, as opposed to high resolution, often on-site, studies where there are concentrations of past human activity in, for instance, Britain. The broad-scale Holocene vegetation histories from the North-west Coast are nonetheless informative. They demonstrate that in the earlier Holocene there were episodes when savanna and prairie vegetation was more extensive in the south-west Washington rain shadow of the Olympic Mountains (Anderson 2009). More open conditions are attested in pollen diagrams at Battleground Lake 8800–3200 cal BC and at Cranberry Lake prior to the deposition of volcanic ash from Mount Mazama *c* 5700 BC (Leopold and Boyd 1999); this is inferred to be the result of more arid climatic episodes which were followed by a wetter late Holocene when woodland expanded. Similarly on the south-eastern coast of Vancouver Island and the islands in Puget Sound there is also some rain-shadow effect from the Olympic range and Vancouver Island ranges giving rise to rather drier and more open woodland of Garry oaks as compared to the temperate rainforest on the west coast. This may explain why the 1778 expedition of Captain Cook (1784) described a virtually continuously forested landscape, since his landfalls in what are now Washington State and British Columbia were confined to the west coast temperate rainforest. In contrast the 1792–4 expedition of Captain Vancouver (1801) explored inner Puget Sound, and the south and east coasts of Vancouver Island where occasional grassy prairies with settlements surrounded by woodland were described; these we now appreciate are the combined result of rain shadow effects and First Nations burning practices (Boyd 1999;

Weiser and Lepofsky 2009). Charcoal from sediment cores on southern Vancouver Island indicate increased levels in the last 2000 years, which do not appear to be climatically related and suggest deliberate burning (Brown and Hebda 2002; 2003).

On the east side of North America there are similar issues on a much larger scale. Pollen analysis shows that west of the Great Lakes the boundary between the continental interior prairie has undergone significant fluctuations, with the prairie advancing to the east between 10,000 and 7500 BP at a time of drier conditions, and then retreating west between 7500 and 500 BP (Webb *et al.* 1984). In this area charcoal occurrence varies through the Holocene and has generally been interpreted as an indicator of natural fires during drier periods, rather than evidence of human agency. In terms of North America as a whole McAndrews (1988) considered that less than 7% of Holocene pollen diagrams register prehistoric agricultural impacts, perhaps reflecting the absence of domestic stock and relatively small-scale crop growing within a more diverse economy in many areas (Wills 1995). Even so, on some sites vegetation changes and evidence of fire can be related to periods of occupation at nearby sites, for instance at Crawford Lake (McAndrews 1988; Clark and Royall 1995). As already noted, early accounts by Europeans provide a good deal of evidence for burning by First Nations people in the eastern part of North America. When burning ceased and bison which grazed the prairies were driven to the verge of extinction by Europeans, then in some areas secondary woodland regeneration from prairie occurred. For instance, after burning was controlled in Illinois the forest margin advanced markedly and it was concluded that some prairie areas would naturally have been wooded if it were not for Native American burning in the last 5000 years (Williams 2003).

The social significance of routes

So far in this chapter we have been concerned with various types of physical evidence for environmental manipulation, some of which provide clues to the mobility practices and routes of Native American communities. The anthropologist Ken Basso (1996), working with the western Apache on a reserve in Arizona, provides a very different philosophical perspective on the issue of landscape routes. The Apache hunters and gatherers of the Great Plains had a mobile way of life, the range of which increased greatly with the acquisition of the horse *c* AD 1650. Basso recounts how the Apache see their past in terms of a well-worn path or trail which was travelled by ancestors and which subsequent generations have travelled ever since. It could be understood with the aid of clues or prompts provided by historical materials, or memorable landscape traces which were sometimes called 'footprints' or 'tracks' that have survived to the present. Descriptive names given to these places conveyed the optimal viewing points from which the ancestor originally gave a place its name, so that people could situate themselves at that point standing in their ancestor's tracks. Descriptive place names served as shorthand anchors for stories which were at the same time moral object lessons guiding people as to social norms. They were reminded about how to behave by events which had taken place at a specific location, hence the title of Basso's book *Wisdom Sits in Places.* The enlightened were seen as moving along a trail of wisdom. In similar terms the English writer Robert Macfarlane (2012, 28), who has drawn on Basso's writing, notes that the Klinchon people of North-west Territories, Canada use the terms knowledge and footprint interchangeably, as do the Tlicho (Dogrib) Athabaskan people of Canada's North-west Territories. Legat (2008) recounts how among the latter, learning takes place through recounting stories and travelling along trails whilst carrying out tasks in places. People describe themselves as following an ancestor's footsteps and using this transmitted knowledge about trails and places to survive (Johnson 2010).

Conclusions

Emerging from this narrative are factors influencing the location and extent of human activity and prairies. These include climatic factors relating to continental interiors and rain shadows, the incidence of wildfire related to climate, the activities of wild grazing animals and other fauna such as beavers, the use of fire by people and their other effects on vegetation. All this is more relevant to the theme of routeways than has generally been supposed because the activities of both grazers and people can be predicted to be more concentrated along the main axes of movement. We have also noted that beaver dams may provide crossing places and beaver lawns attract settlement. This is analogous to the synergistic relationships which have been identified by Smith's (1995) studies of early agriculture in North America where natural disturbance factors in river valleys are also associated with anthropogenic disturbance related to agriculture. People used, and brought into domestication, the plants of disturbed riverine environments. It may also be that the disturbed and nitrogen-rich conditions along routeways brought about the conditions for closer relationships between people and some plants; many of those utilised are, after all, plants of the woodland edge (Turner *et al.* 2003). These processes can be seen as examples of niche construction or co-evolutionary relationships between a series of factors, each of which has some degree of agency (Smith 2011). This situation is very much as foreshadowed in Chapter 1. It no longer seems particularly profitable to search for a single causative factor such as: climate, people, animals etc, when the more realistic and interesting story relates to the interactive agency between a range of factors.

Of the people of the North-west Coast White (1999, 46) has written: 'far from being creatures of their environment the Indians had shaped their world and made it what it was when the whites arrived'. This chapter has emphasised that they did so in the context of a range of interacting influences, or active agents, the existence of which undermines long standing, but artificial, distinctions between human and natural factors. Examples from the Willamette Valley (Fig. 2.11) and the bead-like string of prairies depicted in Douglas's map of the Cowlitz to Nisqually trail (Fig. 2.2) show more clearly than anything the relationship between human environmental manipulation and trails. If we could visualise the pre-European contact prairie/woodland edge at a large scale we may predict the existence of corridors of prairie extending into woodland along major axial routes. We may also predict that patterns of movement will have occurred between more open grassy areas and those with abundant woodland edge resources. These are working hypotheses suggested by the North American evidence; they remain to be tested by developing a linear approach to environmental archaeology which is more focused on routeways than the present site-based emphasis.

Nor should we see people's activity in the landscape in purely economic and utilitarian terms. Basso's Apache studies serve to emphasise the role of perception and storytelling, just as, in many communities no doubt, song, dance and art contributed to understandings of place. Those who have explored phenomenological approaches in archaeology (Chapter 1) emphasise geological and topographic features: rock outcrops, upland peaks, coastal features etc. Such landscape features would have been much more prominent in an open landscape than in one which was densely wooded. Less widely appreciated is that in forests venerable trees, traces of woodland management, cleared areas, food plants and places where animals gathered, are likely to have figured prominently in storytelling and mythology. This highlights the imperative of building a palaeoenvironmental dimension into phenomenological studies.

Oral history would have served to perpetuate patterns of movement through the landscape and the activities conducted, so that the ways in which people have modified plant communities would have helped provide, not just sustenance, but prompts as to the correct way of life. Traces left by people are likely to have created readable routes, along which were favoured resources, some naturally there, some enhanced by human agency. More subtle signatures would also have played a part. Animals would deposit faeces rich in both nutrients and the seeds of their favourite food plants such as huckleberries. The same would apply, probably to a lesser extent due to gut chemistry, to human defecation along regularly used routes. This may have given rise to linear occurrences of those of their favourite food plants which remain viable after passage through the gut and which are suitable for local growing conditions. This would have reinforced the value of using that route during subsequent foraging expeditions. Other factors could include deliberate landscape modification, *e.g.* by burning along corridors (Lewis and Ferguson 1999), the effects of sharing routes with wild and domestic animals, and the more subtle effects of trampling, disturbance and defecation. Each would have contributed, both to the readability of a route, and its capacity to provide food in transit. It may also be speculated that, because of the botanical and other traces along a route, these could have provided readable evidence for the frequency and seasonality of use. The physical evidence of trail-marking blazes and the plants present may also have afforded visual clues as to the particular social groups involved and the bounds of their territories since Hastorf (1999) demonstrates that food preference is a key medium by which people establish their social identity.

3. Niche construction and place making: hunter-gatherer routeways in North-west Europe

Introduction

This chapter picks up the theme of hunter-gather environmental relationships and the evidence which these can provide for routeways from a European perspective. Identification of hunter-gatherer routeways is inevitably speculative. Here we look at sources of evidence for the structuring of landscape by movement from ecology, environmental archaeology, isotopic evidence and material culture. A new direction in environmental archaeology is advocated, with an emphasis on movement and connectivity between sites. This builds on the increasing availability of landscape-scale environmental reconstructions in areas where evidence from multiple sites facilitates the creation of a spatial picture. That aligns with Jordan's (2003a; 2003b) advocacy for increased consideration of the ways in which hunter-gatherer social symbolism transformed landscapes. Evidence presented in Chapter 2 for the ways routeways became defined in North American landscapes suggests how we can approach these questions in much more challenging earlier contexts in Europe.

Environmental archaeologists have written a great deal about vegetation disturbance and burning by hunter-gatherers, yet there has been surprisingly little consideration that these changes would often have been concentrated along axes of movement. In part this reflects an almost unconscious tendency by archaeologists to think in terms of a static site-focused picture. In this chapter a deliberately broad view of disturbance factors is taken, on the premise that since all actions are articulated by movement then the traces they create will be concentrated on the routes used. This chapter is mainly concerned with dry ground situations and how movement may affect woodland and other terrestrial environments. The locations of the British sites discussed are shown on Figure 3.1. Chapter 4 reviews the smaller-scale clues provided by footprints and Chapter 6 evidence for wetland trackways, but those latter are almost all associated with later farming communities. In considering patterns of movement we must keep in mind that terrestrial evidence is only part of the picture of mobility; in prehistory many trips will have been undertaken by boat on rivers and the sea, which would have served as arteries of connectivity, rather than the obstacles sometimes imagined (Chapter 9). Where there was a heavy dependence on boat transport, dryland routes may be short, interrupted stretches originally connected by boat, such that the dry ground parts may appear counter-intuitive in today's very different environmental context, or to those unfamiliar with moving around by boat.

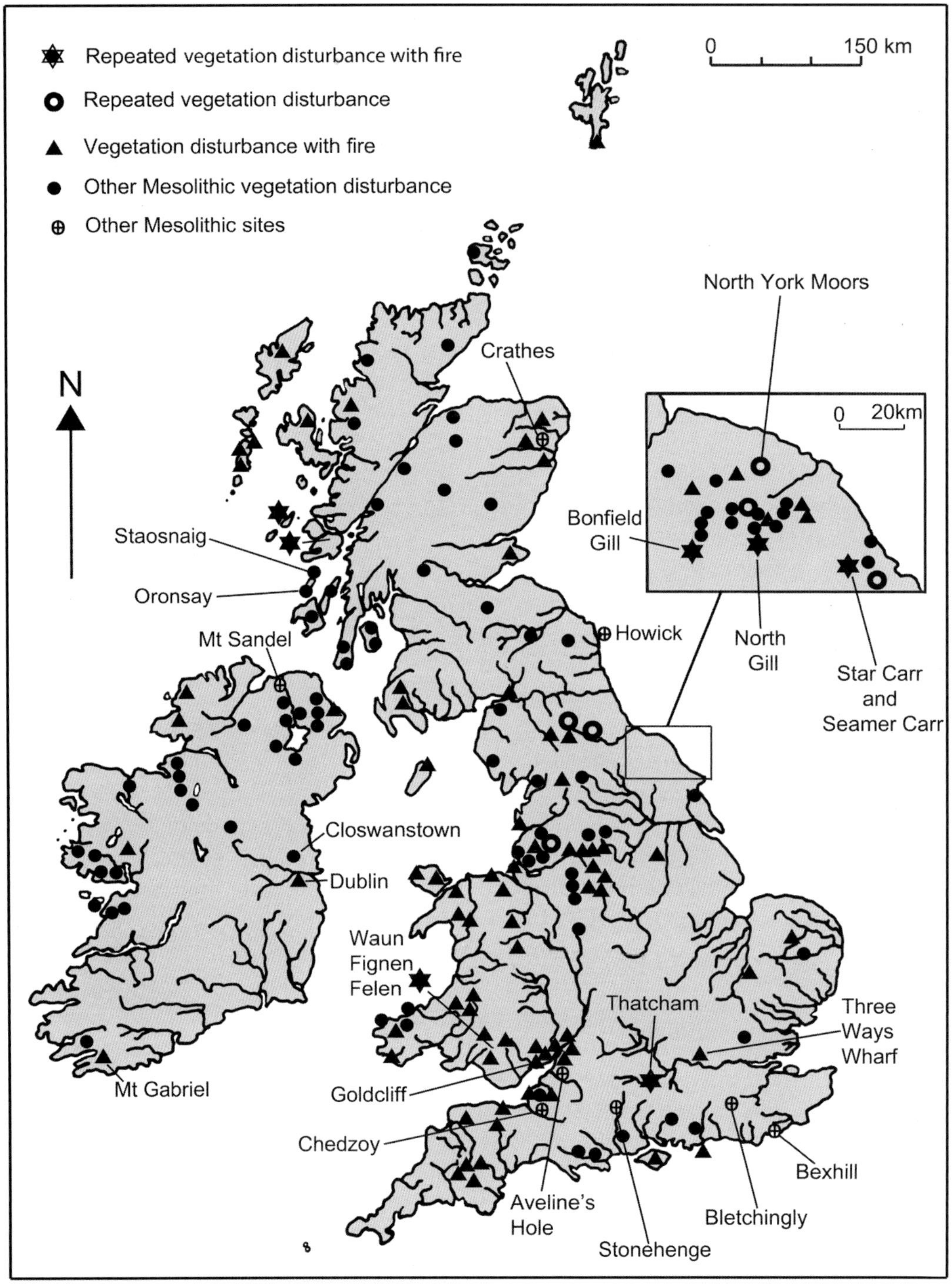

Figure 3.1 Mesolithic sites in Britain and evidence of vegetation disturbance. See on-line digital Supplementary Appendix 3.1 for a gazetteer with details of sites shown (graphic: J. Foster).

Anthropological perspectives

Insights into the range of ways in which landscapes can be structured by movement may be derived from situations where anthropologists have mapped the paths of recent hunter-gatherers. For the Nunamiut of Alaska, Binford (1983) has mapped the annual seasonal round of a group (Fig. 3.2a). He also looked at the contrasting activity areas represented in different landscape facets, and the ways in which residential areas shift due to resource depletion at *c* 10 year intervals and at stages of an individual's life course. Such studies emphasise the considerable areas over which hunter-gatherer activity ranges, both over a year and a lifetime. The household patterns of movement of Skolt Lapps in Finland in the year 1938 (Fig. 3.2b) are very much focused on branching radii from the focal point of the winter camp. Figure 3.2c is derived from the Kutchin of Alaska where two separate traplines in places cross (Nelson 1973). It is notable that three out of four cabin sites are at points where lines of the same trail join. For the Inuit, their country was perceived as a mesh of interconnecting lines rather than a continuous surface (Ingold 2007). Both maps a and c illustrate situations in which the main effects on the environment are focused on seasonal or temporary dwelling sites. Figure 3.2d demonstrates a contrasting pattern created by the migration paths of Swedish mountain Lapps who intercept reindeer at various points along the migratory routes (Ingold 1986, fig. 7.7). There was inevitably great diversity in the forms of hunter-gatherer mobility and settlement. As Broadbent (2010) has demonstrated, in the harsh environment of northern Sweden, periodic changes in economy and patterns of movement can be related to cyclical changes in the abundance of coastal sea mammals and inland caribou. The capacity for such adaptability is a key part of the resilience of hunter-gatherer communities. The duration of stays and the distances travelled between camps and during the course of a year will vary greatly. In the course of a single foraging day it is recorded that the Anbarra of Australia travelled 20 km or more before returning to base camp (Meehan 1982). Of 161 mobile hunter-gatherer groups documented worldwide by Binford (2001; 2006), mean annual mobility distance among those most dependent on animals was 548 km, those most dependent on plants 373 km, and those dependent on aquatic resources 289 km. What is clear from anthropological perspectives is that we cannot reduce patterns of movement to ecological or economic necessity (Politis 2006). Jordan's (2003a) work among the Siberian Khanty shows how space was socially encultured by placement of animal bones, artefacts, shrines and carvings on trees in special places. Settlements were always located upstream of burials, usually 0.8–1 km and some 10% of the landscape regarded as sacred was not visited.

In Australian aboriginal thought (Munn 1970; 1973) 'objects or features of the landscape engender subjects in the form of living persons and are the congealed embodiments of the past creative activity of ancestral beings' (Ingold 1986, 139; Munn 1970; 1973). The Walbiri of Australia see the life of a person as the sum of that person's tracks, the total inscriptions of lifetime's movement. Here again they imagine country as a network of lines (Chatwin 1989; Ingold 2007). Footprints left by humans and animals provide clues to the location of resources and the movements and activities of other groups (Ingold and Vergunst 2008). Often there may be no meaningful distinction between the movements of people and animals. In this context the agency, in creating a linear space, lies not with people or animals, but through their mutual interaction. People must often have followed routes favoured by animals. As Olwig (2008) observes, hunter-gatherers develop a 'feel' for the land which is mediated as much by the feet of the animals they follow as their own and this is also

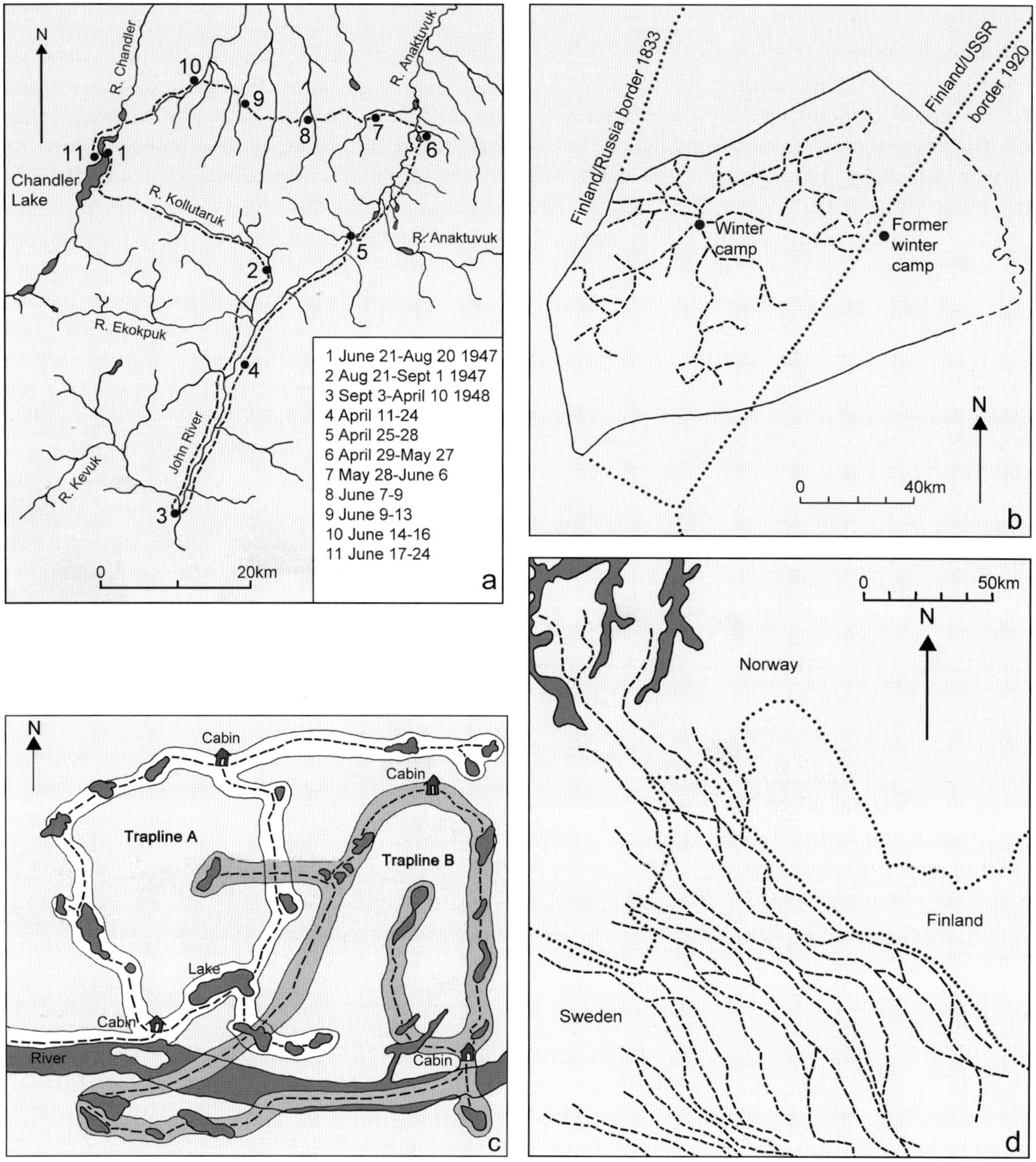

Figure 3.2 Ethnographic evidence for patterns of movement by non-agricultural communities: (a) Nunamiut of Alaska (after Binford 1983, fig. 49); (b) Skolt Lapps (after Nickel 1948); (c) Kutchin traplines and camps in Alaska (after Nelson 1973); (d) Migration paths of Swedish mountain Lapps (after Manker 1953) (graphics: J. Foster).

true, to an even greater extent, of pastoralists who walk with animals (Jordan 2003a; Lorimer 2006). Hunter-gatherers normally share territory with other visiting groups; meeting with them provides a means of acquiring information about resources, exchanging materials and, most crucially for long-term survival, genetically different sexual partners. It is likely, therefore, that the most important intersections of paths where group meetings occurred acquired a special significance. It will be argued in Chapter 5 that these places of deep-seated ancestral significance may, much later, have become elaborated and celebrated as monuments.

Topographic factors and 'natural routeways'

In landscapes of dramatic topography, extensive water bodies, or restricted river crossing places, it may be possible to predict the existence of 'natural' routeways, a concept which will be introduced using Pleistocene and early Holocene examples. Creswell Crags, Derbyshire is a prominent gorge in limestone (Fig. 3.3). The flanking cliffs have many caves, some occupied by Neanderthals (*c* 60–36 ka BP) and later by Creswellian hunters (*c* 12.5 ka BP), who were responsible for art in some caves in the Upper Palaeolithic (Bahn and Pettitt 2009). In each case the evidence suggests short-term visits during which both people and animals made use of the natural routeway through the gorge. Cheddar Gorge in Somerset provides a comparable example; it is a routeway from the lowlands of the Somerset Levels and Axe Valley to Mendip uplands. Caves, such as Goughs Cave, were occupied in the Upper Palaeolithic (Jacobi and Currant 2011) and there is also a very early Mesolithic burial. On the north side of Mendip, in the gorge at Burrington Combe is the location of Aveline's Hole, a cave occupied in the Upper Palaeolithic and containing the largest Mesolithic cemetery in Britain, with evidence for Mesolithic geometric art (Mullin and Wilson 2004; Schulting 2005). The associations with art and, in both the Somerset cases, with burial, hints that the 'natural' gorge routeways were not purely significant as hunting, or stopping places, but they accrued particular social significance.

In the late glacial of northern Europe topographic features inherited from the glaciers both channelled movement and created contexts where animals could be intercepted. The tunnel valley at Stellmoor, Germany creates a natural crossing point, providing the perfect place for interception hunting of reindeer by Ahrensburgian hunters (Terberger 2006). Topography inherited from glaciation can serve as an important influence on patterns of movement in the Holocene. On the Bothnian coast of northern Sweden, eskers (ridges formed by sub-glacial drainage) were favoured dry locations for Sami settlement and provided the axial routes of coastal-inland movement, defining economic zones which are reflected in the distributions of artefact types (Broadbent 2010).

Figure 3.3 Creswell Crags, UK, a limestone gorge with many caves occupied in the Middle and Upper Palaeolithic on a 'natural' routeway. The lake is a recent feature (photo: M. Bell).

The wildwood, disturbance factors and routeways

The Mesolithic has often been portrayed as a period in which people were puppets of the environment affected by the rapid environmental and sea level changes which resulted, for instance, in the inundation of Doggerland between Britain and North-west Europe. There was once an assumption of virtually continuous closed woodland in North-west Europe during the Holocene climatic optimum *c* 9000–4000 cal BC, such that a squirrel could traverse the country from end to end without needing to touch the ground (Fox 1932). That view, maintained until recently, is now being significantly revised in the light of ongoing ecological debates, particularly given the increasing number and geographical coverage of palaeoenvironmental studies. These suggest a distinctly more variable and patchy picture than previously imagined. Drawing on experience of conservation ecology, Vera (2000) argued that the natural wildwood of the early Holocene was more open and park-like than previously imagined. His observations suggested that light-demanding oak and hazel, which are important components of the pollen record at this time, could not have regenerated in closed woodland. He proposed that grazing animals, such as deer and aurochs, created more open areas and that patches of thorny scrub acted as nurseries for woodland regeneration on a cyclical basis. Vera's theories have been significant in promulgating a more dynamic view of woodland history and highlighting the significant role of grazing animals. However, as a universal model of temperate zone wildwood history the Vera model has been subject to critique from multiple palaeoecological perspectives. In Ireland, it is now clear that the main potential grazers, red deer and aurochs, were not present in the first half of the Holocene, yet pollen diagrams show oak and hazel were abundant and must therefore have been able to regenerate (Mitchell 2005). The same applies on Danish islands, after wild grazing animals became locally extinct, probably as a result of Mesolithic hunting. Trees preserved below peat, for instance in the East Anglian Fenland, or in coastal submerged forests, generally have very tall straight trunks (Fig. 3.4) showing that they grew in dense closed woodland, rather than the low branched trees familiar to us today from the park-like landscape which Vera envisaged.

Figure 3.4 Submerged forest tree, Goldcliff, Wales which was associated with charcoal (photo: E. Sacre).

In critiquing Vera we must avoid throwing the baby out with the bath water. It is increasingly clear, from a range of modern analogue studies, that pollen analysis consistently overestimates the proportion of woodland cover, trees being so much more productive of pollen than grasses and herbs. Grazed semi-open woodland in Denmark produces tree pollen values as high as 85–96% of the total pollen count

Figure 3.5 Multiple trees felled by a storm Brighton, Sussex, 16 October 1987 (photo: M. Bell).

(Buchwald and Svenning 2005). Pollen analysis in present-day parklands in the Netherlands shows trees are greatly overrepresented by comparison with grasses and herbs, the pollen productivity of which is suppressed by grazing (Groenman-van Waateringe 1993). A more realistic evaluation of the degree of openness and the effects of grazing animals can be obtained from a multi-proxy approach, both in the case of modern analogues and palaeoenvironmental studies. Non-pollen palynomorphs, such as fungal spore types associated with animal dung or fire, are increasingly used to assess disturbance factors within woodland (Buchwald and Svenning 2005). Modern analogue comparisons of pollen and beetle evidence have been successfully employed to evaluate the extent of disturbance and openness in grazed and ungrazed woodland (Smith *et al.* 2010; Hill 2015). In earlier Holocene contexts, beetles also indicate the presence of grazing animals and openings within woodland, although not generally on the scale which the Vera hypothesis would lead us to expect (Whitehouse and Smith 2004; 2010).

A central weakness of the Vera hypothesis is the emphasis placed on a single disturbance factor: grazers. In the wildwood a range of factors would have contributed to the creation of sub-climax plant communities, these include people, fire (wild and anthropogenic), storms, floods, disease etc (Brown, T. 1997; Svenning 2002; Bell and Walker 2005). A further significant problem is the implicit assumption of the park-like model that disturbance was somehow evenly distributed across the landscape. This is not what the palaeoecological record suggests, nor is it what we would predict from consideration of any of the disturbance factors. The effects of storms (Fig. 3.5) and floods will be focused on particular topographic locations, i.e. those susceptible to high winds, or low-lying areas subject to flood. Fire will be more frequent in areas of dry woodland, those in which pine and birch dominate, and topographic locations susceptible to frequent lightning occurrence. Opening due to pests and tree diseases will often be of patchy occurrence and on average probably more frequent along routes of animal and human movement. The activities of beavers will be concentrated in low lying areas with streams and water bodies. Many finds of bones and beaver-gnawed wood (Fig. 3.6) show they were widespread in riverine environments in Britain during prehistory (Coles and Orme 1983; Coles 2006). Beavers fell trees and create dams in streams and wetlands which can lead to extensive flooding. When beaver ponds fill with sediment, fertile beaver lawns are created. The activities of beavers will create more open tracts in river valleys; their encouragement of flooding must often have made the floodplain difficult to traverse, whereas felling activities could have created more open, easily traversed areas along the floodplain edges. Dams, which in some instances have been shown to persist for millennia, would have made ideal crossing places as noted in Chapter 2 (Coles 2006). Grazing animals play a pivotal role as ecosystem engineers in nutrient biochemical cycling and distribution, depending on the density of grazers. The effects of grazing animals will have been concentrated near waterbodies, in areas of good grazing or browse, or in the case of deer, good rutting grounds. Locations which were warmer and sheltered in the most hostile periods of the year will also have been favoured. Pertinent to the present context, the effects of animals will also have been concentrated on the paths and routeways which link frequented places. They help spread seeds and concentrate edible food sources along animal tracks. Likewise, the activities of hunters and gatherers will also be concentrated in many of the same areas. The various disturbance factors can seldom be envisaged as operating individually. Warren *et al.* (2014) have highlighted the applicability of Niche Construction Theory to the role of people in studies of the Mesolithic landscape of Ireland. Openings created in woodland by one factor, storms for instance, are likely to have been perpetuated by another, such as grazing animals (Buckland and Edwards 1984), or people attracted to an open area and resources of the woodland edge which it offered. These mutual attractions are especially pertinent to the way in which patterns were created within landscapes. It follows that the effects of human agency will often mimic 'natural' disturbance factors (Smith 2011), indeed are integral to them in ways which make the extent of 'natural' or human agency difficult or impossible to disentangle.

Figure 3.6 Neolithic beaver-gnawed wood from Shapwick Heath, Somerset, UK (photo: M. Bell).

Contrary to Vera, we should envisage more of a mosaic environment (Pickett and White 1985; Forman 1995) with extensive areas densely wooded, but other areas, in which disturbance factors were concentrated, being more open and park-like, as well as areas of scrub and grassland (Svenning 2002; Fyfe 2006). An additional factor affecting the openness of landscape is geology. Svennings' (2002) palaeoecological survey of vegetation openness in North-west Europe shows that areas with shallow soils, e.g. chalkland and sandy soils, had more open and patchy woodland with grassland. The hypothesised mosaic environment gains a measure of support from recent palaeoecological studies on the English chalklands. On Cranborne Chase in Dorset the evidence suggests thin Holocene soils and only a partial tree cover during at least the first part of the Mesolithic (French *et al.* 2007). Conversely in another chalk landscape, the South Downs, palaeoecological evidence indicates woodland during the Mesolithic and this was also the case with some sites selected for monument construction during the Neolithic (Thomas 1982). It remains open to question whether these emerging spatial contrasts are the result of superficial geology, soils, the social behaviour of people, or animals, or a combination of these. In the chalk landscape of the Upper Kennet valley, Wiltshire the Encountering Avebury Project (Gillings *et al.* 2008) modelled the susceptibility of landscape topography with a tree-throw windfall risk map (Fig. 3.7a). This in turn generated a predictive map of the locations of possibly clearer areas and hypothetical paths were drawn between these clearings (Fig. 3.7b). This map highlights the susceptibility to tree throw of particular topographic contexts: (i) the steep escarpment on the north side of the Vale of Pewsey on the bottom edge of the map; (ii) to a lesser extent the escarpment formed by the present Ridgeway route to the north of Avebury; (iii) the susceptibility of valley sides to tree throw is also highlighted.

It is clear from the foregoing discussion that closed woodland was more extensive in the early Holocene than a literal interpretation of the Vera hypothesis would suggest and that is confirmed by analysis of the pollen data in some areas, such as Wales (Fyfe 2006). Nonetheless, a range of disturbance factors would have created some open disturbed and park-like areas. In the more open areas routes may have been quite fluid and changeable. The creation of more open areas will rarely have been attributable to one disturbance factor and more often would be the result of interactions between a combination of factors, each exhibiting, to various degrees, agency in influencing change in other factors. The key point for archaeologists, which has not so far emerged clearly from the extensive literature on the nature of early Holocene woodland, is that to a significant extent, particularly in wooded areas, the mosaics created by disturbance factors would have exhibited linear patterns along frequented routes of animal and human movement just as we saw in the American North-west Coast examples, the Douglas Map and Willamette Valley (Chapter 2).

Woodland manipulation and management

In addition to the effects of mobility itself we need to consider European evidence for the deliberate modification of vegetation and its possible role in the structuration of landscape by routeways. Trees may have been consciously modified and thus enter the cultural sphere by virtue of modification, or because they were perceived in particular ways (Jones and Cloke 2002). Examples of Culturally Modified Trees (CMT) have been referred to from North America and there is similar evidence from the Siberian Khanty (Jordan 2003a). Aboriginal Australians also modified trees, removing bark for canoes, containers, shields and shelters; some trees were carved or decorated (Victoria State Government 2008a). In Denmark Mesolithic coastal sites have recently produced vast numbers of even-sized long straight poles which clearly formed part of fishtraps: at the site of Ronas

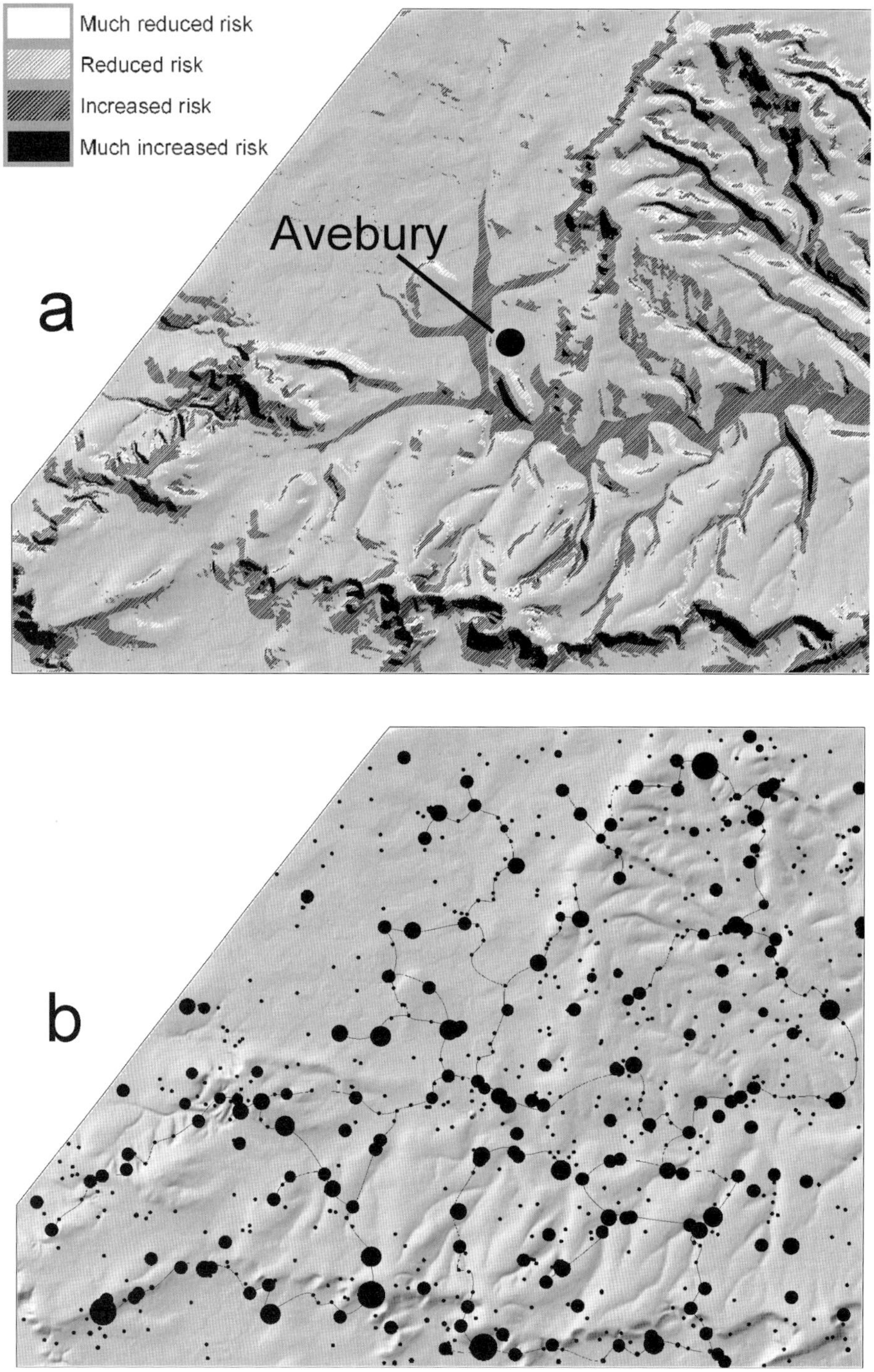

Figure 3.7 The Avebury area, Wiltshire, UK, showing (a) the modelled susceptibility of topography to tree-throw; (b) predicted pre-Neolithic clearances and pathways based on (a) (image: courtesy of Dr D. Wheatley and Dr M. Gillings; Gillings et al. *2008).*

Skov there were 100,000 such rods, and at Tybrind Vig there were 7000 (Andersen 2013). Such numbers of poles imply the deliberate management of hazel coppice in order to provide suitable materials, an inference supported by the high hazel pollen values which are found at coastal sites (Christensen 1997b). Fixed fishtraps and the concomitant long-term management of woodland resources raises important issues of the ownership of resources by hunter-gatherers; ownership is recorded in relation to these intensified practices in a North American context (Stewart 1977, 20). Similarly, very finely-made basketry fishtraps at several sites, including Magrethes Naes on the Storebaelt, Denmark and Ageröd V, Sweden (Pedersen *et al.* 1997; Larsson 1983), imply careful management of willow and hazel in order to produce the perfect materials for such fine work, analogous indeed to the husbanding of basketry resources seen ethnohistorically in North America (Chapter 2). Mesolithic fish traps recently found in Ireland in the River Liffy, Dublin and at Clowanstown have both been interpreted as evidence of the husbandry of wood resources (McQuade and O'Donnell 2007; Fitzgerald 2007). Warren *et al.* (2014), however, highlight the problems in establishing the extent to which formal coppice cycles, as opposed to adventitious coppicing (selecting suitable rods), were practised in the Mesolithic. From the perspective of the development of landscape structures and routes, the distinction between regular and adventitious coppice is not so significant because places where either was practised are likely to have produced readable signs in the landscape.

The broad spectrum revolution and niche construction

People have often husbanded plants and animals which do not differ from wild taxa (Ingold 1996) and they do not necessarily recognise the distinction which westerners make between wild and domestic, nor the concept of nature being transformed by people. Barker (2006) argues that in many parts of the world hunter-gatherers adopted forms of behaviour presaging animal and plant husbandry long before domestication occurred. These often took the form of a developing co-evolutionary process with mutualistic benefits. The development of these relationships, involving a widening range of plant and animal resources, occurred in many parts of the world in the late glacial and initial Holocene, a transformation labelled by Flannery (1969) the Broad Spectrum Revolution (BSR). Such an expansion of the range of resources seems often to have been linked to a process of induced environmental change or Niche Construction which significantly increased the abundance of these resources (Smith 2011; Zeder 2012). In some areas plants and animals which formed part of the BSR became domesticated, in others they did not. The BSR is often linked to more sedentary, less mobile, communities, in part perhaps because, as populations grew, there was less opportunity to move around and greater need to intensify the use of available resources. Zeder (2012, 259) observes that 'greater permanence of human presence opens up new opportunities for plants and animals to move into new anthropogenic environments and develop a range of relationships with humans'. The social implications of this are likewise significant. Increased concentration on local resources strengthens the community's role as a source of environmental knowledge and the social ramifications of this may help to explain the flowering of art, ritual and symbolism which is a characteristic of immediately pre-agricultural and early farming communities in the Near East (Zeder 2012).

Hunter-gatherer plant use

Plant foods would have made a significant contribution to Mesolithic diet. Arguably, human physiology generally requires something like 30–40% of plant food (Zvelebil 1994). Clarke (1976) observed that archaeologists tend to foreground the role of hunting, giving insufficient emphasis to plant resources, of which there are several hundred potentially edible and useful plants in the deciduous forests of North-west Europe (Mears and Hillman 2007). Zvelebil (1994) assembled the evidence for wild plant use in Mesolithic Europe and Bishop *et al.* (2015) have demonstrated systematic and intensive plant exploitation in Scotland. Hazelnuts are the most frequently represented plant resources in Mesolithic Europe; hazel was widespread in many areas and its charred nuts are frequently preserved. The settlement of Staosnaig, Colonsay, Scotland had a pit, dated to 6700 cal BC, containing a large cache of hazelnuts, together with evidence for other plant resources such as lesser celandine tubers and crab apples (Mithen 2000). Significant hazelnut exploitation is likewise demonstrated on various sites in Ireland (McComb 2009) and the Duvensee, Germany (Holst 2010). Charred tubers and evidence for the use of white waterlily (*Nymphaea alba*) seeds occurs on sites in Ireland (Warren *et al.* 2014). At the waterlogged site of Tybrind Vig, Denmark use of a wide range of plant resources is documented (Andersen 2013).

Zvelebil (1994) noted that antler mattocks are widely found in riverine and coastal contexts and he argues these were used for digging of roots and tubers. As noted in Chapter 2, from North American evidence, the very digging of roots and tubers encourages their reproduction and growth. In a similar way, fruit trees can be seen as weedy colonisers which will have flourished in disturbed habitats, to which they may also have been transplanted. Many species with weedy tendencies are plants of the woodland edge which will have been spread along routeways and flourished in these more open, disturbed and nutrient enriched environments. Rackham (1986, 279) notes a number of species characteristic of trackway edges, some fertilised by dung, or by mineral material washed from a road surface, others germinated in water seeps in hollow ways.

Hunter-gatherer vegetation disturbance in Britain

This topic is significant because, as the North American evidence in Chapter 2 demonstrated, environmental disturbance is one of the ways in which early routeways are likely to have become marked. There is significant evidence for Mesolithic disturbance in Britain but specific instances have not generally been related to the topic of routeways. Zvelebil (1994) identified 105 sites with possible evidence of Mesolithic vegetational disturbance, fire being implicated at 29 of them. A predominantly upland moorland distribution and later Mesolithic date also seemed apparent (Simmons 1996; Innes *et al.* 2013). This carried the possible implication that vegetation disturbance was an evolutionary development arising from increased population density, sedentism and environmental manipulation. Sites have continued to be identified and a subsequent survey prepared as background for this book, and presented in detail in Supplementary Appendix 3.1, lists 195 sites which are mapped in Figure 3.1. The evidence is predominantly derived from pollen and charcoal but other palaeoenvironmental evidence contributes in a smaller number of cases (Fig. 3.8a). The new sites change the distribution significantly, with many more sites in

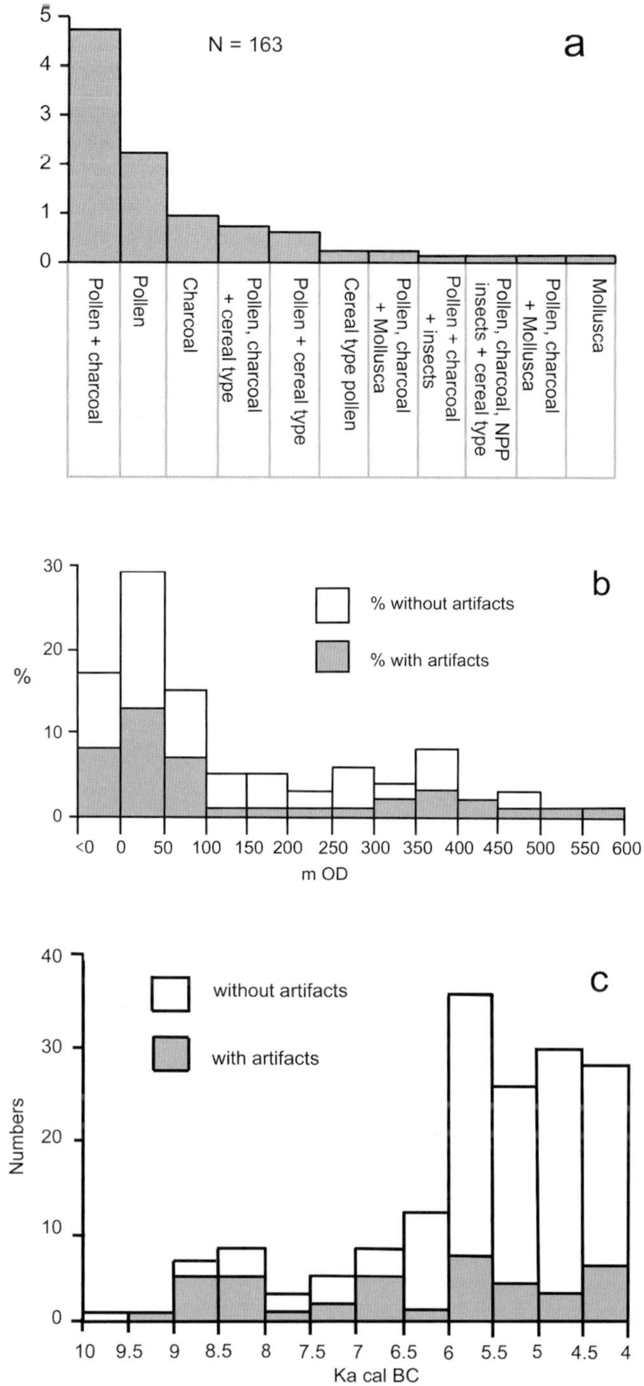

Figure 3.8 Mesolithic vegetation disturbance in Britain. (a) sources of evidence (b) heights above sea level of vegetation disturbance (c) radiocarbon dates of vegetation disturbance (graphic J. Foster; source: Supplementary Appendix 3.1).

southern Britain and Wales. No longer is there a mainly upland distribution. 46% of sites are below 50 m OD and 17% below 0 OD (Fig. 3.8b), emphasising the importance of the coasts to Mesolithic communities, especially if one considers that the majority of coastal sites must now be permanently submerged. There are more radiocarbon-dated sites in the last 2 millennia of the period (56%; Fig. 3.8c). That is further increased to 63% if we include sites which are not radiocarbon-dated but where disturbance clearly predates the elm decline, although not all of these are necessarily pre-farming. However, if we look at the sites where there is a clear association with human agency in the form of artefacts, there is a more even distribution across the period. Earlier arguments that the changes are of late Mesolithic inception are challenged because some of the clearest evidence of human involvement, at sites like Star Carr and Thatcham as outlined below, occurs in the first 1–2 millennia of the Holocene.

The strength of the evidence for human agency as a cause of disturbance varies greatly. In a small number of cases (4%) it is based solely on the presence of cereal-type pollen which could be attributed to certain wild grasses, or precocious farmers (Edwards 1988). In a much larger number of cases there are only slight changes of vegetation e.g. woodland reductions and the presence of species of disturbed conditions which could be the result of a range of disturbance factors. It would probably be fair to say that there has been too much readiness to attribute disturbances to people without considering alternatives. Also problematic is establishing the contribution of fire, whether deliberate or natural wildfire. Fire determines the character of vegetation in many parts of the world, interfacing with human culture in myriad ways (Pyne 2012) from its earliest occurrences associated with archaeological sites in Africa *c* 1,500,000 years ago (Gowlett 2016). The occurrence of natural wildfire provides a proxy record of drier climatic periods in many areas (Harrison and Sanchez Goni 2010; Marlon *et al.* 2013). Some episodes of charcoal occurrence in Britain seem to correlate with periods of drier conditions, for instance in late glacial Scotland (Edwards *et al.* 2000; Tipping 1996). As dating precision increases due to Bayesian statistical modelling of sequences, so the possibility of correlating disturbance phases with distinctive climatic episodes attested from other sources increase and Wicks and Mithen (2014) have recently argued that disturbance phases in woodlands of the Inner Hebrides correlate with climatic episodes. It is also probable that a proportion of the charcoal occurrences relate to campfires as opposed to the deliberate modification of vegetation. Even so, there are many cases of marked charcoal occurrence in pollen sequences of Mesolithic date, both on sites with artefacts and in off-site contexts. Often this accompanies other forms of vegetation disturbance such as a reduction of trees and peaks of plants of open conditions. It is generally agreed that the temperate woodlands of Western Europe, and Britain in particular, are very difficult to ignite (Brown 1977; Rackham 1988; Pyne 2012). Mellars (1976) has highlighted the role of fire among hunter-gatherers generally and in terms of the significant archaeological evidence from Britain, arguing that it created favourable conditions for game especially deer. Clarke (1976) also highlighted the role of fire in the encouragement of wild plant foods. Particularly telling is that in several moorland sequences charcoal occurrence virtually ceases with the appearance of the first farmers c 4000 cal BC and this is without obvious climatic explanation (Edwards 1988). There is a marked concentration of sites in upland moors especially the North York Moors and Pennines (Fig. 3.1) where some episodes of charcoal deposition are associated with artefacts below later blanket peats. The evidence is most persuasive on sites such as North Gill and Bonfield Gill, Yorkshire where there were repeated episodes of charcoal occurrence

(Simmons 1996) and those where pollen and charcoal evidence is supplemented by additional sources such as non-pollen palynomorphs indicating the presence of grazing animals (Ryan and Blackford 2010; Innes *et al.* 2010; 2013). Evidence for Mesolithic vegetation disturbance in Ireland has recently been reviewed by Warren *et al.* (2014); out of 33 palaeoenvironmental sequences they found evidence of disturbance at 16. On two sites this was as early as the eighth millennium cal BC but instances were more frequent late in the Mesolithic, with 16 sites having disturbance evidence after 6000 cal BC; one site, Mount Gabriel, showed evidence for multiple episodes of burning. Deer and aurochs being absent in Mesolithic Ireland, anthropogenic disturbance was more likely for the encouragement of wild plant resources.

Star Carr

One of the most pronounced concentrations of Mesolithic sites is in the Vale of Pickering, Yorkshire around a former lake. Here the Mesolithic site at Star Carr has long been recognised as the most significant of this period in Britain. Occupation occurred at the beginning of the Mesolithic between 9300–8500 cal BC in open grassland with birch scrub and reeds fringing a lake shore. Beavers were already active in a landscape undergoing gradual woodland colonisation (Milner *et al.* 2018). However, the vegetation was impacted by repeated burning of vegetation including reedswamp on the lake shore (Mellars and Dark 1998). Phases without burning and with reduced activity suggest occupation was not continuous. Evidence for a distinct seasonal focus is not apparent because animals were killed in all seasons. There is evidence of dwelling structures on the dryland and at the lake edge successive timber platforms were constructed and ritual deposition of artefacts took place, including frontlets of deer antler. These activities attest to common patterns of deposition and belief over some 800 years at what was clearly a focal and socially significant point repeatedly returned to over centuries. Flint sources suggest patterns of connectivity to the coast 10–20 km to the east and also south to the Yorkshire Wolds.

Case study: Kennet valley

At about the same time in the early Holocene in southern Britain the Kennet Valley, a tributary of the Thames, has a concentration of Mesolithic activity and disturbance along a 25 km stretch of the valley around Thatcham (Fig. 3.9); key aspects of the evidence from these sites is summarised in Table 3.1. The sites are on the floodplain and the edge of the river terrace (Wymer 1962; Froom 2012). Activity is concentrated in the first two millennia of the Holocene, with some sites occupied in the later Mesolithic. Activity occurred as woodland became established on the valley sides, initially birch and pine, then hazel. However, trees remained patchy. The floodplain was more open and colonisation by woodland seems to have taken place more slowly than in surrounding areas, apparently due to a combination of disturbance factors. Several of the occupation sites began on dry ground but became subject to increasing seasonal flooding, though activity continued despite increased wetness, suggesting occupation was seasonal. The flooding can be attributed to the activities of beavers, represented by bones on several sites, and they will have contributed to the maintenance of open areas (Coles 2006). Deer bones and boar predominate on many sites of this period and grazing by animals moving up the river valley may have helped maintain more open conditions. Particularly significant is evidence from Thatcham reedbeds for repeated episodes of burning, including that of reedswamp, suggesting that Mesolithic communities were deliberately creating more open conditions to favour grazing animals and the plants of the woodland edge (Barnett 2009). Thus, a series of niche-constructing factors were in play: the river flooding, animals and human agency interacting together to create a linear corridor of more open conditions and productive environments attractive to human communities. The corridor can be seen as part of a natural axis of movement between the Thames valley to the east and the River Avon and the Severn Estuary to the west. That the later Neolithic centre at Avebury lies mid-way along this hypothetical axis may be further evidence of its longer term significance.

A notable feature of these sites, both the Vale of Pickering, the Kennet and also other tributaries of the Thames such as the Colne is that the concentration of early Mesolithic activity occurs near Long-Blade sites of the late glacial or initial Holocene (Fig. 3.9; Conneller and Schadla-Hall 2003; Froom and Cook 2005; Lewis and Rackham 2011). This may indicate that patterns of river valley focused mobility were originally established in the Pleistocene–Holocene transition, at a time when horse and reindeer roamed very different open landscapes. Such patterns may then have been perpetuated by following existing trails enhanced by human and animal niche-constructors, as woodland encroached in the first millennia of the Holocene. Such a natural corridor of mobility may be seen as analogous in a way to the case of human niche construction documented from botanical and historical sources in North America such as the Willamette Valley (Chapter 2).

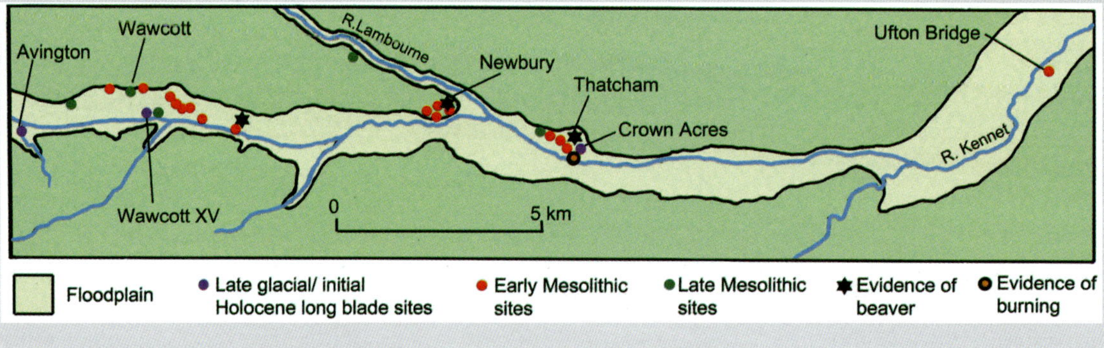

Figure 3.9 Kennet Valley showing the location of Late glacial and Mesolithic sites (graphic: J. Foster).

Table 3.1 Mesolithic sites in the Kennet Valley outlining date, environmental evidence and its sources.

Site	*Date cal BC*	*Environment*	*Sediments*	*Pollen*	*Mollusca*	*Flooding*	*Beaver*	*Reference*
Victoria Park, Newbury	8430–8240	floodplain open, some woodland	✓		✓	✓	✓	Bell *et al.* in prep.
Faraday Road, Newbury	8600–8300	grassland / sedge floodplain some trees	✓		✓	✓	✓	Ellis *et al.* 2003
Thatcham Sewage Works	8500–7200	pine hazel grass herbs partly open	✓	✓				Healy *et al.* 1992
Thatcham Wymer sites	10,300–6550	pine, birch grasses	✓	✓		✓	✓	Wymer 1962
Thatcham reedbeds	9150–7520	floodplain sedge and grass, dry ground birch, pine then hazel	✓	✓	✓	✓	✓	Barnett 2009
Ufton Bridge	9300–7380	pine birch partly open floodplain	✓	✓	✓	✓		Barnett 2009

Case study: A Welsh model of river valley based mobility

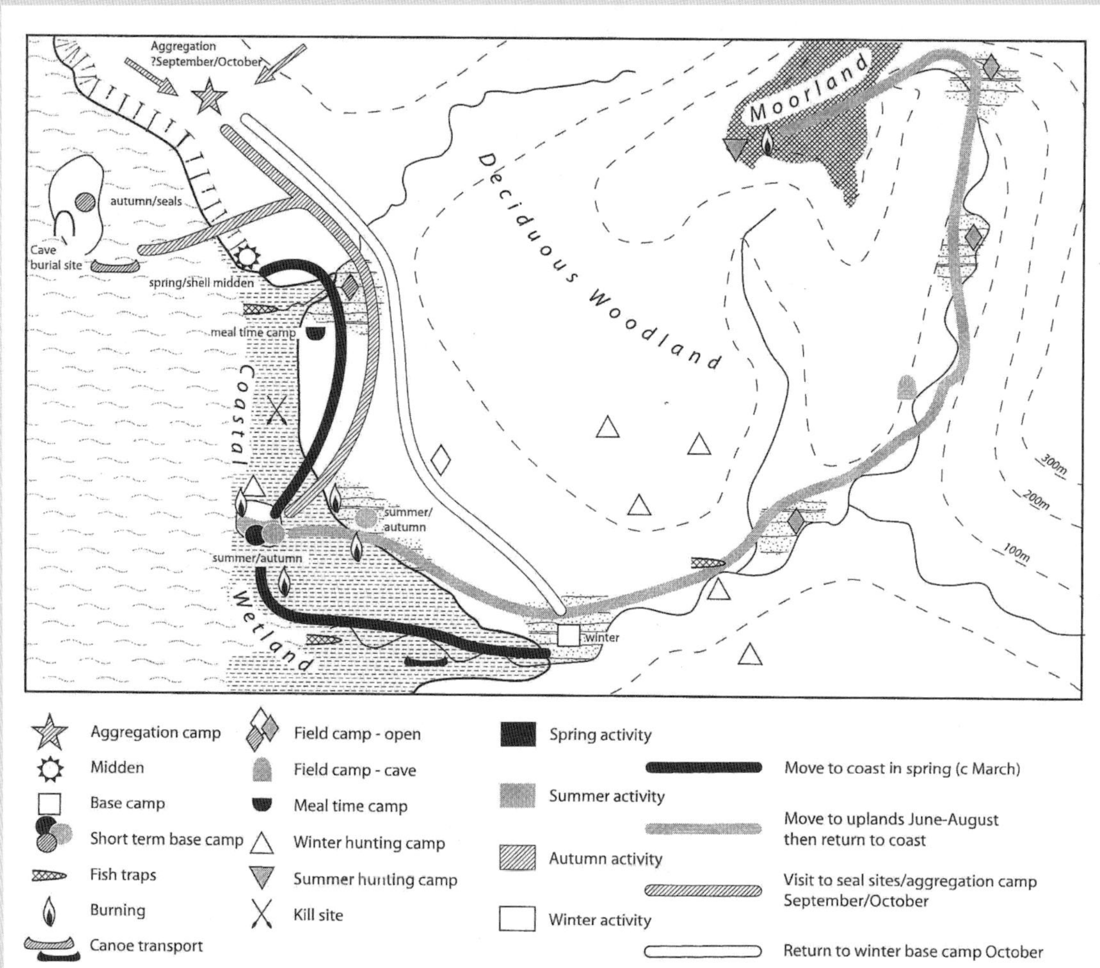

Figure 3.10 Hypothetical model of Mesolithic annual movement based on evidence from Wales (after Bell 2007a; graphic M. Matthews).

A hypothetical model of coastal and river valley based mobility is shown in Figure 3.10 based on evidence from sites in Wales (Bell 2007a). Here coastal locations with submerged forests have produce evidence of burning, from both on- and off-site locations (Fig. 3.4). Goldcliff is an example of the former. On the edge of a bedrock island surrounded by wetland are several small and short-lived seasonal camps (Bell 2007a). The model envisages episodes of activity on the coast to gather plant resources, hunt and fish at times of eel and salmon runs. Seasonality evidence shows this was particularly in the autumn, with shorter visits at other times of year. The model postulates the existence of aggregation camps, represented by major artefact concentrations, at times of particular abundance in autumn. The possibility of winter base camps in sheltered locations such as estuary heads is suggested and in summer movement upriver to the uplands is attested. Caves in the Wye valley were used and seashell beads attest to a coastal connection (Barton and Roberts 2015). People went up-valley to exposed moorlands, probably following deer in the summer, where, at sites such as Waun-Fignen-Felen, there is evidence for the repeated burning of vegetation over several millennia from 6000 cal BC, a classic case of a regularly visited 'persistent place' (Smith and Cloutman 1988; Barton *et al.* 1995). The evidence for burning in both coastal locations, favoured in autumn, and in uplands, utilised in summer, means that Mesolithic communities were engaged in niche construction at both ends of their hypothetical annual territories. Such a coastal/inland model makes sense in the topographic context of Wales with an extensive central upland and a radial pattern of rivers flowing to a generally narrow coastal plain. If this model is anything like reality then the drowning of coastal lowlands as a result of sea-level rise, which inundated the Severn Estuary, Cardigan and Liverpool Bays, as Figure 3.11 shows, would have had the effect of reducing annual river catchment-based territories by an average of 46% (Bell 2007a, 335).

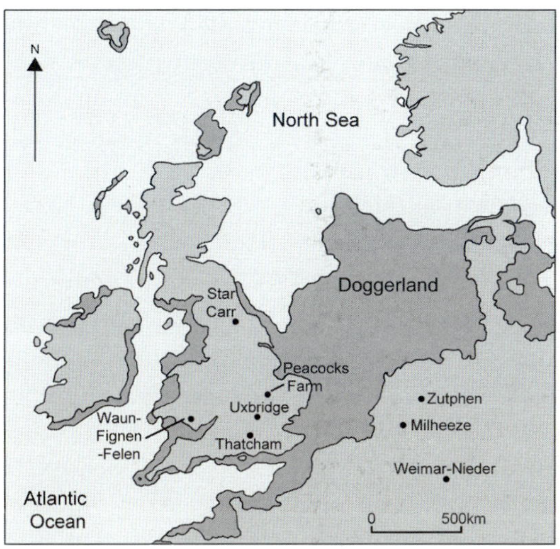

Figure 3.11 Early Mesolithic sites in Britain and North-west Europe with evidence for fire and vegetation disturbance. The coast is shown as it was at the beginning of the Holocene c *9500 cal BC (graphic J. Foster; coast after Shennan* et al. *2000).*

Continental Europe

Until recently, the evidence for Mesolithic vegetation disturbance and burning in Britain was in marked contrast to that from continental Europe, where it is still much more limited, but instances are increasingly recognised (Fig. 3.11). Some charcoal occurrences in late glacial Allerød contexts have been interpreted as a result of lightning strikes (Behre 1988). However, at the Upper Palaeolithic site at Milheeze, Netherlands, a reduction in birch/pine forest is associated with pine charcoal and is contemporary with an adjacent site dated 11,200–10,600 cal BC (Bos and Janssen 1996), appearing to suggest an association with human activity. A little later at Zutphen, Netherlands, an early Mesolithic site *c* 9100–8500 cal BC was in open grassland which seems to have been maintained by burning (Groenewoudt *et al.* 2001). Particularly revealing is the site of Weimar-Niederweimar, Germany in the valley of the River Lahn, a Rhine tributary, where there are two episodes of woodland reduction (*c.* 8470 and 8350 cal BC) at Mesolithic sites on the river terraces; both were associated with charcoal and the occurrence of plants of disturbed and nutrient-rich conditions (Bos and Urz 2003). The association on this site of charcoal with disturbed, nutrient-rich conditions hints at a natural routeway of animal movement along these river terraces, as was suggested in the case of the Kennet. The Dutch and German evidence for vegetation disturbance seems to be more marked in the early Mesolithic, as woodland cover was developing, and perhaps as communities burnt to retard woodland succession in areas of key resources, which may often have been along routeways.

In Denmark there has been extensive research on the Mesolithic, including many substantial shell midden sites, the oldest dating from *c* 5600 cal BC and some in use for 700–1000 years (Milner 2002). In such cases a significant effect on the environment might be anticipated. Yet in terms of clear evidence for burning, or vegetation disturbance, there is limited evidence from palynology. Only at one site, the inland site at Ringkloster, does a reduction of woodland occur at the time of the site's occupation *c* 4400 cal BC (Rasmussen 1995). Submerged forest trees in the Storebaelt also produced two charred trees, which might be attributed to human activity (Christiansen 1997a, 40). At Hassing Huse, Jutland, two elm declines, slight hazel rises and slight charcoal peaks in the mid and late fifth millennium cal BC occur (Andersen 1993; Andersen and Rasmussen 1993), but the authors regard the evidence for Mesolithic disturbance as extremely weak. Paradoxically, however, as already noted above, Denmark has yielded the most convincing evidence for the manipulation of woodland by hazel coppicing to produce vast numbers of straight poles and some fine basketry for fishtraps (Christiansen 1997b). At Tybrind Vig the pollen diagrams show a landscape of primeval forest with a dense canopy and it is clear from submerged trees that the forest extended to the shore (Andersen 2013). The pollen of terrestrial herbs was very limited and it was concluded that openings in the forest were few, and that 'the forest was without visible trace of human interference'. However, the plant macrofossil evidence suggests a different picture: 42% of the utilised wood was of hazel and the charcoal is dominated by hazel and other understorey species. This indicates that people had affected the composition of the lime forest, to encourage hazel which provided nuts and poles for fish traps. Evidence for burning occurs in association with a small number of Mesolithic sites in Norway, Sweden and Finland (Welinder 1990; Regnell *et al.* 1995; Berglund *et al.* 1996, 167).

Across Europe some notable and paradoxical contrasts emerge in terms of Mesolithic environmental disturbance. In what is now mainland Europe, the evidence seems to be mainly early in the Mesolithic and to relate particularly to the stages of vegetation succession prior to the development of climax woodland. In that sense it is analogous to the early Holocene British sites at Star Carr and Thatcham. At this time Britain was linked to continental Europe. In the later Mesolithic, after the drowning of Doggerland (Fig. 3.11) *c* 7000 cal BC, which must have had a severely disruptive effect on patterns of Mesolithic connectivity and environmental knowledge (Leary 2015), there is very little evidence for vegetation disturbance, even in parts of Denmark and Schleswig Holstein where sites are abundant, rich and long-lived. Overall there is far more evidence for Mesolithic vegetation disturbance in the British Isles than continental Europe.

Mobility and sedentism

The emphasis here has been on the identification of patterns of movement in the Mesolithic, but it has been argued that some groups were sedentary or semi-sedentary, especially in rich coastal environments capable of supporting year-round occupation. The later Mesolithic shell middens of Denmark are vast accumulations of shells and cultural material which demonstrate a wide range of resource utilisation. Some may represent year-round activity (Andersen 2000), associated perhaps with the ownership of fixed resources, such as the fish-traps and managed woodland and also burials in some middens. However, studies of shell collection seasonality have revealed concentrations at certain times of year and clear evidence for year-round occupation remains elusive (Milner 2005). Identification of seasonality and sedentism is a challenging topic and, in common with many of the themes discussed here, rests generally, not on a single source of evidence, but on evaluating a range of sources in a comparative way. The middens may have served as base camps, with sections of the community decamping elsewhere for periods to exploit specific resources in logistical (i.e. task-specific) camps. The shell middens on the tiny Scottish island of Oronsay provide biological evidence for occupation of the island in all seasons of the year (Mellars 2004), although that seems counter-intuitive on a small and exposed island and Mithen (2000) has argued for shorter visits at various times through the year. Sedentism has also been proposed where there are substantial pit houses, as for instance at Mount Sandel, Ireland (Woodman 1985) and Howick, Northumberland (Waddington 2007). However, recent analysis of date sequences from pit house sites suggests that, rather than continuous occupation, they are characterised by repeated visits with interruptions over extended timescales (Mithen and Wicks 2018).

Some recent British excavations have uncovered extensive areas of Mesolithic landscape, revealing multiple discrete artefact scatters with little between, which would seem to suggest recurrent, relatively short-term, activity over extended timescales. Had activity been more continuous one would have expected a much more blurred picture and substantially greater quantities of artefacts than characterise most sites. Discrete clusters are found at some of the Vale of Pickering sites, such as Seamer Carr (Conneller and Schadla-Hall 2003), a site complex at Bletchingly, Surrey where activity occurred over 4000 years (Jones 2013), the Goldcliff intertidal sites occupied for 1000 years (Bell 2007a), and at Bexhill, Sussex there is a complex of over 200 discrete artefact scatters on the edge of coastal wetland (M. Donnelly, pers. comm.). Discrete artefact clusters are also seen in the Kennet valley sites at Thatcham and Wawcott (Froom 2012). Where various comparative sources of seasonality evidence are available, as at Goldcliff, Wales, they tend to point to activity mainly concentrated in one time of year, but with evidence for occasional visits at other times of year (Bell 2007a). In other words the general picture seems to be of seasonal and logistical (task specific) mobility. The discrete multiple artefact scatters and the pit houses with episodic use, point to recurrent patters of activity which have often been rationalised in terms of the attractive ecological resources of the place. Such long-term foci may, however, make more sense as places on frequented routes marked by animal and human passage imprinted on landscapes by linear niche construction. Sometimes, as perhaps at Star Carr, these places may have become significant for social reasons as much as for the ecological resources they provided.

Artefact areas and 'monuments'

Much of the evidence considered so far takes the form of what may often have been unconscious traces of human activity and movement in the landscape. Mesolithic communities also left traces of their activities in the form of lithic and artefact scatters which would often have been obvious to others moving along the same trackways. Sometimes there is evidence that artefacts were deliberately buried and hidden in tree-throw pits before people moved on, although this might relate to the deliberate caching of materials which could be reused on a subsequent visit (Evans *et al.* 1999). Lithic artefacts with distinct episodes of reuse, for instance retouch through patination on previously worked flints, indicate that this occurred. The association between Mesolithic artefacts and tree-throw pits is seen also in Belgium where the inference drawn is that people were attracted to tree throw areas (Crombé 1993). Animal footprints on the edge of tree throw features in Belgium are noted by Langohr (1993) as highlighting the attractiveness of these areas to grazing animals.

In coastal areas the focal points of activity often left obvious traces in the form of shell middens. These would have stood out on account of their smell which would have been obvious at significant distance; their white colour; the trails and disturbance by scavenging animals they attracted; and the distinctive plant communities associated with the geochemical signature created by such concentrations of calcium and nitrogen-rich midden material. This would have been enhanced by a halo of plants growing from seeds deposited in faeces. It has been argued that middens should be seen as Mesolithic monuments playing a social role in delineating landscape (Chatterton 2006). This inference is strengthened by middens, such as those on Oronsay, which contain individual human bones (Mellars 1987; Richards and Sheridan 2000), and some Scottish middens which have Neolithic tombs built on them (Pollard 1996). An association between shell middens and Neolithic tombs is also found on Guernsey (Sabre 2005). Shell middens are numerous and often very substantial round the coast of Denmark and are likely to have acted as important way marks for those travelling by water. The soils created by concentrations of shells in otherwise non-calcareous environments are also likely to have attracted early farmers to the sites of Mesolithic middens because they would have been fertile places for garden agriculture, as is suggested by evidence for later Neolithic cultivation on middens at Northton, Harris and Prestatyn, north Wales (Gregory *et al.* 2005; Bell 2007a).

However one interprets the midden evidence, it has become increasingly clear that Mesolithic communities did create structures which are akin to monuments and there is growing evidence that these related to routes. Three or four substantial post pits found in the Stonehenge car park have been dated to 8800–6590 cal BC. It is at least possible that the relationship to the much later Stonehenge (Chapter 5) is not coincidental and that the posts mark the special significance of a place which was much later monumentalised, Mesolithic activity having left some tangible trace marking the significance of the place. That possibility becomes less fanciful if we posit a path or route, or a convergence of paths, which led successive generations of people and animals to the same point, maintaining a grassy clearing to which myths and legends and genealogies became attached. Allen and Gardiner (2002) cite several other instances of post holes and/or pits dated to the Mesolithic on sites which later became significant foci of Neolithic activity, some of which are discussed in Chapter 5. A particularly interesting example is at Crathes, Aberdeenshire, Scotland where 12 pits have been excavated forming an alignment over 50 m; at least two seem to have held wooden posts (Murray *et al.* 2009). They have radiocarbon dates centring on the period 7000–6800 cal BC and appear to have been dug in open hazel and birch woodland over several hundred years, with some activity continuing to the Neolithic. The surrounding area remained important in the Neolithic with two rectangular buildings at Warren Field and Balbridie nearby.

Isotopes and mobility

So far this chapter has concentrated on the traces that peoples' activities and movements leave on the landscape. A complementary perspective may be obtained from the evidence which movement leaves in the stable isotopes incorporated in the bodies of people and animals (Sealy 2001). Carbon ($\delta^{13}C$) and Nitrogen ($\delta^{15}N$) isotopes provide evidence of diet and trophic level (position in the food chain). Using this evidence, it is possible to estimate the proportion of marine/terrestrial food in diet from the carbon isotopes. Nitrogen isotopes can produce evidence of what contributed to that diet as between, for instance, primary producers and top carnivores. In the Bristol Channel, UK analysis of early Mesolithic burials in Aveline's Hole, a site 100 km from the coast at the time of its occupation before sea level rise, shows no evidence for the consumption of marine resources, indicating that the people did not spend extended periods on the coast (Schulting 2005). Burials further to the west on the south Wales coast mostly show a high reliance on marine foods, indicating that people spent most of the year on the coast (Schulting and Richards 2002). Significantly, there are one or two individuals with little evidence of marine diet and they may have spent most of their time at inland locations, although they were buried near the coast. In Denmark some individuals from inland locations have been shown to obtain a significant proportion of their diet from marine resources indicating that most of the year was spent on the coast (Fischer 2003). On Oronsay, Scotland human bones from shell middens have an isotopic composition indicating that most of the year was spent in coastal locations (Richards and Mellars 1998). Midden sites on the islands of Teviec and Hoedic, Brittany have male burials who consumed higher proportions of marine protein than younger adult females, who, it was inferred, had spent the early part of their lives in inland locations (Schulting and Richards 2001; Schulting 2003). From this it was deduced that some females moved to coastal locations on marriage indicating patrilocal post-marital residence. In many areas of North-west Europe, it is striking that, following the advent of farming, most burials, even those close to the coast, show little or no evidence for the consumption of marine resources, pointing to a marked and rapid change in economy and patterns of movement (Richards *et al.* 2003). There is, however, continuing debate concerning the interpretation of this evidence and its implications for understanding the process of Neolithisation (Milner *et al.* 2004). Oxygen isotopes could also be employed to identify individuals who derive from climatically contrasting areas and strontium isotopes those who originate from areas of contrasting geology.

Material culture and movement

Material culture, particularly the geological sourcing of lithic materials, can also provide clues to patterns of hunter-gatherer movement (Conneller and Schadla-Hall 2003; Preston and Schorle 2013). However, in this case we have to consider the question of whether it is the people that are moving, or just materials being passed on from one community to another by a process known as down the line exchange.

In northern Fennoscandia patterns of Mesolithic mobility can be established from the occurrence of lithic raw materials that originate on the coast but are found on sites up to 100 km inland (Manninen 2009). That indicates a similar extent of mobility to that recorded for recent East Sami people in the same area (Fig. 3.2d). In the Mesolithic, movements of raw materials suggest movements of tens, sometimes hundreds of kilometres. The Dutch sites at Hardinxveld derived some lithic raw materials from distances of between 50 km and 150 km (Louwe Kooijmans 2001a; 2001b). Lithic raw materials from the Mesolithic site of Chedzoy in the Somerset Levels include flint obtained from chalk outcrops and Greensand chert, both indicating connections to Cretaceous strata 40–50 km to the east (Norman 2003). In north-west England, Preston (2013) has used lithic raw materials, artefact typology and the sequence of lithic artefact production (*châines operatoires*) to challenge existing simple models of east-west movement and others suggesting distinct coastal and inland social groups. He argues that the evidence points to a more complex and extensive regional pattern of mobility, with the Pennine upland as a key node. On the Inner Hebrides islands of Scotland there is a developing picture of task specific settlements on several islands, with mobility between relatively distant groups of islands (Mithen 2000; Wicks *et al.* 2014). Wickham-Jones (2005) has identified the use of bloodstone from the island of Rhum on sites up to 40 km away, whilst mudstone from Staffin on the island of Skye was used on sites up to 120 km distant.

Conclusions

Tilley (1994, 206), in writing about phenomenology, noted that Mesolithic sites would often have been situated on paths. We have noted a whole range of ways in which the places and paths used by hunter-gatherers would have left visible traces. Some might be seen as accidental, others deliberate, although for people who would have been so adept at the skills of tracking that distinction must often be questionable. What to us are ephemeral or equivocal traces are likely to have been very obvious to them. This range of traces, together with the presence of natural routes produced by topography would all have contributed to the creation of persistent places used over extended timescales (Barton *et al.* 1995).

A range of disturbance factors will have created openings within woodland and thus contributed to the patch dynamic, e.g. storms, floods, the activities of animals, disease etc. Open areas, however caused, are likely to have attracted both animals and people. It will often have been easier for people to maintain existing openings than to create them anew (Simmons 1996). It could thus be argued that debates about causation are, to an extent, academic because in most cases cleared areas, including those along routes, will reflect the interaction of animals, natural disturbance factors and people, which in combination had agency in helping to determine subsequent actions and routes through landscape. Burning may often have taken place for social reasons as much as resource enhancement (Lillie 2015).

The ethnohistorical record refers to ease of movement as one reason for burning by aboriginal Australians (Jones 1969). Fire may sometimes have served as a symbol of presence and territoriality (Simmons 1996), perhaps marking, or laying claim to an area, as much as for any direct economic benefits. What cannot at this stage be directly demonstrated is that the distribution of those clearings was determined in part by routeways, although this seems a likely proposition which would repay investigation of the spatial distribution of disturbance whether linear, or curvilinear, patterning related for instance to topographically 'natural routes'.

As noted, multiple sources of evidence can contribute to the identification of patterns of movement by hunter-gatherers including: artefact typology and raw material sourcing; groupings of sites on natural topographic routes; and isotopic studies. Further sources at a more intimate scale will be discussed in Chapter 4 on footprints. Geographical Information Systems and Agent Based

Modelling provide ways of integrating these various strands of evidence in models of Mesolithic settlement, as has been done for the Southern Hebrides (Woodman 2000; Lake 2000). Fieldwork around the Federsee, Germany adopted a least-cost path approach to the study of Mesolithic movement; in that case it was concluded that there was a need to integrate other factors than functional least cost, including social factors concerned with the significance of particular places (Star and Harris 2009). Inevitably the emphasis here has been on paths which may have engendered long term routeway continuity. However, the converse is equally interesting, where factors led to the closure of routeways. Sometimes the cause would have been environmental, such as the radical changes consequent on the inundation of Doggerland by sea level rise. Other more local changes are described, for instance, by Jordan (2003a), writing of Khanty ethnohistorical practices in Siberia, where paths leading to burial places were symbolically closed with saplings after feasts to remember the dead.

A key point is that the effects of human activity such as disturbance and the associated plants, animals, chemistry etc will be concentrated around settlement areas as archaeologists generally recognise; less routinely recognised is their enhancement along routeways. This is perhaps most evident in the case of plant resources, but routeways and associated more open grassy areas will also attract many animals which will act to reinforce the distinctive character of the route. In the case of both plants and animals the manipulation is a mutual process, people affecting and being affected by the floral and faunal patterns which people, animals and natural disturbance factors have helped to create. As people settle down, the way they construct their environment may be predicted to change from circular/net-like, linear/branched or multi-stranded patterns, identified in Figure 3.2a and c–d, to patterns with a stronger focus on individual sites, more like Figure 3.2b. Settling down has other consequences; it concentrates things in a central place and in the routes to and from that place. Places and routes become enriched biologically and chemically. Parasites and disease organisms will likewise become concentrated and, as routes become more formal and frequently used, so that transmission of these organisms and that of exotic species from place to place will increase.

There does seem to be clear evidence that hunter-gatherers in Europe played an important role in niche construction which involved the manipulation of plant resources. In some cases, the active agency is not with humans alone but in the interactions between people, animals, plants and fire, all of which contribute to the conditions which bring about change. The evidence for niche construction, and the intensified use of plant resources and burning, fits a larger pattern of the Broad Spectrum Revolution, which in some areas of the world, such as the Mediterranean and south west Asia and east Asia, is well-attested from the Upper Palaeolithic and initial Holocene when conditions ameliorated after the Last Glacial Maximum (Barker 2006). Many of the situations considered in this chapter such as environmental changes associated with settlement areas, burnt patches and areas of modified vegetation, will have been articulated together by patterns by movement. All of these contribute to what the writer has described as the structuration of landscape by antecedent conditions (Bell 2007a). The legacy traces of these movements will often be one of the most significant factors in that landscape, and will sometimes have long-lasting influences on subsequent activity.

4. Footprints of people and animals as evidence of mobility

Introduction

Investigation of patterns of prehistoric movement is necessarily multi-scalar, encompassing everything from a footstep to an intercontinental journey. The human footprint is a key piece of evidence and a topic which readily captures the imagination of the public (Fig. 4.1). People feel that they can relate to the individual in a way that is more difficult to achieve with artefacts such as pottery and metalwork, because we are observing the actions of a living flesh-covered body. Obvious as the popular appeal may be, the question remains: can this, rather unusual, evidence supplement and go beyond what we can learn from other sources? We often have some idea of the age-group structure of a population from burials, but it is frequently clear that that only a sample of the population is represented in the burial record and footprints provide an alternative way of addressing demographic questions. They also tell us about human ecology at a specific place, and in a defined environment and sedimentary context and individual human actions and sometimes patterns of social and family interaction. Ingold (2011) contended that movement is the heart of perception, being in the world and how we make ourselves. Walking is a way of knowing the world which Ingold and Vergunst (2008) refer to as 'tactile feet first engagement with the world'. Those who habitually go barefoot have much greater dexterity with their feet and are bound to have an enhanced perception of the terrain and substrate over which they move.

Ingold (2011) goes so far as to say that a function of modern footwear has been to separate locomotion from cognition so that people are no longer able to think through their feet. Thus, traces of movement, however fragmentary and difficult to interpret, should lie at the heart of the archaeological project. In some situations we may be able to glimpse multiple fragments of past footprint trails and thus to identify the locations of lost or buried settlements. Previously, when the discoveries of footprints in archaeological contexts were very few, these would have seemed rather pious expectations, however, in the last few years discoveries have proliferated, especially in British coastal environments, greatly encouraging the belief that this is a form of evidence which is far more common than previously supposed, and the vast majority of footprint finds probably go unrecognised. As the number of finds increases, so does our capacity to generalise about them in a way not previously possible.

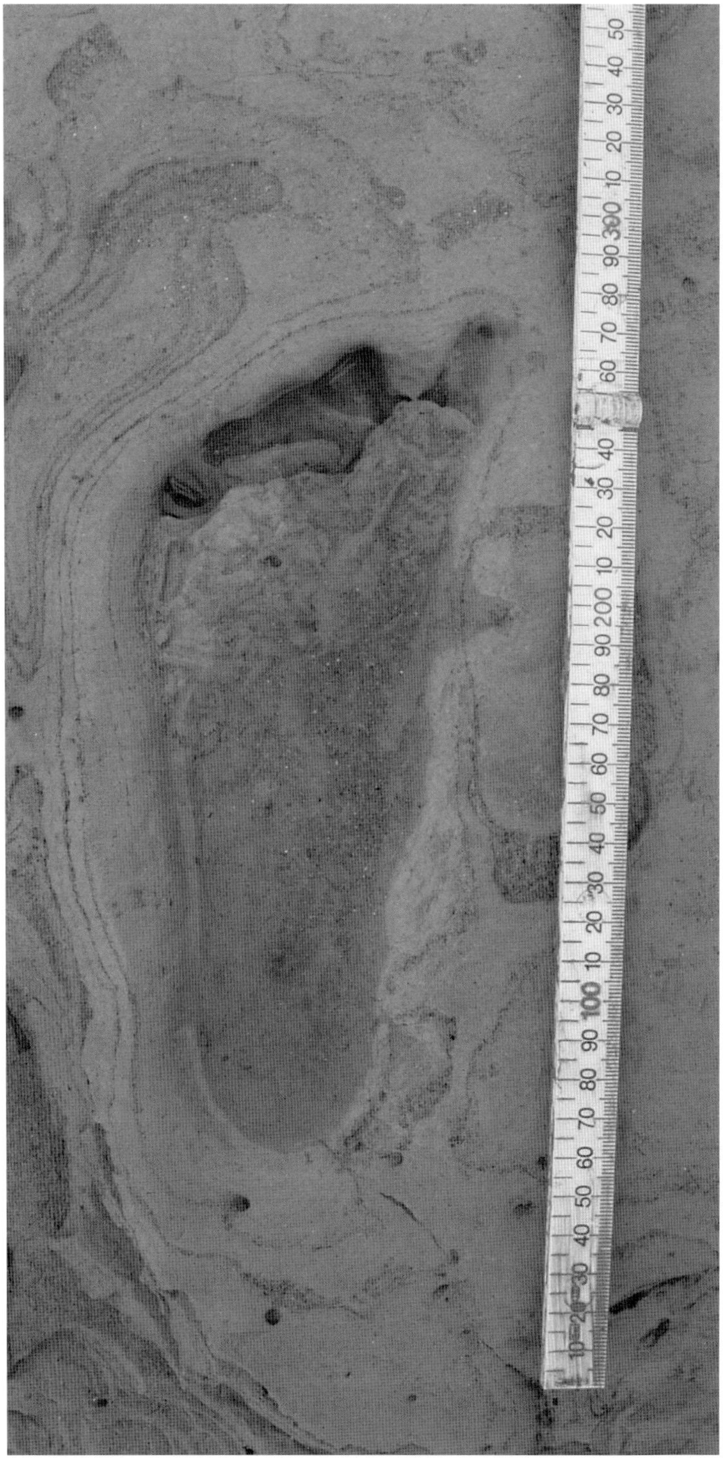

Figure 4.1 Goldcliff East, Wales: the Mesolithic footprint-track of a young person aged 10–12 (photo: M. Bell).

Methodologies for footprint investigation are still being developed. Some previous studies have been insufficiently critical and many, whilst identifying past human and animal footprints, have not made a great deal of the evidence they can provide. Fortunately a recent monograph by Bennett and Morse (2014) makes a substantial contribution to the field of archaeological footprint recording and analysis by putting it in a more solid scientific basis. That book has its main emphasis on Palaeolithic human footprints and their implications for human anatomical evolution. This chapter, whilst briefly reviewing the Palaeolithic evidence, has a particular emphasis on Holocene human and animal footprints and the social implications of the evidence they provide. Bennett and Morse (2014) refer to some 44 sites and partial surveys have also been published by Lockley *et al.* (2008) and Willey *et al.* 2009 for the Americas. Figure 4.2 shows the location and sedimentary context of 93 sites of which 79 have human footprints and 39 animal footprints. Supplementary Table 4.1 provides details of all sites. The list of sites is certainly not comprehensive; discoveries are now regularly being made and many are not yet fully reported.

Trace fossils

Footprints are a form of trace fossil known as ichnofossils after *Ichnos*: Greek for trace. They are the marks made by the activities of organisms in sediment, as distinct from fossils of body parts, such as bones, shells etc. The field of ichnology has grown and a journal *Ichnos* devoted to trace fossils was founded in 1990. It has published many papers on footprints in geological deposits, especially those of dinosaurs. Only since 2008, however, have Quaternary animal and human footprints started to feature in the journal, highlighting the developing nature of this field. The investigation of footprints draws on a remarkable range of disciplinary fields, including palaeobiology, archaeology, physical anthropology, forensic science, natural history and ecology, ergonomics and biomechanics, but there has been limited exchange of information between these separate fields (Allen *et al.* 2003, 55).

The study of dinosaur prints is particularly instructive, in terms of comparison with the trajectory of archaeological research (Lockley and Hunt 1995; Lockley and Meyer 2000). The first examples were described in Scotland in 1828, although incorrectly identified at the time as tortoise prints (Pemberton *et al.* 2008). Rapidly growing interest in dinosaurs in the 1970s was followed by something of a revolution in the study of dinosaur tracks from the last two decades of the twentieth century, when it became clear that the evidence they provided significantly supplemented that from fossils of body parts. The new evidence concerned herding, social activity, the ecological and sedimentary environments in which the animals were actually living. For work on human footprints the field of forensic science, the use of footprints in criminal investigations, has been especially influential. Robbins (1985) developed this field of study in the United States and contributed to a number of related archaeological investigations including the earliest human footprints at Laetoli, Tanzania and footprints in a number of North American caves; however, her work was not, as we shall see, without controversy.

Formation processes and terminology

As with any form of archaeological evidence particular attention needs to be given to the formation processes of footprint-tracks, a topic pioneered by Allen (J. 1997). His careful observation of present-day examples was followed by laboratory experiments employing, for instance, punches

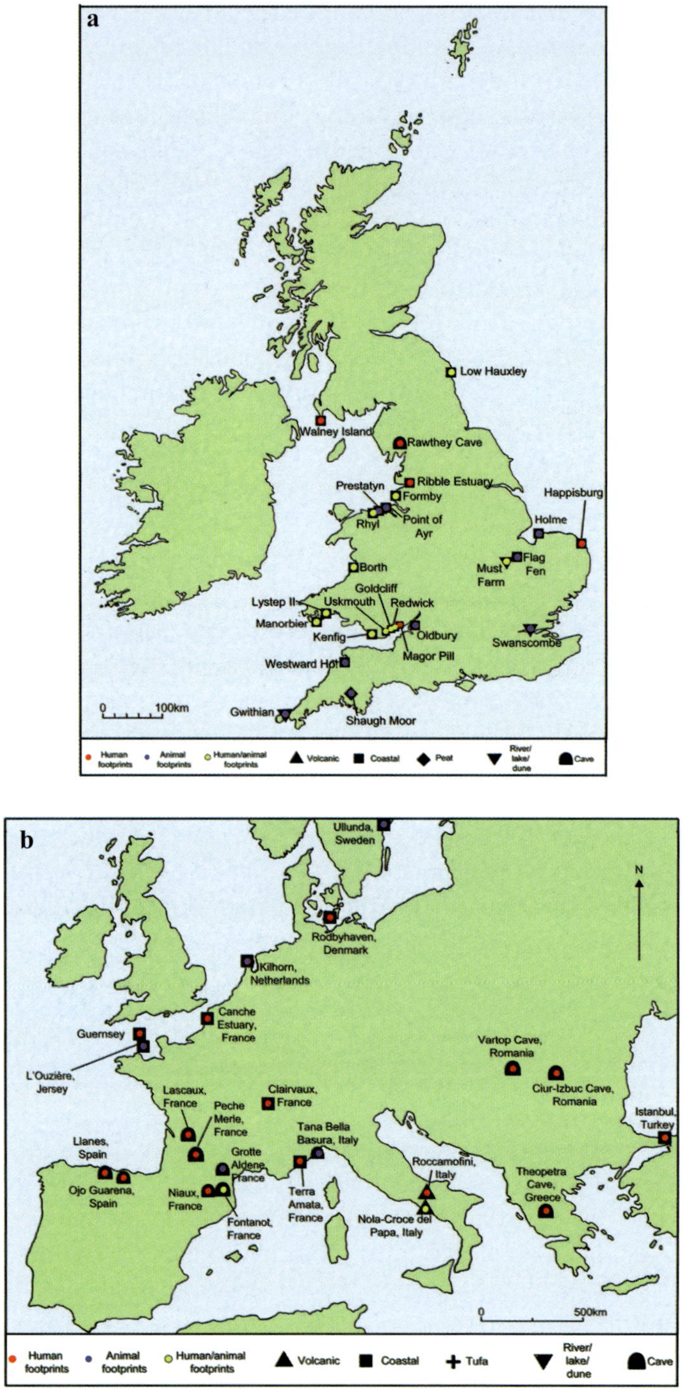

Figure 4.2 The geographical distribution of Pliocene, Pleistocene and Holocene human and animal footprint-tracks in: (a) the British Isles, (b) Europe, (c) Africa, Asia and Australasia, (d) the Americas (graphics: J. Foster; sources in text and Supplementary Table 4.1).

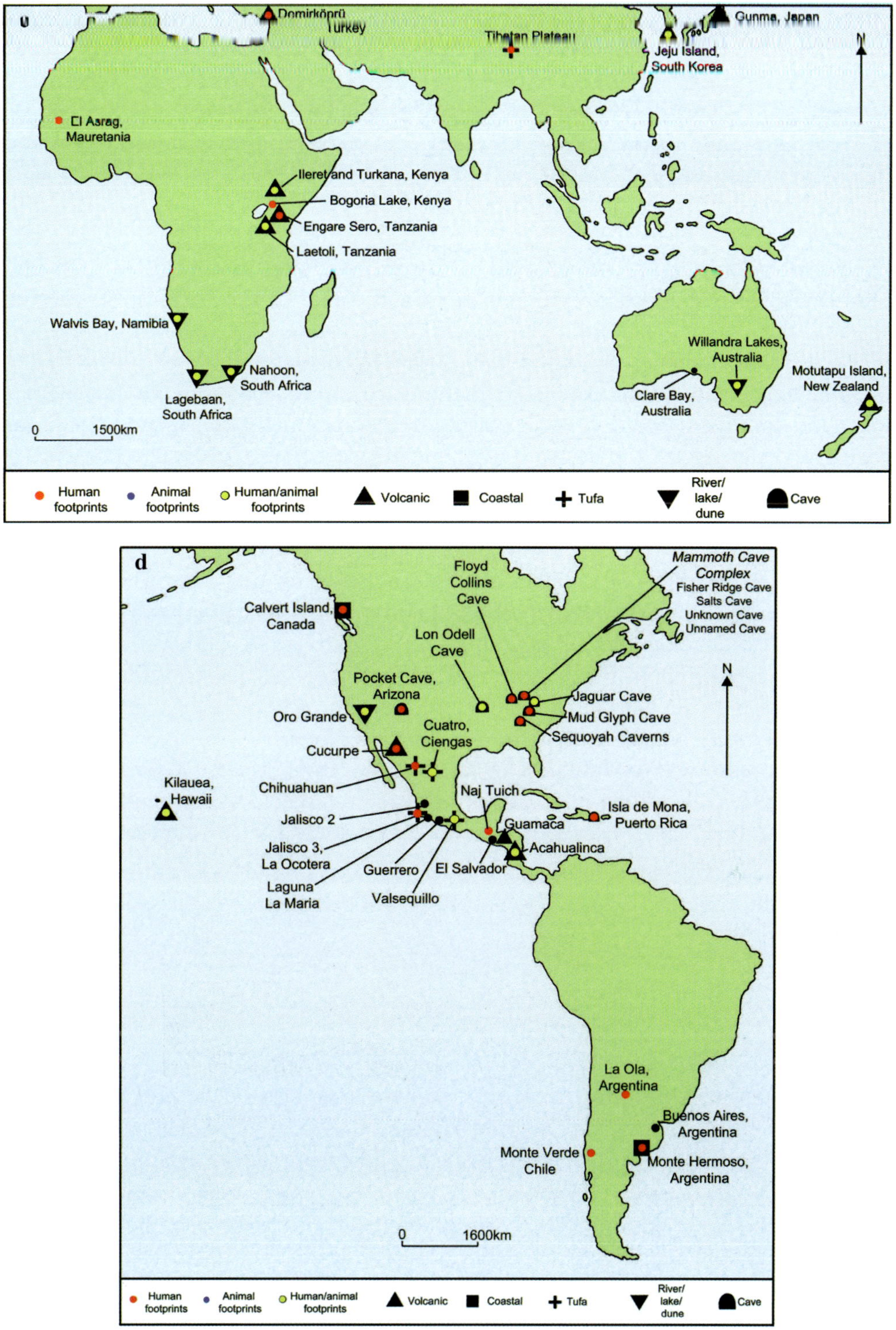

Figure 4.2 (Continued).

and laminated plasticine to refine understanding of footprint impact on sediment. The effects will differ according to the type and consistency of the sediment, the same foot leaving very different traces under contrasting conditions (Bromley 1996; Allen, J. 1997; Bennett and Morse 2014; Gatesy and Falkingham 2017). The ideal conditions for preservation are, relatively fine grained sediments, moderately moist conditions, a change of lithology (e.g. clays or silts to sand), low rate of disturbance, rapid sedimentation, and minimal compaction (Ashley 2003). In the case of minerogenic sediment, water content is a particularly important factor. A footprint in very wet sediment may leave little more than a disturbed oval crater. Sediment which is yielding but not too sloppy may produce a faithful print, if it is then allowed to dry and becomes rapidly filled with sediment of a contrasting type, for instance a print in silt filled with sand, or volcanic ash. Footprints can be particularly well preserved in alternating laminated bands, for instance, where the footprint is made in relatively fine grained silts or clays and then buried by sand laminations (Fig. 4.1).

When laminated sediments are exposed by natural erosion, most commonly marine or deflation (wind blow), the contrasting particle size, compaction and cementation may result in lamination surfaces being revealed. If the footprint fill is of a coarser particle size, or less compact, it will be differentially eroded away so that the original footprint is exposed. In other cases, a footprint-track is revealed by differential fill, as for instance when a footprint in peat is filled with minerogenic silt (Figure 4.14c–e). Footprint horizons may also be cemented by secondary carbonate deposition; this contributed to preservation of footprints at Willandra Lakes, and Clare Bay, Australia (Webb *et al.* 2006; Belperio and Fotheringham 1990). Deposition of secondary carbonates in some contexts is associated with the development of microbial mats (Marty *et al.* 2009). Given the exigencies of preservation we need to consider to what extent the footprints preserved are a representative sample. If people or animals were present at a site at a time when conditions were unfavourable for preservation they may be unrepresented or underrepresented.

Based on observation of modern tracks and simple experiments, a terminology has been developed by Allen (J. 1997; Allen *et al.* 2003) which is applicable in the range of fields to which footprint-tracks contribute. The following key terms are illustrated, where appropriate, on Figure 4.3.

Footprint-tracks, the individual traces made by people and animals. The entire three-dimensional body of sediment disturbed by the descent of the foot and subsequent sedimentary processes (Fig. 4.3a–c).

Trail, a succession of footprint-tracks representing the axis of movement.

Stride length, the complete cycle of limb movements, e.g. from left heel to left heel of successive prints.

Cadence, the rate or timing of footfalls.

Footprint, the surface which was in contact with the underside of the foot.

Undertraces are formed when the footprint pushes into and distorts the sediments below producing deformations which are particularly clear where sediments are laminated (Fig. 4.3a–c).

Shaft is created by the impressed foot, which, when withdrawn, becomes filled with sediment.

Overtraces are the distorted, often layered sediments which blanket the uneven surface created by a footprint-track.

Marginal ridges, sediment squashed up in a rim around the print.

Marginal folds, emplacement of a foot produces small-scale folding of adjacent sediment.

Footprint assemblage, the complete sample from a site.

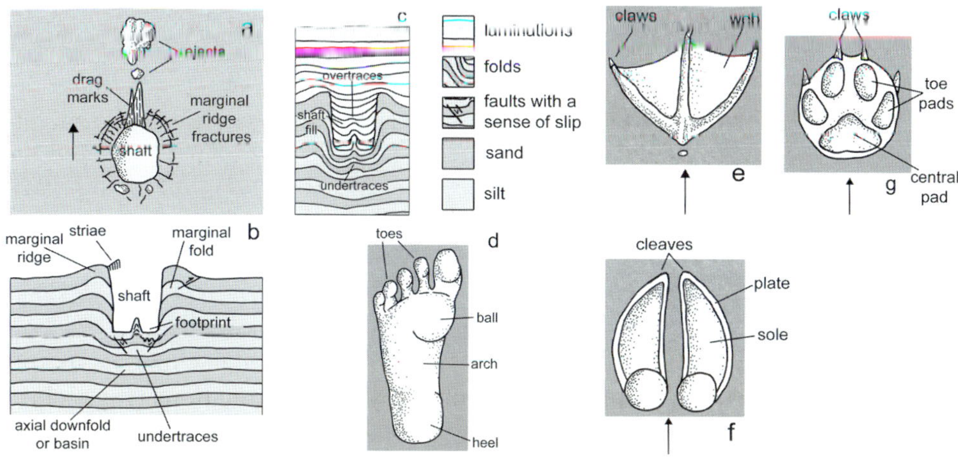

Figure 4.3 Terminology used in the description of human and animal footprint-tracks: (a) plan view, (b and c) cross sections, (d) human, (e) bird, (f) generalised cloven hoof, (g) dog. Arrows show direction of movement (graphic: J. Foster after J.R.L. Allen).

Recording methodology

Footprint-tracks are three-dimensional forms, the detail of which is important in understanding of the formation processes of the evidence and identification of the maker. Conventional archaeological planning is important in putting the evidence in a wider context. However, the challenges of recording are greatly exacerbated in intertidal contexts in which discoveries may only be exposed for an hour or two. Previous studies have used conventional photography, tracing and have taken casts using plaster of Paris or dental alginate. Today's more sophisticated approaches include photogrammetry, and portable optical laser scanning; the range of methods and their relative strengths and limitations are outlined by Bennett and Morse (2014). Where exposures are brief 3-D photogrammetric recording (often called 'structure from motion') provides major advantages (Plets *et al.* 2012; De Reu *et al.* 2013). Distinctive targets at measured intervals are placed around the subject and multiple photographs are taken from all angles in a surrounding hemisphere. These are then digitally stitched together to make a 3D model of the surface which can be rotated, lit and viewed from any position and converted to a contour diagram. Figure 4.4 shows this method being used in the Severn Estuary by Dr Kirsten Barr (2018).

Dating and timing

A wide range of dating techniques can be applied to sediments with footprint-tracks. In some cases dates will be derived from stratigraphic contexts in relation to units of known date. For Pleistocene contexts before about 50 ka BP dates are often provided by Potassium-Argon in volcanic situations and by Uranium Series where there is calcareous sediment, bones or shells; where the latter are present Amino Acid Racemisation can also be used as a dating technique. In the case of quartz or feldspar-rich minerogenic sediments with footprints, Optically Stimulated Luminescence (OSL; Duller 2008) dating is appropriate. From *c* 50 ka BP radiocarbon dating is the most widely employed technique used to date footprints in peat, or where organic material

Figure 4.4 Photogrammetric recording of Mesolithic human footprints at Goldcliff East, Wales: (a) targets along a trail, (b) 3D model of footprint-tracks (a. photo: M. Bell, b. photo: courtesy of K. Barr).

is incorporated in minerogenic sediments, although care is needed to avoid reworked organics. There are a few Holocene sites where dendrochronology has contributed. There is also a growing number of sites where footprints occur within occupation contexts on archaeological sites which provide artefacts, or other evidence for date. In these cases the footprints may make a particularly direct contribution to the understanding of human activity patterns, or, in the case of animal footprints, patterns of husbandry. In the case of some laminated minerogenic sediments which are seasonally banded the footprints provide evidence for the seasonality of human or animal activity at that site.

Identification and interpretation

The information which can be obtained from a footprint-track is dependent on the faithfulness and detail of the footprint impression. Where these are excellent and there is a trail of footprints much can be deduced. In the case of poorly preserved, or indistinct footprint-tracks (often over- or under-traces), the first issue is to identify specific features which indicate whether they were made by humans, or other animals. Some previous studies have been insufficiently rigorous in this regard. Claimed footprints at Valseqvillo, Mexico (Huddart *et al.* 2008) were said to indicate human colonisation of the Americas *c* 40 ka BP, which is about 25,000 years before the earliest well-attested other evidence. Published photos show no characteristic features of human footprints and they are now thought to be the teeth marks of a quarry machine. If marks form a trail with a distinct bipedal gait (alternating right/left), and no other bipeds (e.g. apes) are known in the area, they are clearly human. Where other bipeds may be present, or in the case of single footprints, there are identification features (Fig. 4.3d), which may be present even in poorly-preserved examples. A prominent big toe is often evident, somewhat splayed from the other toes which are often less clear. Other features are the presence of a distinct heel and a curving arch between the heel and the ball of the foot with a relatively straight outer edge to the foot. These and other features are more fully outlined in anatomical terms by Bennett *et al.* (2010, 69) and Bennett and Morse (2014). Even poorly preserved over-traces, or under-traces, have value if they are clearly the product of human bipedal gait. They demonstrate that people were present in the area, the number of trails providing some idea of the intensity of activity. The directions in which they are moving may give clues to the resources being used, and axes of movement, for instance towards the sea or a lake or they may converge on activity areas.

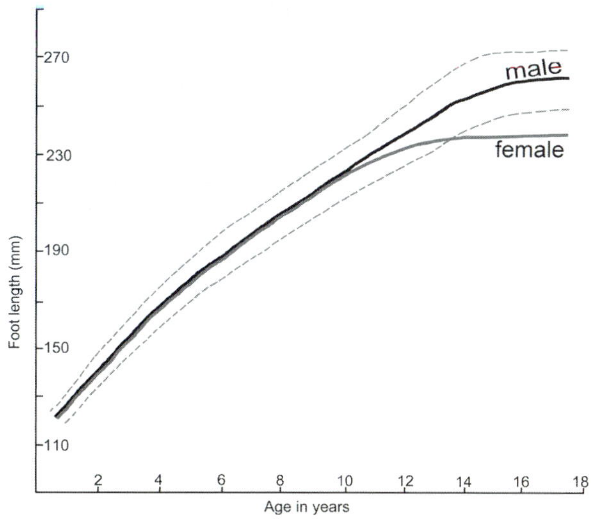

Figure 4.5 Graph of the relationship between age and foot size in modern human populations. Broken line error margin for males (graphic: J. Foster after Bennett and Morse 2014, fig. 6.7).

When a footprint is sufficiently well-preserved for its size to be measured, then it is possible to make an estimate of the age of an individual. Calculations are based on the relationship between foot size and age in modern human populations. Figure 4.5 shows a graph of this relationship. For a given age female feet are on average up to 1 cm smaller. Up to about age 14 in girls and 16 in boys there is a steady increase in foot size with age, so footprints provide a reasonable estimation of age, although from the age of about 10 there is an overlap between children and a proportion of smaller adult foot sizes (Bennett and Morse 2014, fig. 6.2). Above the teens we may simply be able to say that an adult is represented. We do need to be aware of course that these figures are based on modern populations; a range of factors particularly genetic and dietary, but perhaps also behavioural, are likely to mean that the relationship between foot size and age differed in prehistory to some degree. To address this problem a number of studies have used comparative footprint data from non-western populations local to the footprint site. For instance, recent aboriginal data was used in the interpretation of the late Pleistocene footprints from Willandra Lakes, Australia (Webb 2007). Davies *et al.* (2014) discuss the effects of habitual movement on skeletal plasticity and the extent to which specific activities and degrees of mobility may be identified in the skeleton. In discussing early Neolithic burials Wysocki and Whittle (2000) have observed that the males show stronger musculature than females, which they interpret as showing they had a more mobile lifestyle accompanying herds.

The topic of ageing is an important one because, as we will see, children are remarkably numerous among prehistoric footprint-tracks. However, many of those children are so young, in several cases below 10, that any likely difference in prehistory in the age/size relationship will not alter the inference that young children were involved, though it may alter their ages by a year or two. It has long been established that there is a relationship between foot length and a person's height and it is generally taken that foot length is 15% of height (Tuttle 2008; Scales 2006; 2007). Formulae are also available which enable calculation of the speed of walk from the foot and stride length (Bennett and Morse 2014, 154).

There are many other attributes which may be determined from footprint-tracks. In the case of Pleistocene examples such as the 3.6 ma year old tracks at Laetoli, Tanzania, there is the question of whether they are made by a fully bipedal hominid and in later Palaeolithic examples there are comparisons between the footprints made by Hominins at different evolutionary stages. There is also the question of footwear and the differences which wearing shoes makes to the structure of the foot. Nearly all prehistoric footprint-tracks are barefoot and distinct from those of shod individuals today. A doctor writing in *The Lancet* on a mission hospital on the Solomon Islands compared shod and unshod footprints (James 1939). Those of the natives were broader than the Europeans, the toes were spread and radiated. The prehistoric examples are similar; they tend to have a more prominent splayed big toe, often with a gap between that and the other toes (Bennett *et al.* 2010). Tuttle *et al.* (1990) compared early hominid footprint-tracks at Laetoli, Tanzania with those of present day Machiguenga people from Peru who have never worn shoes; they show many similar features which, it was argued, showed that the Laetoli individuals exhibited many features of modern human gait.

More problematic is attribution of sex. On average female feet are smaller than those of males but there is significant overlap between small male and large female foot sizes (Bennett and Morse 2014, fig. 6.2). Whilst we may be reasonably confident that especially large footprints are mostly those of adult males, smaller examples cannot be confidently assigned, nor does it seem to be possible to make a clear distinction between females and a small proportion of males with small feet and adolescents. Some writers have taken a far more optimistic view. Robbins (1985) maintained that an individual's footprints were almost unique and could be used to determine sex, ethnicity etc. The most ambitious of Robbins' claims have been successfully challenged as pseudo-scientific since her death in 1987 (Tuttle 2008). This had significant implications for the role of forensic expert opinion in criminal justice, because convicts were on death row awaiting execution on the basis of her footprint interpretations and were subsequently released (Tuttle 2008).

As generally in palaeoenvironmental studies, the pitfalls of a single 'expert' view are reduced if a critical and questioning approach to scientific evidence and a multiproxy approach is employed, whereby evidence from a range of sources is compared in a self-reflexive (i.e. critical) way by a group of researchers of varied disciplinary backgrounds. This in any case reflects the collaborative nature of many archaeological projects (Hodder 1999; Bell 2015). Significant extensions of this approach have been the involvement by Pastoors *et al.* (2015) of Bushmen trackers in interpretation of Palaeolithic footprint-tracks in French caves and Webb's (2007) work with Pintubi Australian Aboriginal communities, brought up without European influence, in the interpretation of the Late Pleistocene Willandra Lakes human footprint trails. With their tracking skills they were able to enhance some archaeological interpretations and challenge others. However, in the Willandra case it would also have been interesting to know if there were any interpretations by which they were less convinced and whether archaeologists modified these accordingly.

Associated animals

Animal footprints are very relevant to our theme given the evidence presented in Chapters 1–3 that humans and animals often follow the same paths and that much human mobility is related to animals in terms of hunting, herding, etc. The identification of animal tracks is aided by some very good guides, mainly created for those with wildlife interests; the best of these cover not only the footprints and trails but also trace fossils such as beak marks, feeding traces, faeces etc. Available

Figure 4.6 Laetoli, Tanzania: footprint-tracks of three hominin individuals crossed by the trail of a three-toed horse Hipparion *and its foal (photo: John Reader).*

guides include coverage of Europe (Bang and Dahlstrøm 2001); European birds (Brown *et al.* 2003); south and east African wildlife (Stuart and Stuart 1998); and North America (Jaeger 1948; Sheldon and Hartson 1999; Halfpenny 2007).

Modern analogies of animal footprint-tracks have been used in the interpretation of sub-fossil examples. Allen (J. 1997) compared the tracks of recent domestic animals in the Severn Estuary and Barr and Bell (2016) compared the footprints of primitive rare breed sheep and cattle with prehistoric examples, especially to establish the age structure of the herds and test the hypothesis of seasonal estuarine saltmarsh utilisation (Bell 2013).

Palaeolithic footprint-tracks on open sites

Volcanic and related contexts

The discovery which put footprints on the archaeological radar was made in 1978 at Laetoli, Tanzania in an expedition led by the renowned archaeologist Mary Leakey (1984). Even so, it was to be another 20–30 years before they started to be more widely recorded; although there had been several earlier discoveries in the Americas and in French caves, these did not generate widespread attention. What made Laetoli so significant was its Potassium-Argon date 3.66 ma years ago and the fact that the prints were made by a fully bipedal Hominin prior to the earliest evidence for stone tool use, currently 3.3 ma (Gowlett 2016). Tuttle (1987) describes this as 'among the most dramatic and scientifically important discoveries of modern palaeoanthropologists'. The footprint-tracks were of three individuals walking in parallel (Fig. 4.6): G1 was the smallest (1.2 m tall); G2 was the tallest (1.8 m); and G3 was 1.4 m tall and walked, apparently deliberately, in the footprints of G2. It was suggested that they walked as a group which may have comprised male, female and offspring (Leakey and Harris 1987). They are generally considered to have been made by *Australopithecus afarensis* (Bennett and Morse 2014, 47) and were walking in an area of falling volcanic ash from an eruption of the Sadiman volcano 35 km to the east. The volcanic tuff is cemented, resistant to erosion and splits along bedding plains exposing footprints (Hay 1987). It was inferred that it was laid down in a few weeks at the end of a dry season and in the early stages of the wet season; grass had been grazed down to stubble and later in the tuff there are traces of raindrop impact. Study of associated animal footprint-tracks supplemented this picture significantly. In all 9500 animal prints were recorded from nine separate exposures of the footprint tuff (Leakey and Harris 1987). In places there were deeply-worn game tracks, some produced by rhinoceros, others by buffalo. Below the human footprint level those of animals represent a restricted range of species which remain in the savannah during the dry season. Above this, in the horizon with human footprint-tracks, there is a wider range of animals suggestive of rainy season migration.

Two Palaeolithic footprint-track sites have been located in the Koobi Fora Formation of Kenya where other early hominid evidence has been found (Bennett and Morse 2014, 53). At Ileret, two distinct footprint-track horizons were identified; they are thought to have been made by *Homo erectus* and were accompanied by animal and bird footprint-tracks. The two levels were bracketed by volcanic tuffs with dates *c* 1.52 ma (Bennett *et al.* 2009). In the same sedimentary formation 45 km away on the shores of Lake Turkana are footprint-tracks in lake margin and fluvial sandy mud sediments, buried by sands and dated to 1.4 ma (Behrensmeyer and Laporte 1981; Bennett and Morse 2014, 58). They include a bipedal hominin trail, large vertebrates including hippopotamus and bovids, as well as birds. Olduvai Gorge, Tanzania, cuts through a major sedimentary sequence made famous by the Leakey family's discoveries of hominin fossils; there are also footprint-tracks in sediments here representing groundwater-fed wetlands between volcanic tuffs dated 1.75 ma. There is much evidence for trample by vertebrates including hippopotamus (Ashley 2003), but no human footprints. Early Palaeolithic footprint-tracks were preserved in volcanic sediments at Roccamonfina, Italy. Three fossil trails are dated by Argon/Argon to 345 ka BP; they were on the surface of a pyroclastic flow (Mietto *et al.* 2003; Avanzini *et al.* 2008). The trails had been recognised by local people prior to scientific investigation and were given the name 'Devil's trails'. The prints are not well preserved; some have traces of toes, but they were fully bipedal, they appear to be adult, and one of them zig-zags down an 80° slope, where there is occasional use of the hands to steady the walker in steep areas. A large collection of 350 human footprints is also reported from Engare Sero, Tanzania; preliminary dating suggests they were made *c* 120 ka BP in wet volcanic ash near the former shore of Lake Natron. They are of children and adults, walking at different speeds, thus not apparently walking together (Hatala *et al.* 2011). The volcanic contexts are particularly important because of their contribution to human anatomical evolution, because they can be dated by U-Series and in some cases, as at Laetoli, the wealth of associated environmental evidence.

Coastal and riverine contexts

Lower Palaeolithic footprint-tracks have recently been found in estuarine sediments at Happisburgh, Norfolk, UK (Ashton *et al.* 2014). This forms part of a sequence of Pleistocene freshwater, terrestrial and marine sediments which are extensively exposed in cliff sections round the coast of Norfolk and north Suffolk and have been well known since the nineteenth century for the animal bone and other biological evidence they contain (Ashton 2017). No human artefacts had been found in these sediments until, in 2000, flint tools were found with the bones of deer and bison at Happisburgh. These deposits are thought to date to 1–0.78 ma and this represents the earliest evidence of Hominins in northern Europe (Parfitt *et al.* 2010). Above gravels containing flint artefacts there were laminated estuarine sediments and it was within these that human footprint-tracks were revealed by intertidal erosion (Fig. 4.7). A recent origin for these prints could be discounted because the sediments were compacted, had a low water content and were too firm for recent prints. The footprints were recorded by multi-image photogrammetry and laser scanning. They survived coastal erosion for less than a month. Five individuals, possibly *Homo antecessor*, both adults and children, were present, suggesting that people were out foraging with no division of labour between sexes and ages (Ashton *et al.* 2014). They walked across mudflats to a tidally influenced river within a wider landscape of coniferous forest with alder in wet areas, and on drier ground, heath and grassland. The plant communities suggest the activity took place at the beginning or end of an interglacial which is thought to correlate with Marine Isotope Stage 21 or 25. A younger coastal campsite dating to *c* 400 ka BP at Terra Amata, Nice, France was associated with a single human footprint in sand (de Lumley 1969). Footprints also occurred in later Hoxnian interglacial sediments in the Thames valley at Swanscombe dated *c* 420 ka BP; these were of animals, probably elephant, rhinoceros, cervid and bovid (Davis 1996). They were in riverine sediments, the Lower Loam, where it was overlain by sand and pebbles. Coastal and riverine sites are important because of the information they provide for early human colonisation at Happisburgh, the utilisation of these environments and the associated biological evidence.

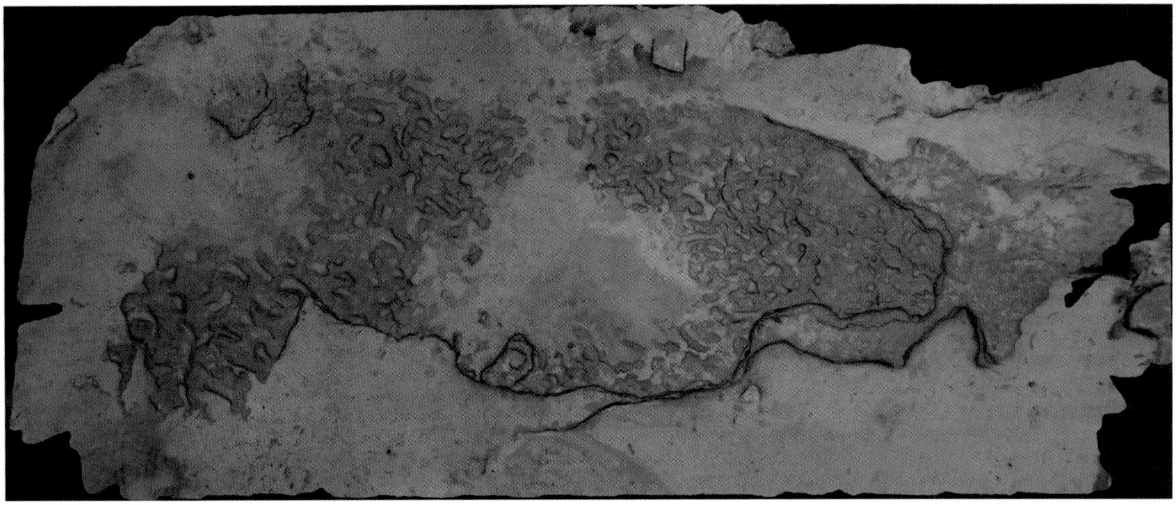

Figure 4.7 Happisburgh, Norfolk, footprint-tracks (image: Sarah Duffy, courtesy of Nick Ashton).

Aeolian contexts

In south Africa coastal wind-blown calcified sands preserve footprints thought to be those of anatomically modern humans at two sites of last glacial date *c* 120 ka BP: at Langebaan, where there are human and hyena trails, and Nahoon where humans are accompanied by other animals and birds (Roberts 2008). Probably the most extensive area of footprints so far recorded is at Willandra Lakes, Australia, a complex of lakes and former wetlands associated with runoff from the last glacial maximum; Optically Stimulated Luminescence provides dates between 23–19 ka BP (Webb *et al.* 2006; Webb 2007). The footprints are in laminated calcareous silts deflated from a lake and buried by aeolian sands (Fig. 4.8). Extensive areas of clay lamination were exposed by wind erosion and trails could be excavated where they were still buried by laminated sediments. In all 563 human prints were identified, made by both adults and children, and forming 23 distinct trails.

Travertine

Finds from the Tibetan Plateau illustrate particularly unusual contexts in which footprint-tracks may be preserved. They were found at a height of 4200 m and provide remarkable evidence for the human presence close to the last glacial maximum at 20 ka BP in an area previously believed to be completely covered in a huge ice sheet (Zhang and Li 2002). The evidence is in the form of foot and hand prints of six individuals, two of them children, in a calcareous travertine laid down by hot springs and dated using Optically Stimulated Luminescence. Nearby was a hearth shown by luminescence dating to be contemporary, important evidence from an area where prehistoric finds are few. Of even greater significance in terms of the history of human colonisation is the finding of human footprints at Monte Verde, Chile dated *c* 14.6 ka BP. Though at the extreme south of the Americas this represents one of the earliest really securely documented sites of human occupation in the continent (Dillehay 1989; Dillehay *et al.* 2008).

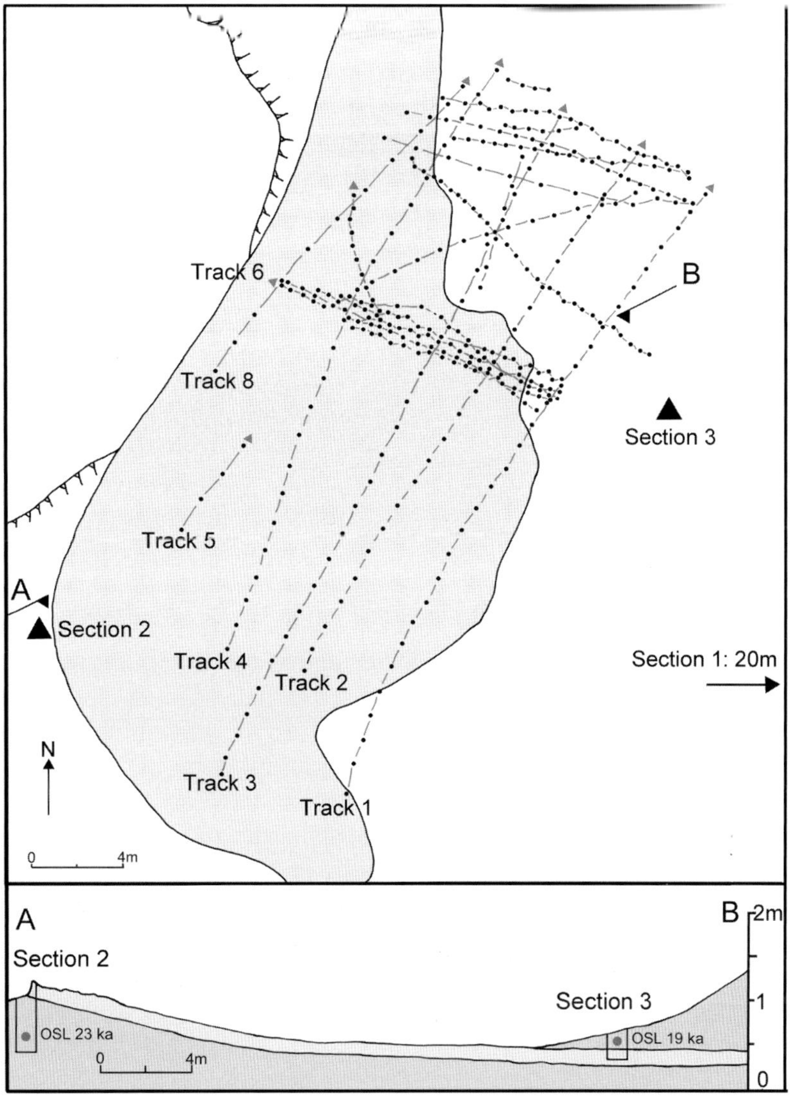

Figure 4.8 Willandra Lakes, Australia: multiple human footprint-tracks dating to 23–19 ka BP (graphic: J. Foster after Webb et al. *2006; Webb 2007).*

Caves

The wider practical and spiritual motivations for people venturing underground have been imaginatively explored by Macfarlane (2019) and it is in caves that people's footprints leave some of their most insightful traces. The earliest examples are Vartop Cave, Romania, dated 62 ka BP (Onac *et al.* 2005), and Theopetra Cave, Greece, dated 48 ka BP (Bennett and Morse 2014, 67). There is also a series of important finds from the floors of French caves. These sites inform, not so much about everyday activities, but those possibly involved in ritual within these caves and, surprisingly perhaps, children are often well represented. Where caves have remained sealed from prehistory to the time of discovery and the finders have had the foresight to protect the traces of activities left on cave floors, remarkably detailed records can be obtained, some from caves with Palaeolithic art. The way people moved around the caves can be established from footprints, invariably barefoot, the remains of torches, carbon smudges on the roof of caves and human faeces. A particularly well-documented example is Chauvet cave, which is exceptionally rich in Palaeolithic art dated 30 ka BP. In this case a good radiocarbon chronology demonstrates that the footprints are later than the art (Clottes 2003). The art was followed by a period of cave bear occupation represented by hollows in which the bears hibernated, many bear bones and footprints. Then *c* 26 ka there followed a human visit represented by 20 footprints forming a trail of 70 m made by a single pre-adolescent of height 1.3 m, with soot-marks on the roof and path showing the passage of the person's torch. The cave has also produced a single wolf or dog print, fossil canid faeces and ibex prints. Reinvestigation of footprint-tracks in four French caves employed the expertise of African Bushmen trackers who advanced more detailed and precise interpretations than had been suggested by conventional analysis (Pastoors *et al.* 2015). In the extensive cave system at Niaux, France, parts of which are richly decorated with art, the trackers identified footprint-tracks forming deliberately made patterns created by a 12 year old girl (Bahn and Vertut 1997, figs 1 and 2). Fontanet cave found in 1972 had been preserved because the entrance became blocked in the Magdalenian period. The front part of the cave is decorated with paintings dated 13–14 ka BP, whilst deeper within the cave are what the trackers identified as the footprint-tracks of 13 individuals, men, women and children aged 3–60. Pech Merle cave, richly decorated with art, had the tracks of five individuals aged 9–50 (Lockley and Meyer 2000). Particularly notable were the trackers' observations at Tuc d'Audoubert where they concluded that the footprint-tracks were those of a man aged 38 and a boy of 14 engaged in obtaining clay to model two bison figurines. At other cave art sites a relationship to animal tracking is demonstrated by the depiction of feet in twisted perspective such that the footprint is depicted at the base of the leg. On this basis it has been argued that art played a part in the acquisition of hunting skills in the young, something potentially critical to survival in the worsening environmental conditions of the last glacial maximum (Mithen 1991). One other cave site should be noted, although it is not associated with rock art. This is Ciur-Izbuc cave, Romania where 400 footprint-tracks were found in 1965; dating cave bear bones has recently revised the date of this site to 36 ka BP (Webb 2014; Webb *et al.* 2014). Other Pleistocene caves with evidence of human footprints are shown on Figure 4.2 with details in Supplementary Table 4.1.

Holocene hunter-gatherer-fishers

Case Study: Mesolithic paths in the Severn Estuary

The Severn Estuary in western Britain has emerged as a particularly significant area for the study of prehistoric footprint-tracks, due to the extensive exposures of sedimentary contexts, especially laminated sediments, that preserve them well. An additional factor is the enormous tidal range of 14.8 m, the second or third greatest in the world, which means that large areas of intertidal sediments are exposed at low tide. The great body of water mobilised by the tides and the effects of wave erosion, especially during storms, cuts sections through the stack of Holocene and earlier sediments exposing impressive sections on the foreshore. Extensive investigation and dating of the sedimentary sequence by Professor J.R.L. Allen (1987; 1997; 2001) has established a secure framework for the many archaeological sites and footprint-tracks which have been discovered in the intertidal exposures. It is the close association of footprint-tracks with excavated sites which justifies rather more detailed treatment of the evidence than that from other areas.

Figure 4.9 Uskmouth, Wales: Mesolithic red deer footprint-tracks (photo: M. Bell).

The first footprint-tracks in the estuary were discovered by the late Mr Derek Upton, an amateur archaeologist and natural historian with a lifetime's intimate knowledge of the estuary and exceptional gifts of observation. He spotted the first footprint-tracks on the foreshore at Uskmouth in 1986 and a team of specialists under Dr Stephen Aldhouse-Green (*et al.* 1992) was assembled for their investigation, leading to further work on the formation and stratigraphic occurrence of footprint-tracks in the Severn Estuary by Allen (J. 1997). The Uskmouth footprints were below reed peat dated *c* 5245 cal BC and an antler mattock found on the foreshore nearby was of a similar date (Aldhouse-Green and Housley 1993). Three trails were identified in laminated sandy silts, two made by adult males, one by a child. Palaeoenvironmental evidence shows they were walking in a saltmarsh/mudflat environment. Red deer footprint-tracks are also very abundant in the laminated sediments (Fig. 4.9). Modelling of the palaeotopography from coring indicated that the nearest dry ground at the time was 4 km to the north or east (Allen 2001), and the nearest known settlements are at Goldcliff 4 km east.

Goldcliff is of special significance because a series of four excavated Mesolithic settlement areas are associated with extensive areas of footprint-tracks and abundant palaeoenvironmental evidence (Bell 2007a). Figure 4.10a shows the geography of the area which is dominated by the remains of a former bedrock island originally 1 × 0.5 km, today much reduced in extent by coastal erosion (Allen 2001). The margins of the island are buried by Pleistocene and Holocene sediments which, in places, reach 12 m and are exposed in section

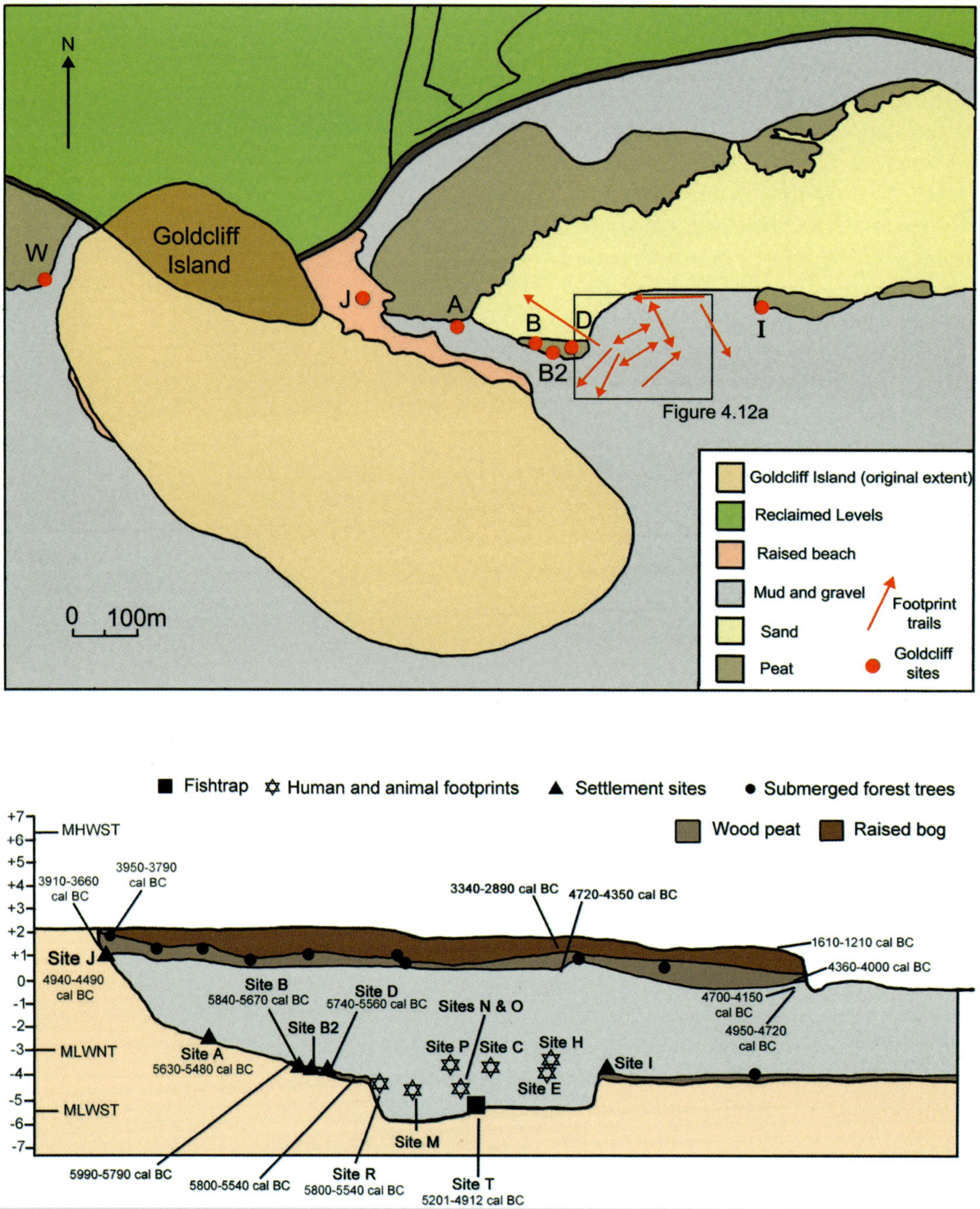

Figure 4.10 Footprint-tracks at Goldcliff East, Wales: (a) landscape context in relation to former island, with intertidal peat exposures, excavated Mesolithic occupation sites, and the axes of Mesolithic movement indicated by footprint-tracks; (b) stratigraphic section (graphics: J. Foster).

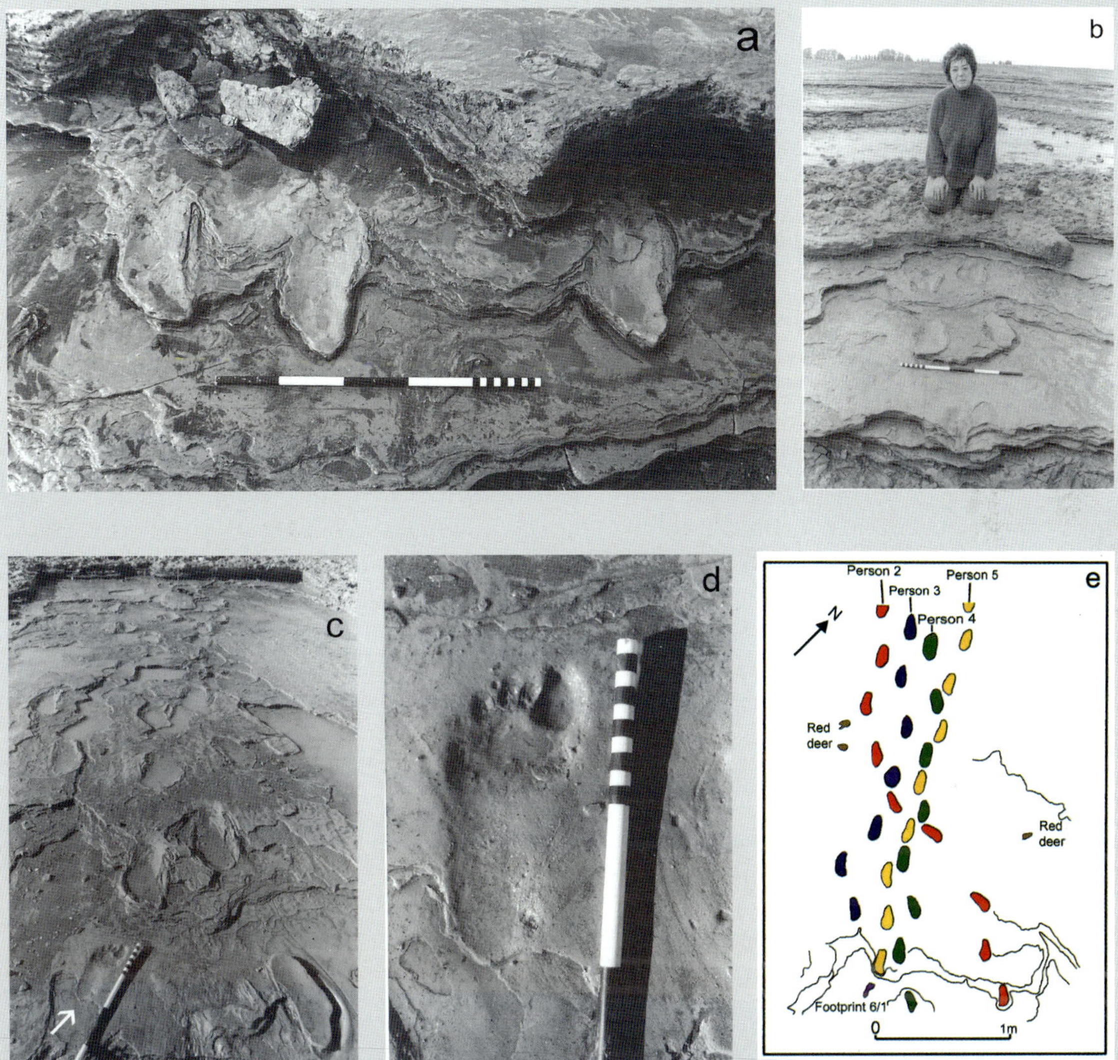

Figure 4.11 Footprint-tracks revealed within laminated sediments: (a–c) by successive stages of excavation at Site E at Goldcliff East, Wales, which revealed the tracks of four individuals walking together; (e) At the end of the excavation, tidal erosion revealed an excellently-preserved footprint (Footprint 6/1) at a lower level, arrowed in (c) and shown in detail in (d) (graphic: J. Foster; photos: E. Sacre).

by coastal erosion as shown in Figure 4.10b. At the base of the sequence above Pleistocene Head is an old land surface with a submerged forest of large oak trees (Fig. 3.4). These were drowned by rising sea levels, and a thin reed peat formed *c* 5700 cal BC. Then, with continuing sea-level rise, 4–6 m of silts were laid down, some parts of which are strongly laminated and have very abundant footprint-tracks. An upper peat started forming 4500 cal BC and a second submerged forest grew from about 4400 cal BC. Most of the footprint-tracks are in the lower part of the laminated sediments, which formed *c* 5500–5200 cal BC. The Mesolithic occupation sites date to 5900–4800 cal BC; each successive occupation was at a higher level on the island edge as people moved upslope in response to sea level rise which buried and preserved the earlier settlements below estuarine silts. The sediments are laminated and comprise fine silts alternating with more sandy bands. It has been shown that these rhythmical coarse and fine bands reflect annual changes in the estuarine tidal regime with the coarse bands being laid down in the cooler and more turbid waters of autumn and winter months and the finer silts in the warmer and calmer waters of spring and summer (Allen 2004; Allen and Haslett 2006). This interpretation has been confirmed, both by sedimentary modelling and the study of pollen in the bands, which has also shown that charcoal, and thus perhaps human activity, was more frequent in the summer months (Dark and Allen 2005). This banding evidence is of great significance for the interpretation of the footprint-tracks because it contributes to understanding the seasonality of activity by showing that the most well-preserved examples were made during occupation in high summer. The footprint-tracks were securely stratified in laminated sediments as demonstrated by a sequence of photographs of Site E (Fig. 4.11): (a) shows initial discovery where they project from a small erosion cliff; (b) and (c) show trails progressively revealed by a technique Dr Rachel Scales developed by careful finger-tip peeling back of laminations over a month to expose four trails comprising 33 over-traces. All were heading out to sea; three were youths aged *c* 12–14, and one aged *c* 16. Scales (2007) demonstrated they almost certainly walked together, no prints overlapped and tellingly the individuals paused, and the stride pattern altered at the same point, indicating that they acted together with common purpose, perhaps hunting, or fowling; red deer prints were found at the same level. As this excavation was ending the tide eroded part of the area exposing an exceptionally well-preserved footprint (Fig. 4.11d–e) of a young person aged 8–9 walking in the finest of summer silts in the opposite direction to the party of four. In a second area (Site C), where footprint-tracks were also observed disappearing into a low erosion cliff, Scales uncovered 165 footprint-tracks of two individuals, one a child aged 3–5, the other aged 10–11 (Fig. 11.1). A year later erosion exposed a really well-preserved footprint in the same area (Fig. 4.1), of a young person aged 10–12.

When all the evidence assembled over 19 years of intertidal survey at this site is put together, this includes 342 recorded footprint-tracks and 21 distinct trails (Barr 2018). These create maps of footprint alignments, although in interpreting this we cannot necessarily assume that paths were straight (Fig. 4.10a). A few lead to excavated settlement sites, others converge on points on the former island edge where we may suppose there were settlements now lost to erosion. Two predominant axes of movement are evident. Low in the laminated sequence (below –4.2 m OD) the footprint-trails have a predominantly north-east to south-west orientation, roughly at right-angles to the axis of the palaeochannel within which they lie (Fig. 4.12). On Site N multiple, roughly parallel trails form a band 16 m wide which could be described as a Mesolithic footpath (Fig. 4.13) perhaps leading from a lost island edge encampment to traps, or a boat landing place in the channel. Indeed, in 2017, V-shaped post and wattle structures identified as fishtraps, and dated 5210–4912 cal BC, were found at Site T on the alignment of this path between Site N and the island edge. It is notable that the trails on Site N mainly lead away from the island and those on Site M predominantly converge on an area at the island edge, the inference being that people followed a circuit, rather than two-way path, although two-way passage is also represented in several cases. A circuit would perhaps be consistent with visiting a number of fishtraps or other stations. Later in the laminated sequence (above –4.2 m OD; Fig. 4.12b) footprint-trail orientation is more varied but predominantly north-west to south-east, which is parallel to the axis of the palaeochannel, probably as a result of walking along its banks, as in the case of the four companions at Site E.

Of the 21 distinct trails 19% were of children under 8, 19% probably children under 10, 43% older children, females or slight males and 19% probably adult males. These proportions may imply that activities within 300–500 m of the campsites were mainly the province of children and females rather than adult males. Children, some as young as 4, were venturing out into the muddy estuary and can be assumed to have contributed to the active life of the community. The Uskmouth footprint-tracks show that children with adults also ventured several kilometres from dryland camps. The most faithfully preserved footprints were clearly made in the fine sediments of spring and summer, although much of the other seasonality evidence indicates that the main period people were present at the occupation sites was late summer and autumn (Bell 2007a, table 18.4); only one site had animal bone evidence of winter occupation. Virtually all the footprints which were clear enough were made by people who were barefoot. Associated with the human footprint-tracks at Goldcliff were those of red and roe deer, the occasional aurochs and many birds.

Recognition of footprint-tracks in the Severn Estuary has been followed by discoveries in similar intertidal contexts round the Welsh coast. At Lydstep, Pembrokeshire, a boar with microliths from a weapon was found in intertidal peat (Leach 1918) and is dated to the final Mesolithic; footprint-tracks of adults, children and deer have recently been found on the peat (Murphy *et al.* 2014). Other Mesolithic footprint-tracks of people and deer were found at Rhyl, North Wales (Bell 2007a) and Porth Eynon, Gower (Philp 2019).

On the other side of the world, around Clare Bay in South Australia, calcified saline coastal lagoonal sediments revealed footprints of adult and child humans, emus, kangaroos and wallabies dated *c* 5000 BP (Belperio and Fotheringham 1990). Footprints were first spotted here by the notable pioneer of aboriginal studies, Daisy Bates, *c* 1914.

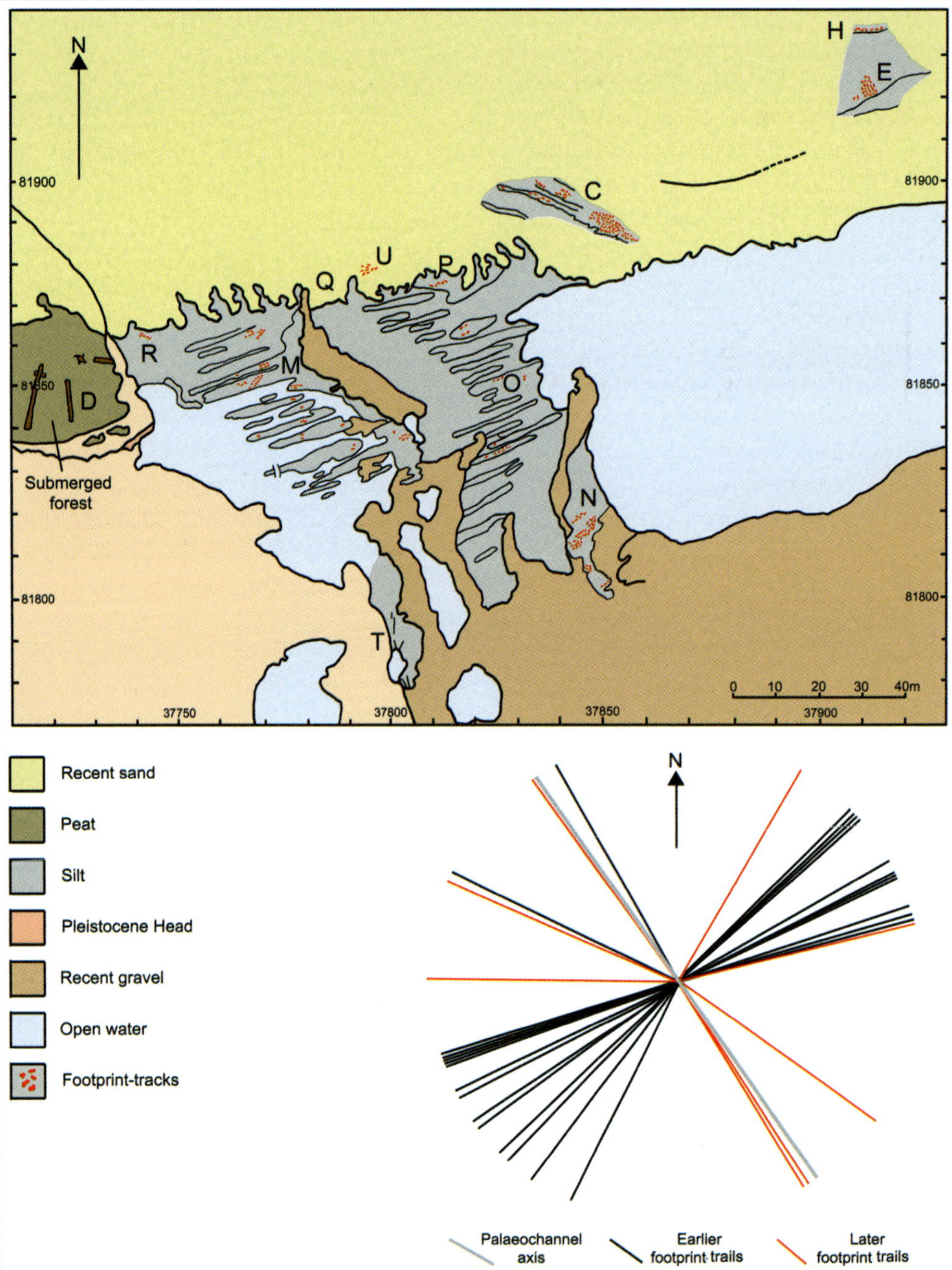

Figure 4.12 (a) Goldcliff East showing sites with Mesolithic footprints in relation to Quaternary sediments; (b) rose diagram showing the orientation of Mesolithic footprint trails (graphics: J. Foster).

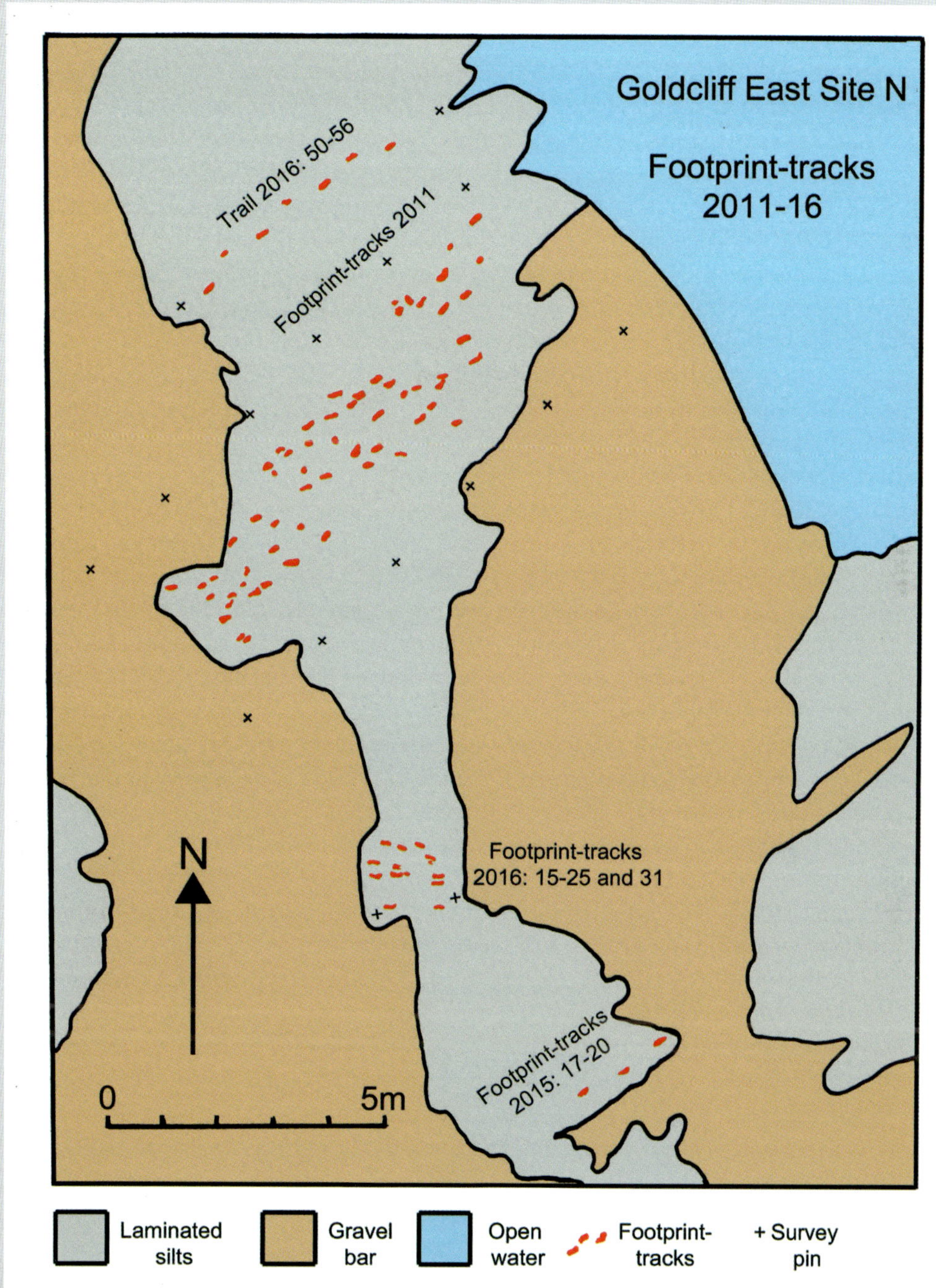

Figure 4.13 Goldcliff Site N showing Mesolithic footpaths (graphic: J. Foster).

Footprint-tracks in later prehistoric contexts

Footprint-tracks from contexts post-dating the introduction of farming provide similar information to the foregoing hunter-gatherer contexts, concerning localised patterns of movement and demographic composition. Additionally there are often footprints of domestic animals which provide insights to patterns of animal husbandry. A site similar to the Mesolithic examples previously discussed comes from the very extensive excavation of coastal environments at Rodbyhavn, Denmark in advance of a major tunnel between Lolland, Denmark and Fehmern, Germany (S. Sørensen, pers. comm. 12.8.15). Many wooden fishtraps have been discovered in former lagoons and next to one of these, dated *c* 3000 cal BC and thus Neolithic Funnel Beaker period, were *c* 20 human footprint-tracks. The footprints were made in gyttja (organic mud) and filled with sand in a storm which preserved the prints but damaged the fishtrap. A further very significant outcome of this major excavation has been to demonstrate that people did not visit this coastal wetland for purely utilitarian activities such as fishing, for there is also evidence for the ritual deposition of artefacts in the late Mesolithic and Neolithic. Similar deposition practices occur across the two periods thus providing new insights to the process of Neolithisation.

Another huge archaeological project which has recently produced footprint-tracks is the Istanbul Metro and associated rail link across the Bosphorus in Turkey (Polat 2013). Not only has this revealed 37 ships, many complete with cargo, of Roman to medieval date but also a Neolithic settlement some 8 m below sea level dated 6500–6000 BC. Associated with this were more than 2000 footprints of people who had walked on wet clay around the settlement, their prints then filling with sand. Animal footprints have not so far been reported. The published human prints are all clearly made by people that wore footwear. This is the earliest evidence the writer has found for shoes specifically related to footprint tracks, although examples in mud brick and other contexts in the Near East are to be expected. One sedimentary context where we might expect footprint evidence is the lake edge contexts of the circum-Alpine zone where so many wonderfully preserved Neolithic and Bronze Age lake villages have been excavated. Surprisingly, the writer has only come across one example with human footprints, the Neolithic site at Clairvaux, France (Pétrequin and Pétrequin 1988, 171).

The greatest concentration of human footprint-tracks from later prehistory has been recorded on the foreshore at Formby, Lancashire, UK, as a result of 25 years of dedicated recording by retired Formby schoolmaster Gordon Roberts (2014; Roberts *et al.* 1996; Huddart *et al.* 1999b; Roberts and Worsley 2008). The evidence extends along 4 km of shore and the best footprints are preserved in laminated silts, interstratified with sand which are seaward of extensive dune sands along the shore. Some of the 219 trails which Roberts reports were excellently preserved. Most people were barefoot, though some had footwear. They occur in two main horizons; the lower horizon has Optically Stimulated Luminescence dates with quite large standard deviations which dates them between 5400–3200 cal BC, i.e. late Mesolithic to early Neolithic. Others are dated from *c* 2000 cal BC and through the Bronze Age. As at Goldcliff children predominate, with smaller numbers of adult females and a few adult males. The latter tend to be associated with red deer prints and this has been interpreted as reflecting hunting activity. Other wild animal prints are roe deer, aurochs, wild boar, with crane, oyster catcher, and many wading birds. Domestic animals are cattle, sheep/goat, small unshod horse and dog/wolf. Palaeoenvironmental evidence shows that at times there were offshore dunes to seaward and the humans and animals were walking on lagoonal mudflats subject to direct marine influence (Gonzalez *et al.* 1997). No artefacts have been found associated with the Formby footprint-tracks. The nearest known occupation sites are Neolithic, *c* 5 km inland around the peatlands of Downholland Moss, and there are also lithic scatters of Mesolithic–Bronze Age date and a Neolithic trackway at Hightown *c* 6 km to the south (Cowell 2008). Pollen analysis from the peatlands of Merseyside has produced a good deal of evidence for burning in the Mesolithic and later (Cowell and Innes 1994), including charcoal in sediments in which footprints occur (Roberts and Worsley 2008). It appears therefore that the Formby footprints relate to people whose settlements were on drier land at some distance and that they were visiting the shore particularly for foraging, hunting and fishing activities; domestic animals were grazed in the coastal environment, but not apparently to the extent discussed below in the Severn Estuary, perhaps because the sandy Formby shore offered less rich grazing than the Severn saltmarshes.

Case Study: seasonal pastoralists in the Severn Estuary

In the Severn Estuary many foreshore sites have produced footprint-tracks of domestic animals; they are particularly found on the surface of a Bronze Age peat which overlies silts containing the Mesolithic footprint-tracks discussed above. The peat is in turn overlain by estuarine silts of the middle Bronze Age to Iron Age (Allen, J. 1997). At the stratigraphic interface between the peats and silts are footprints, six sites with buildings and trackways interpreted as seasonal settlements of pastoralists who grazed animals on the saltmarsh during the spring and summer period of lush vegetation growth and reduced tidal range. The most detailed study of footprint-tracks from one of these settlements has been at Redwick, where they were associated with four rectangular buildings dated to the middle Bronze Age (Fig. 4.14a–b; Bell 2013). Ten human footprint-tracks were present and one of these was very well preserved where a child aged *c* 7 had trodden a thin layer of silt into the peat (Fig. 4.14c). In all five were of children aged under 11 and two may have been adults. The presence of young children is significant, especially given the specialist pastoral nature of this seasonal settlement, once again it suggests the active engagement of children in the activities of the community. The rectangular Bronze Age buildings are on slight natural rises of peat partly encircled by shallow curvilinear depressions. The floors of these depressions are covered by animal footprint-tracks, mostly cattle (71%), with some sheep/goat (16%), and a few pigs (Fig. 4.14d–e). The presence of footprints and bones of neonatal calves and lambs in the first three months of life indicates activity in spring and summer. Isotopic evidence from animal teeth strengthens the case for saltmarsh grazing for part of the year (Britton *et al.* 2008; Britton and Müldner 2013).

Figure 4.14f is a reconstruction of the setting of one building constructed from palaeoenvironmental evidence. A cow is being driven by the 7 year old along the depression to the building for milking. Tidal incursion occurs along the depression and there are saltmarsh plants on its margins and to seaward and this provides grazing. Landward is extensive raised bog and beyond, about 4 km away, the rising land where the main settlements of this community probably lay.

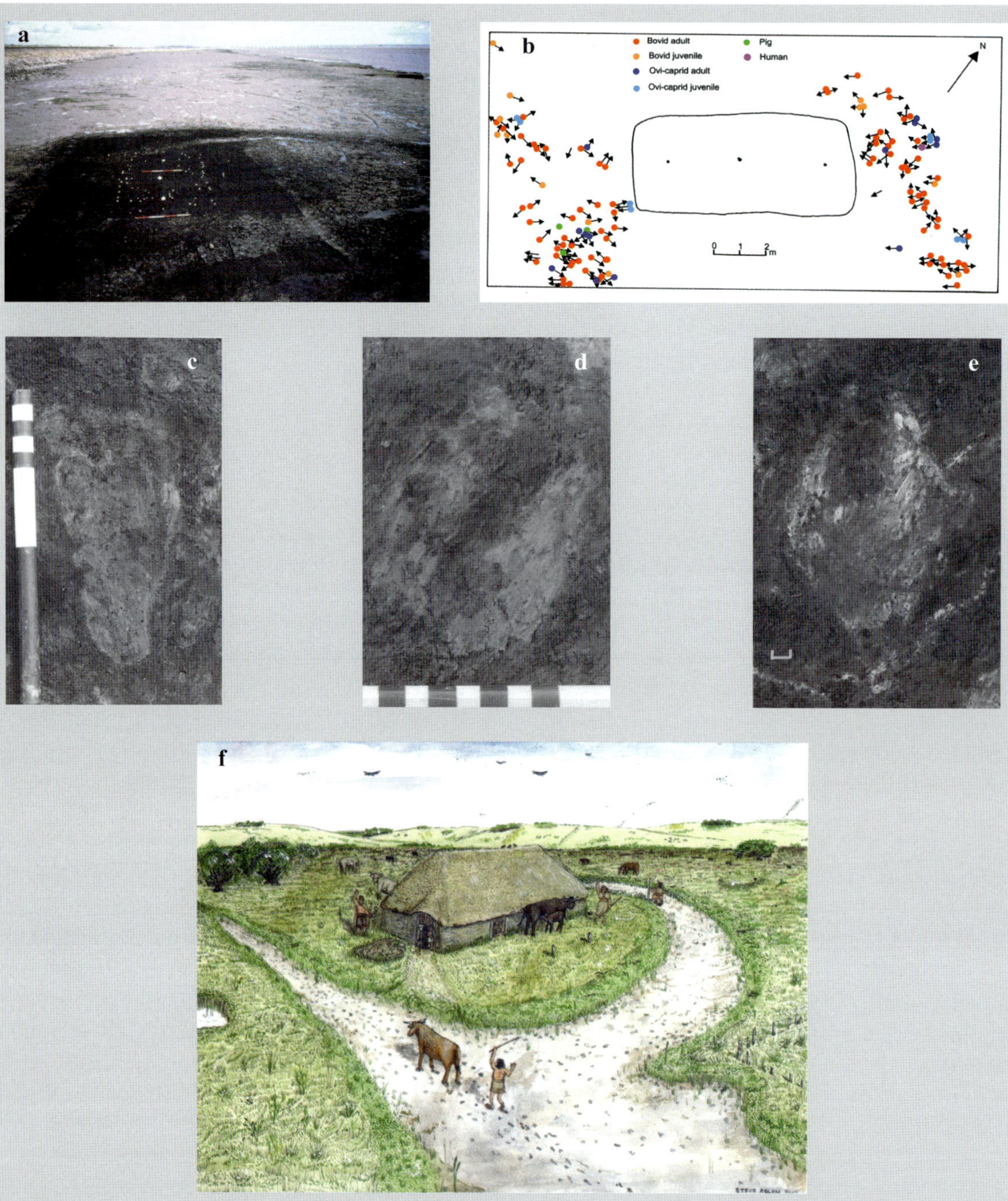

Figure 4.14 Middle Bronze Age seasonal settlement at Redwick, Wales: (a) Building with animal footprint-tracks in flanking palaeochannels; (b) directions of animal movement in relation to buildings; (c) footprint of a child aged c 7; (d–e) cattle footprints; (f) reconstruction of scene around the building (photos: E. Sacre; graphic: J. Foster; reconstruction: S. Allen).

Six kilometres west is an Iron Age site at Goldcliff with eight rectangular buildings, again occupied in the summer months; some were associated with the footprint-tracks of cattle (Bell *et al.* 2000). One such building housed subdivisions interpreted as animal stalls and the inference that the buildings were shared by people and animals is confirmed by the presence of cattle lice (*Damalina bovis*) and human fleas (*Pulex irritans*) in the flooring. The widespread occurrence of cattle footprint-tracks on the surface of the Bronze Age peat is testimony to the extensive nature of seasonal animal husbandry on the coastal wetlands from the middle Bronze Age to the Iron Age.

Other later prehistoric examples

There is evidence for similar practices more widely. On the north shore of the Bristol Channel at Kenfig, Wales the surface of an intertidal peat exposure revealed by a storm was marked by some rather indistinctly preserved footprint-tracks, some forming trails and thus apparently human; they were optical laser scanned by Bennett *et al.* (2010) and dated to 3700–2200 cal BC.

Further north in west Wales at Borth is probably the finest intertidal exposures of submerged forest in Britain. Marine erosion is eating into the margins of the vast Borth Bog at the mouth of the Dovey Estuary. A storm early in 2012 revealed footprint-tracks on the peat surface dating between *c* 2000–1000 cal BC (BBC 2012). They were of people ranging in age from a 4 year old child to adults, with cattle, sheep/goat and possibly bear.

In the East Anglian Fenland footprints of animals including cattle and people have been found in the edge of a palaeochannel, some pre-dating a wonderfully preserved settlement at Must Farm (see Fig. 6.11) dated 900–800 BC and some just outside the palisade which surrounded the settlement (M. Knight, pers. comm.). Nearby at Fengate/Flag Fen, animal footprints were found at the end of a droveway where it reached the wetland, consistent with use of the droveway to move animals from enclosed dryland fields to open wetland seasonal grazing (Pryor 1998a). In the Netherlands at Kolhorn a context of the single grave Neolithic culture had cattle hoof prints, which were in soft clay of the maritime zone with a fill of darker sediment. They mainly defined a path across a channel and also radiated from a freshwater well in the settlement. The animal bone assemblage is cattle-dominated but there was also much evidence of fowling and fishing. No buildings were found, despite favourable preservation conditions, and the site is interpreted as seasonal (Bakels and Zeiler 2005).

A significant Swedish footprint site is at Ullunda, in an area on the Baltic still subject to huge isostatic land rise following the melting of the ice sheets (Bell and Walker 2005, 117); consequently this site, originally just above the beach on a fjord, is now well inland. The site is a Bronze Age long house dated 1400–1200 BC (Price 1995). On a clay surface below a gravel spread several hundred unshod horse hoofprints were recorded; they appeared to be engaged in the very specific activity of importing the gravel to the site as hard standing. They provide the earliest evidence for the horse in Sweden and demonstrate the involvement of horses in apparently everyday, rather than purely high status, activities, as had previously been supposed.

A site at Walvis Bay, Namibia highlights the potential of coastal sediments in other parts of the world. Here the footprints date to AD 450–1450. Thousands of probable ovicaprid prints are represented and the animals were apparently being driven as a flock. Overlying human prints are interpreted as those engaged in driving them and significantly some are of children as young as 5.

Footprint-tracks are rather rarely reported from inland locations, perhaps because excavators have not considered the possibility of their occurrence. However, there are some examples and significantly they tend to be associated with the coaxial field systems and drove roads of Bronze Age date which are more fully discussed in Chapter 8. A muddy ditch which formed part of the Shaugh Moor Bronze Age reave field system on Dartmoor revealed many footprint-tracks of cattle, sheep, the occasional horse and badger (Smith *et al.* 1981), welcome evidence of the animals living on acid moorlands where bones rarely survive.

Very different contexts for footprint preservation are represented by later prehistoric volcanic deposits. The most spectacular of these is at Avellino, near Vesuvius, Italy where thousands of human footprint-tracks are associated with a Bronze Age settlement at Affagola; many were apparently footprints of those fleeing an eruption *c* 1830 cal BC, the ash from which preserved the settlement and footprints (Livadie 2002; di Vitro *et al.* 2009). At Jeju island off the south coast of the Korean Peninsula, human footprint-tracks were found in shoreline sediments stratigraphically related to volcanic deposits. They were originally dated to the Pleistocene (Kim *et al.* 2009) but a critical review of the stratigraphic context and redating using a combination of radiocarbon, U-Series and OSL has shown that the footprint horizon is coeval with a volcanic eruption dated *c* 1750 cal BC (Sohn *et al.* 2015). Paradoxically it was the apparent Pleistocene age of the Jeju site, and its implications for human colonisation of Korea, which stimulated a conference bringing together much of the existing work on footprints (Lockley *et al.* 2008). Almost 2000 human footprint-tracks are reported at Kilavea, Hawaii and are interpreted as people fleeing a volcanic eruption in the eighteenth century AD (Mayor and Sarjeant 2001).

Footprint-tracks in the Americas

The earliest and most significant find from the Americas at Monte Verde II, Chile has already been noted in the Palaeolithic section which, given the site's date *c* 14 ka BP, is of great significance in terms of the human colonisation of the Americas. Colonisation is generally considered to have taken place when low sea level linked Beringia to Alaska with people moving south of Alaska around 14 ka BP. The recent discovery of 29 human footprints, of three individuals including a child, on Calvert Island, coastal British Columbia has a radiocarbon date of 13 ka BP and demonstrates coastal adaptation and a possible coastal colonisation route into the Americas (McLaren *et al.* 2018). Also of significance in demonstrating early human presence are footprint-tracks at Cuatro Ciengas, Mexico within tufa (calcareous spring) deposits dated 10,550 BP and the later 7240 BP (Felstead *et al.* 2014). The earlier group demonstrates human presence in that area *c* 1500 years before the earliest other dated evidence, which comes from coprolites (fossilised faeces).

At Monte Hermoso, near Buenos Aires, Argentina footprint-tracks are associated with fresh-water pools with a fill of silty clay laminae intercalated with wind blown sands, which as a result of erosion are now exposed in the intertidal zone, though there was no marine influence at the time of their formation (Politis and Bayon 1995; Aramayo 2009). In these lagoon silts 420 human footprint-tracks have been recorded; they were predominantly of children, with some probable women's prints and infrequent prints of adult males. Associated plant and animal remains are dated around 5900 cal BC. Footprint-tracks of birds and artiodactyls (even toed ungulates) are also present. Just 200 m away is the site of La Olla, where there are bones associated with the processing and consumption of sea mammals dated 6200–5500 cal BC; the site is considered to be associated with the nearby footprint-tracks. These well stratified and dated sites have resolved debates about the early coastal settlement of this area. Around Lake Managua, Nicaragua, footprints were revealed in a quarry in 1874; and were once considered as evidence for the very early colonisation of the Americas. They have now been redated to *c* 4800 cal BC. They were in volcanic mud which was covered by volcanic ash. A total of 322 human footprint-tracks, some of exceptional quality, were found together with those of deer, opossum and bird. The humans represented 15–16 individuals in a series of sub-parallel lines, some forming trample paths and certainly indicating defined axes of movement. It was suggested that the people were fleeing from a volcanic eruption to the nearby lake shore (Bryan 1973; Lockley *et al.* 2009). On a terrace of the Mojave River, California footprints dated *c* 3000 BC were found in silty clays underlying a later excavated settlement. There were 54 footprints of five individuals and several animal tracks forming a pathway; three of the individuals were immature. A few artefacts were found at the footprint horizon suggesting nearby occupation (Rector 1979; Rector *et al.* 1983).

North American caves have, like the Palaeolithic caves of France, been productive of footprint evidence and some of them have particularly dry (and/or saline) conditions which also preserve remarkable organic evidence such as footwear, plant remains and coprolites. One of the most productive is Salts Cave, Kentucky (Watson 1969). Two individuals left adult size prints in dust where they walked along the passage and back leaving the cave walls with soot marks from their torches. That cave also contained a child's mummified body and bundles of wood faggots. The people had left faeces in which 45% of the plants were of varieties cultivated by these early woodland communities, with dates from the cave between 1400 and 270 cal BC. Jaguar Cave, Tennessee has produced the footprint-tracks of nine cave explorers, the number based on metrical analysis (Watson *et al.* 2005). They were identified as males and possibly some female; most were barefoot but one was shod. Their route was also marked by carbon smudges from torches on the cave roof and radiocarbon dating of this carbon showed they were formed *c* 3000 cal BC. Ritual practice in caves is illustrated by middle American cases, such as Naj Tunich, Guatemala where Maya art, hieroglyphs and associated footprints on the cave floor date to the eighth century AD (Brady and Stone 1986). Extensive cave systems on Isla de Mona, Puerto Rico are decorated with pre-Columbian art *c* AD 1200–1500 below which are footprints of the artists in calcite crusts on the cave floor (J. Cooper, pers. comm.).

Footprints: perceptual and symbolic aspects

Just as we may be able to predict the location of a campsite 7000 years ago from patterns of footprint-tracks one wonders what people then made of the footprints they observed as they moved around. Observing animal footprints would have been an important part of tracking skills. Reading human footprints as evidence of the composition, routes and activities of other groups would have been an equally important part of life. It is a reasonable proposition that adolescent males living in a sparsely populated Palaeolithic or Mesolithic world were a great deal better at identifying the footprints of young females (and vice versa) than we are today. People would naturally think about and rationalise the footprints they observed whether fresh or fossil.

Fossil footprints in many parts of the world have provided a basis for mythology and beliefs (Mayor and Sarjeant 2001). One example is the Algerian legend of a giant Roc bird which may derive from observation of dinosaur footprints. A fascinating insight to the perception of footprints comes from a !Kung Bushmen cave painting in Mokhali cave, Lesotho (Ellenberger *et al.* 2007). It is thought to have been made in the first two decades of the nineteenth century and depicts a footprint and an image which is interpreted as the print maker, a large bird-like bipedal creature like an ornithopod dinosaur. Tellingly a dinosaur skeleton is exposed in the cave and their remains are frequent in the area. It is supposed that the !Kung, renowned trackers, observed the footprint and imagined the creature which produced it, which we now know had been extinct for some 65 million years. Equally tantalising is a case reported from the Weaber Range in Australia. Here aboriginal communities depicted bird tracks and images in their art. One of them appears to be a precise anatomical representation of the footprint of a bird which is thought to have become extinct either just before, or just after, the human colonisation of Australia (Ouzman *et al.* 2002). Although undated, the artistic representation is not thought to be anything like that old. The favoured interpretation is that it represents a creature from aboriginal Dreamtime mythology. Oddly the possibility does not appear to have been considered that Aboriginals had observed sub-fossil footprints of the extinct bird and that these contributed to the mythology.

The perception of footprints by prehistoric people is also illustrated by footprints in the Bronze Age rock art of Scandinavia (Fig. 11.2; Nimura 2016). A recent study of their orientations suggests they may relate to solar observance and are part of the wider cosmological scheme of Bronze Age Scandinavia which involved ships, horses and the passage of the sun, all significantly symbols of mobility (Chapter 9; Skogland *et al.* 2017). Assuming the footprints are lifesize, some appear to represent children or adolescents (Bradley 2009). A surprising number appear to be shod. This is the reverse of the actual footprint-tracks where shod prehistoric examples are very rare. Some lines of footprints are seen as walking away from barrows as at Järrestad, Sweden where Bradley (2000) suggests these are the footprints of the dead moving from cairns across the liminal coast to the sea.

In more recent medieval Christian times marks on rocks interpreted as the supposed footprints of saints were venerated and in contrast others were supposed to be the cloved hooves of the devil (Mayor and Sarjeant 2001; Walsham 2010). One wonders if some of these ideas were based on religious rationalisation of observed sub-fossil footprints. It was noted above that Lower Palaeolithic footprints at Roccamonfina, Italy which, when originally observed, were interpreted by local people as 'devils trails'. In terms of perception most recent and evocative of all are the bare footprints in concrete of Jewish slave labour workers involved during World War II in the construction of the Atlantic defensive wall in France (Roberts 2010). They represent a permanent ichnofossil record of one of the darkest episodes in recent human history.

Conclusions

When footprint evidence began to be found more frequently on archaeological sites a comment made several times to the writer by archaeologists was that they were fascinating, but what could

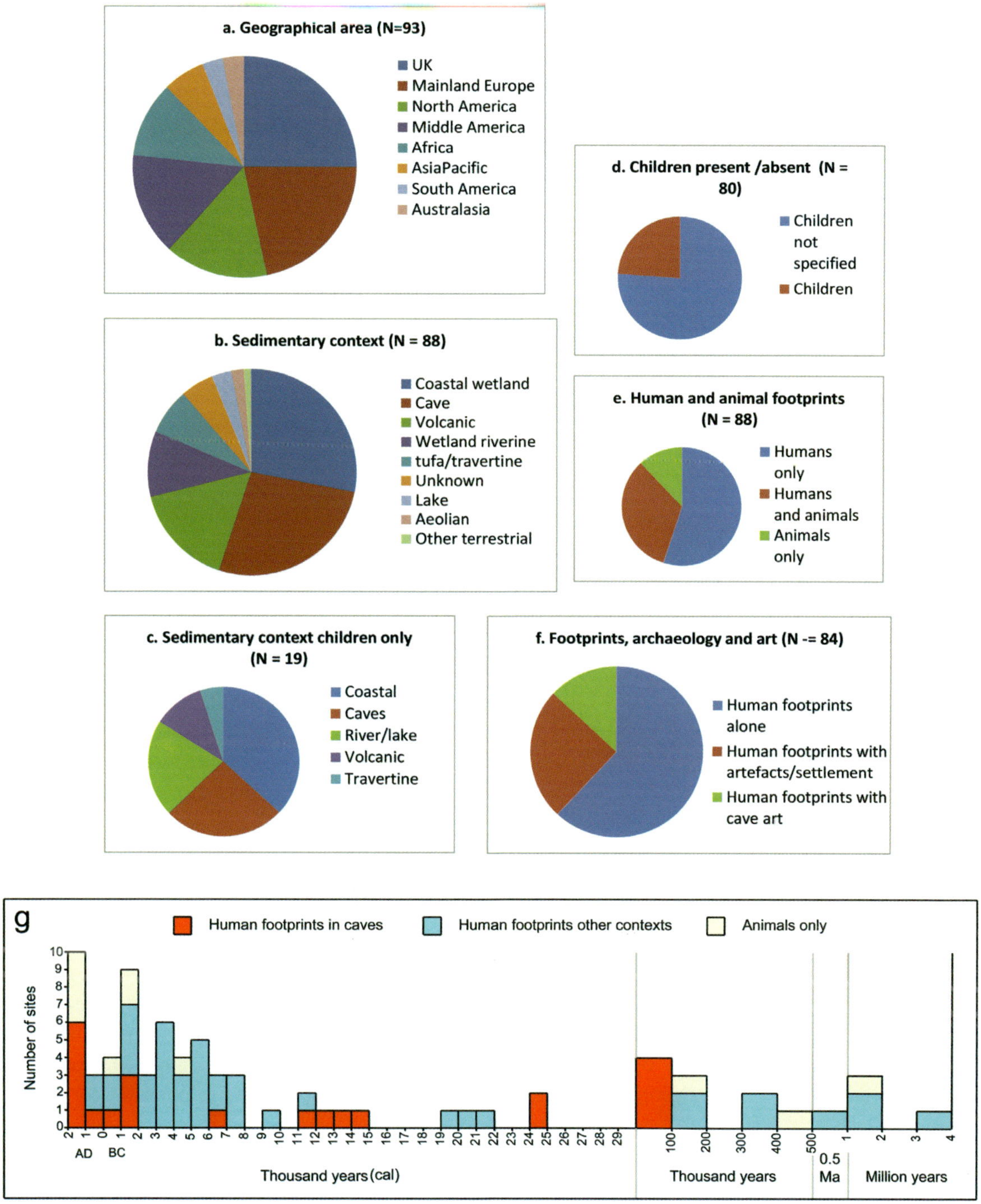

Figure 4.15 The emerging picture of human and animal footprint track evidence in terms of (a) geography; (b–c) sedimentary context; (d) age groups represented; (e) human and animal footprints; (f) archaeological context; (g) dates of footprints; note changes of scale marked by vertical lines. Based on Supplementary Appendix 4.1.

they actually tell us? The question perhaps tells us more about archaeologist's preoccupation with pottery, flints and buildings than it does about the limitations of footprint evidence. It is up to us as scientists to tease out what they can reveal and the objective of this chapter has been to show that they can contribute to a very wide range of archaeological questions. The chapter has also included as many different examples as possible with the objective of demonstrating the increasingly widespread nature of this form of evidence and types of context in which it occurs. Figure 4.15 (based on Supplementary Appendix 4.1) picks out some of the key patterns in the footprint data presently available including: (a) the geographical spread of footprints; (b and c) the sedimentary contexts in which they occur; (d) the occurrence of children; (e) occurrence of human and animals; (f) associations with art and archaeology; (g) footprint occurrence through time. Note the changes of scale marked by vertical lines.

It follows from the range of sedimentary contexts that sub-fossil footprints are likely to be far more frequent than the published literature indicates, a point also made by Bennett and Morse (2014). That point is also demonstrated by a dramatic increase in the number of finds in recent years. Of the British finds mapped on Figure 4.2a, seven out of 18 have been reported in the last 6 years, almost entirely as a result of an increasing focus on the archaeology of the intertidal zone. There must, however, also be many terrestrial contexts where the evidence will be found. Lockley and Hunt (1995), writing about dinosaur tracks, make the point that footprints ought to be far more abundant than bones; a creature has one skeleton but creates myriad footprints during a lifetime. A active individual human has been estimated to make 224 million footprints in a lifetime (McLaren *et al.* 2018). Of course, only a minute proportion is likely to escape the destroying hand of time.

So many of the cases touched on in this chapter highlight the importance of a geoarchaeological approach, which has been so well illustrated by the work of Allen (J. 1997) and Bennett and Morse (2014). These show why it is important to pay particular attention to the sedimentary and geomorphic context of the footprints, the formation processes of the record and of course the palaeoenvironmental evidence, to which animal footprint-tracks make their own distinctive contribution.

Just as we need to be aware of the possibilities of this relatively new form of evidence so we need to be alert to its limitations and the problems of interpretation. It was noted above that footprints occur in specific sedimentary contexts, so we need to consider to what extent these patterns are representative of the totality of activity. Some apparently represent people fleeing the catastrophic event of a volcanic eruption. Those are not everyday activities but provide a snapshot of humans coping with environmental disaster. In other situations, such as the laminated silts at Formby, the Severn Estuary, and Monte Hermoso, we are dealing with something much more like the palimpsest pattern created by everyday activities. However, the evidence may be more likely to be preserved at particular seasons, or events resulting, for instance, in the change from fine to coarse sediments at the transition from summer to autumn. We have therefore to consider to what extent the footprint-track record can be considered representative.

In the study of footprint-tracks we are particularly constrained by the quality of the record. The vast majority of cases are poorly preserved and in these cases it is important that interpretation is not pushed beyond what the evidence will allow. It may for instance be clear from the overall shape of the print that it is human and if it is part of a bipedal trail then that may provide confirmation.

We can infer from this that people were present and in what direction they were moving. Only if the dimensions of the foot can be accurately determined can we say anything about the age and height of the individual, the speed of walk etc.

There is obviously a level of uncertainty concerning the age of those represented. Present estimates of age as quoted from many studies earlier in this chapter are based on averages of modern population without consideration of the standard deviations of those averages, or how modern populations may have differed in terms of growth rates, height or physique from those of the past. We can, however, compare stature determined from footprints with that derived from skeletal remains. Scales (2006) did this for Mesolithic examples and found that the range of heights were similar. More problematic are uncertainties in establishing the sex of footprint makers. Identifying large males is perhaps possible with greater confidence, but distinguishing small males and females is problematic. Indeed Tuttle (2008), with an eye to the forensic and criminal applications of footprints, argues that, without knowledge of the range of statures and habitual footwear of an individual, it is probably not possible to establish age and sex. As footprint evidence starts to be found and used more in archaeology the uncertainties and problems of interpretation clearly need to be addressed. It remains to be established whether heightened observational skills, such as those applied by Bushmen trackers to interpretation of footprint-tracks in French Palaeolithic caves (Pastoors *et al.* 2015), can facilitate more precise determination of age, sex and activity patterns than currently achieved by western metrical scientific methods.

Perhaps the most striking aspects of many of the sites noted in this chapter are the proportions of children represented by footprint-tracks. This is particularly marked in the case of the Severn Estuary, Formby, Monte Hermoso and the French Palaeolithic caves. It may be contended that this evidence is put in question by the uncertainties of establishing the ages of footprint makers. Whilst that may introduce an element of doubt in some cases it does not undermine the overall point because in many of the cases the footprints are so small that they are highly likely to belong to young children and we have noted many cases involving children under 10 and some as young as 3 or 4. This is important evidence because it casts light on a segment of the population which is generally very hard to recognise in the archaeological record, except in death (Moore and Scott 1997). Chamberlain (1997) makes the point that children (defined in his case as under 18) would have formed about 50% of prehistoric communities and the tendency for them to be ignored in archaeological discourse he sees as a result of an academic male hegemony.

The footprint evidence shows children actively engaged, particularly in the activities of hunting and gathering communities. It demonstrates that a range of age groups were involved in the seasonal camps of Mesolithic communities of the Severn Estuary, in the foraging parties at Formby and in the seasonal pastoralist camp at Bronze Age Redwick. Nor of course should we ignore the possibility that the presence of children in some of these situations might reflect, not everyday activities, but special circumstances such as initiation rites, or rites of passage. Van de Noort and O'Sullivan (2006) have made the interesting suggestion that seasonal pastoral camps may have been significant contexts for adolescent rites of passage. This might also be an explanation for the predominance of child prints deep within some cave systems.

The involvement of children in activities is what we would expect from the ethnographic record which shows some as young as 3–5 as fully participating members in many activities of hunter-gatherer groups (Hawkes *et al.* 1995; Zeller 1987; Bird-David 2005; Politis 2005). Meehan

(1982) records that shellfish collection among the Anbarra aboriginals of Australia was largely done by woman, often accompanied by girls and children as young as 2–4; it often incorporated family time and opportunities for play (Cläassen 1991). Turner (2005) writing about the First Nations communities of the Canadian North West Coast discusses the opportunities which logistical harvesting expeditions provided for children to observe, practise skills and learn the techniques needed for their future survival. In the Mesolithic cemeteries of Teviec and Hoedic, Brittany some child burials had rich grave goods and elaborate ritual (Cunliffe 2001a). Likewise Mesolithic and early Neolithic child burials from Ukraine show that they were buried with status objects and there was no evidence of differential access to food resources (Lillie 1997, 225). This supports the footprint evidence that they were fully active participating members of their communities. Zeder (2012, 258) emphasises that the Broad Spectrum Revolution and the intensified process of niche construction which took place in the late glacial and initial Holocene, led to an increased role for women and children in the exploitation of a wider range of resources at a range of times of year from one site. Around such places, one supposes, a more clearly defined radial network of paths would develop.

Footprint evidence from Palaeolithic sites, most notably from Laetoli, has made an important contribution to the study of hominid evolution and especially the development of bipedalism. It also contributes to the study of social behaviour, as illustrated for instance by the three Laeotoli individuals apparently walking together. There is also the evidence for patterns of human colonisation. Some footprints represent the earliest evidence for human presence in particular areas, for example the Tibetan plateau at a time when it was thought to be ice covered. Others are the Calvert Island evidence for human colonisation of the Americas down the coastal so-called 'kelp corridor' or the Monte Verde examples from Chile, so soon after the apparent date of the colonisation of the Americas via Alaska.

In many of the examples reviewed in this chapter footprints contribute to an understanding of the ecological context of particular deposits by providing evidence for the range of animals present. In this respect the footprint evidence is complementary to that provided by animal bones. Footprints have the advantage of providing information about animals actually present at a site, whereas bones may represent food brought from elsewhere. The ecological aspect was particularly well illustrated by Laetoli where a stratigraphic sequence of contrasting assemblages of animal footprint-tracks established the seasonality of human activity.

Human footprint-tracks have also contributed to the investigation of seasonality in the British Mesolithic at Goldcliff by demonstrating that the most well-preserved examples were made in the height of summer when the finest sediments were deposited. This is the most conclusive of a wide range of seasonality indicators from these sites (Bell 2007a). As regards later agricultural communities, footprint-tracks contribute to an understanding of the animal economy and particularly the evidence for seasonal saltmarsh grazing in the Severn Estuary during the later Bronze Age and Iron Age. We have also noted instances in which human footprint-tracks provide evidence of particular patterns or axes of movement such as paths which have been identified at Lake Managua, Mojave and Goldcliff. The convergence points of trails indicate the location of lost, or buried, campsites or activity areas. It is clear from the sites which have been studied in the greatest detail, that when we have the opportunity to use footprint-tracks as part of a multi-proxy approach to interpretation, comparing animal and human footprint-tracks, sedimentary, pollen, macrobotanical, bones, molluscan and other evidence, then interpretations are notably rich and more robust.

Bennett and Morse (2014) whilst highlighting the importance of footprint-tracks in the Pliocene and Pleistocene to studies of human evolution, a particular strength of their book, imply that footprint studies are less significant in Holocene contexts to which they add 'local colour'. It is hopefully by now clear from the foregoing examples that they offer far more in terms of evidence for social interaction, axes of movement, seasonality, interaction with animals, demographic composition of groups and the role of children in society, and even aspects of human perception etc. They reveal the small-scale patterns of movement by which people structure their daily lives and their relationships with animals.

5. Early farmers: mobility, site location and antecedent activities

Introduction

In the previous chapter we explored how the footprints of individuals can contribute to understanding of small-scale patterns of mobility. That is a form of evidence so individual, fleeting and personal that everybody feels they can relate to it. This chapter begins with a case study illustrating how the exceptional preservation of one individual, the Ice Man, can enhance our understanding of mobility in many ways. This is followed by evidence from skeletal and isotopic sources for human and animal movement as preludes to the body of the chapter which focuses on the first farmers in the British Isles, beginning with site-based and environmental evidence for their patterns of movement, and then briefly considering artefacts. Linking themes between the introductory sketches and the body of the chapter are the importance of remembering that we are dealing with people rather than sites, environments and things, and that to understand people's patterns of movement we must necessarily integrate multiple forms of evidence and work on diverse spatial scales.

Case Study: the Ice Man

The so-called Ice Man or 'Otzi' from the high Alps of the Austrian–Italian border illustrates the wide range of evidence for movement which may be preserved under deep frozen conditions in glacier ice (Spindler 1993; Fleckinger and Steiner 1999). This formed in prehistory and is now melting for the first time in 5000 years as a result of global warming. The body was found at a height of 3210 m and dates to 3365–3040 cal BC. His clothing and artefacts, which were superbly adapted to the harsh conditions, provide many clues to his origins. Charcoals, wood and pollen evidence shows him passing through a sequence of Alpine vegetation zones with mixed oak and hornbeam woods at lower elevations, then mature spruce forest pollen from higher elevations to the snowfields above the woodlands where he was found. He originated in an agricultural settlement; spikelets of cereals were found in his clothing, he had eaten cereals and the evidence suggests he was in the valleys as little as 12 hours before his death, which appears to have occurred in late spring or early summer. He carried a very early copper axe, for which the closest parallels are the Remedello culture of northern Italy, where carvings on menhirs also show similar axes. Pollen evidence and wood from his tools suggest he came from the Senales valley to the south, where there are Neolithic settlements (Bortenschlager and Oeggl 2000). Some of the plants present are also found to the south, the moss *Neckera* and hop hornbeam (*Ostrya carpenifolia*) (Dickson *et al.* 1996). Well before the Ice Man, *c* 4300 cal BC, people were already engaged in niche construction at lower elevations around the tree line, *c* 2760 m. Heather with dwarf birch was being burnt and replaced by meadows typical of Alpine grazing land. This is thought to reflect the activities of transhumant pastoralists, who even today take their animals across high Alpine passes to exploit seasonal pastures to the north. Spindler suggests that it was transhumance which brought the Ice Man to the High Alps. However, it is questionable whether he was directly involved in animal herding; had he been one would have expected abundant animal hairs in his clothing.

Also in the Swiss Alps melting ice patches are revealing organic artefacts such as clothing, shoes, bows and arrows as evidence of Alpine hunting expeditions some 5000 years ago (Suter *et al.* 2005). In the north Italian Alps at Monte Bego, Neolithic and Bronze Age rock art most frequently depicts horned cattle (Fig. 5.1) in a valley leading to Alpine passes between Italy and France. Above the tree line in those high mountain pastures animals are likely to have grazed for about 3 months of the year (Bradley 2009; Huet 2017); the sojourn of their herders would have provided opportunities for social interaction and exchange with groups from across the pass, which might help to explain the very many depictions of weapons. What we may be glimpsing in the case of the Ice Man and the Monte Bego art are the early origins of patterns of Mediterranean transhumance from lowlands in winter to upland pastures in spring and summer. Such practices are well-documented aspects of the economy in Classical, medieval and up to recent times (Walsh 2013). It may be questioned whether at this early date grazing was in such short supply as to necessitate transhumance. Two points should be made in response to this. Firstly, that extensive natural meadows would have existed above the tree line and could have been extended by burning.

Figure 5.1 Monte Bego, Italy: prehistoric rock art with two cattle drawing a plough (photo: M. Bell).

Secondly, transhumant expeditions may well have been as much for social as stock management objectives; they presented opportunities to meet with people, and exchange goods, genes and knowledge with those from the other side of the mountains. This point is also pertinent to other areas where mobility with animals occurred, including Neolithic to Iron Age Britain.

Skeletal, isotopic and DNA evidence for Neolithic mobility

Human skeletal evidence from the Neolithic enables us to identify the movement of individuals. Male burials from the Parc le Breos, a megalithic tomb in south Wales dated *c* 3800–3400 cal BC, exhibited musculoskeletal stress markers evidencing extensive leg exercise, suggestive perhaps of transhumant herding (Whittle and Wysocki 1998). Isotopic evidence indicated the diet of these people was mainly terrestrial and meat-based with no significant use of marine resources despite their availability within 1 km of the tomb. Isotopic evidence can also provide more specific evidence of patterns of mobility, as in the case of burials from a henge-like ritual site at Monkton-up-Wimborne on Cranborne Chase (Budd *et al.* 2003). These included a male aged 30 and a young girl who had previously lived in an area where there was a high lead content in the water; the nearest source is on Mendip, 34 km distant.

Later Neolithic cremated humans from Stonehenge also exhibit isotopic evidence for non-local origins, possibly including west Wales from whence the site's famous bluestones originated (Snoeck *et al.* 2018). In the final Neolithic, evidence for animals such as cattle and pigs being driven to the great henges of Wessex at Avebury, Durrington Walls by Stonehenge, Marden and Mount Pleasant, from diverse isotopically distinct areas across the British Isles, points to previously unsuspected pan-British connectivity (Chan *et al.* 2016; Madgwick *et al.* 2019). Only relatively small proportions of these animals had been raised locally and they are interpreted as having been driven considerable distances for gatherings at particular times of year, in the case of the Durrington Walls cattle in late autumn and winter (Viner *et al.* 2010; Parker Pearson 2012).

Isotopic evidence is likewise transforming our understanding of mobility patterns involving the first metal users in Britain who were buried with Beakers. The earliest and richest of these, dated 2470–2280 cal BC, is the Amesbury Archer from Wiltshire, who was buried 5 km east of Stonehenge (Evans *et al.* 2006; Fitzpatrick 2011). This adult male was accompanied by Beakers, barbed and tanged arrowheads, copper daggers, a boar's tusk, a tiny stone for metal testing, two gold earrings and archers wrist guards. Strontium and oxygen isotope analysis of his teeth indicate mainland European origins; the Alps were initially proposed but subsequent work has suggested perhaps the Middle Rhine (Parker Pearson 2012). Three other Beaker burials near Stonehenge, known as the Boscombe Bowmen, all appeared to have originated from an area of older Palaeozoic rocks, probably in the west of the British Isles (Evans *et al.* 2006). Isotopic analysis involving 360 burials of this period indicates that something like half grew up in an area different from that in which they were buried; however, a very small proportion (*c* 2%) produced detectable isotopic evidence of origins outside Britain (Jay *et al.* 2011; Parker Pearson *et al.* 2019).

The emerging picture seems consistent with significant pastoral mobility in the earlier Neolithic developing into longer distance ceremonially-motivated mobility in the final Neolithic henge period, with significant human mobility among early metal users. This is consistent with the paucity of evidence for permanent settlements through much of the Neolithic (Whittle 1997). Current opinion sees Beaker communities as semi-mobile cattle pastoralists, and micro-residues in Beaker pots indicate that some contained milk products (Soberl *et al.* 2009). However, in parts of southern Britain at least there is evidence for arable activity and changed patterns of animal husbandry in the Beaker period (Allen and Maltby 2012), so perhaps we may be seeing a degree of spatial variation and specialism.

Work on ancient DNA (aDNA) is radically changing our understanding of patterns of connection between populations and movements associated with the spread of farming. The current view among most writers is that both indigenous and incoming groups made a significant contribution to Neolithisation, both being fundamentally changed in cultural terms in the process (Cummings and Harris 2011; Garrow and Sturt 2011). The early Neolithic Linearbandkeramik groups of Central Europe have been shown to have a connection with the Starçevo gene pool of the Balkans, which could point to population migration rather than the adoption of farming by indigenous communities of Mesolithic origin (Bickle 2016). Other studies infer that the female line of the present-day population of Europe is largely of pre-Neolithic genetic origin (Haak *et al.* 2005), with the implication that overall acculturation (adoption of farming by indigenous groups) was more significant than population replacement. That the latter took place in certain periods is indicated by evidence from the Beaker period in Britain where the latest aDNA studies point to an almost total population replacement over a few centuries, as noted in Chapter 1 (Olalde *et al.* 2018). The scene is clearly set for aDNA studies to totally transform our understanding of prehistoric mobility in the coming years.

Neolithic landscapes in Britain

Growing evidence for the mobility of people and animals in Neolithic Britain requires a focused investigation of the routeways involved. With the advent of the Neolithic we have many more sites, including the creation of monuments, some in alignments. These are potential clues as to how the early farmers moved around, although the interpretation of that evidence is frequently contested. Recent studies of the Neolithic have also given a greater emphasis to mobility by comparison with the prehistoric periods considered in other chapters. Two edited books in particular explore this theme: *Prehistoric Journeys* (Cummings and Johnston 2007) contains several Neolithic case studies and *Moving on in Neolithic Studies* (Leary and Kador 2016) is entirely focused on the period. Both are case study based, rather than focused on the general theme of how we can identify past routeways; they are also generally site and artefact-based studies rather than having the palaeoenvironmental dimension which is emphasised here.

The Neolithic is marked by the appearance of domesticated plants and animals, pottery and, within two or three centuries, the appearance of monuments. The chronology of the beginning of the Neolithic has recently been greatly refined due to the increasing availability of radiocarbon dates, the critical examination of the significance and validity of dates, and the Bayesian statistical analysis of sequences of dates on stratified sites (Bayliss and Whittle 2007; Whittle *et al.* 2011). This indicates that farming first arrived in the British Isles around 4100 cal BC and rapidly spread to the west, reaching Ireland and Scotland around 3800 cal BC. The first long barrows were constructed from *c* 3800 and by 3700 there was a peak of causewayed enclosure construction. Between 3800 and 3700 cal BC the first rectangular buildings are represented; in Britain, and even more so in Ireland, rectangular buildings are very much concentrated in the early centuries of the Neolithic and later there are fewer domestic structures. Despite the rather dramatic changes which this chronology implies, it is generally considered that in many areas of the British Isles the Neolithic way of life was not totally settled, hence the paucity of domestic structures. In Ireland and Scotland where more structures occur, some with significant evidence of crop growing, there seems to have been less emphasis on mobile pastoralism (Cooney 2000), reminding us that regional variability is to be expected.

In much of the British Isles the Neolithic seems to involve mobility focused on animal husbandry (Thomas 1991; Bradley 1993; Whittle 1997; Pollard 2000b). Neolithic communities grew crops, but to what extent remains a matter of debate (Fairbairn 2000; Stevens and Fuller 2015). There is often evidence that crop growing was combined with the continued use of wild plant resources. There is little physical evidence of their fields. However, long accepted models of shifting Neolithic agriculture have recently been called into question as a result of the detailed analysis of weed seeds associated with crops, both in Central Europe and the British Isles, which points to longer established fields in which fertility was maintained by manuring (Bogaard 2004; Bogaard and Jones 2007). The agricultural activities of Neolithic communities have mainly to be inferred from pollen evidence for clearings and cultivation (Simmons and Tooley 1981). There has been a tendency to treat environmental and cultural evidence as separate categories in the Neolithic, whereas understanding requires an integrated perspective in which the woodland and other vegetation forms part of the 'architecture', as it were, of sites and landscapes (Noble 2017). Many pollen diagrams show distinct vegetation changes, a reduction in trees, and an increase in herbs and plants of disturbed ground soon after 4000 cal BC. Sometimes these changes are associated with small numbers of cereal-type pollen grains, which are generally interpreted as reflecting the activities of the first farmers. This interpretation is complicated by several factors. One is the increasing evidence for Mesolithic vegetation disturbance and burning discussed in Chapter 3. Another is the pronounced decline in elm pollen which occurs *c* 3800 cal BC across a wide area of western Europe. This was once interpreted as the result of Neolithic farmers using elm leaf fodder for animals; today it is widely considered to be an early outbreak of Dutch Elm Disease, although early farmers' movements and practices may have contributed to the spread of the disease (Parker *et al.* 2002). The fact that the elm decline occurs at about the same time as the onset of the Neolithic means that these parts of pollen sequences tend to be subject to particularly close sampling, with the consequent likelihood of detecting more small-scale vegetation changes than have been recognised at other times by less high-resolution sampling. A further issue is that plants of disturbed ground have tended to be interpreted in terms of areas used for agriculture, although as emphasised in Chapters 2 and 3, they are often associated with the woodland edge and concentrated along routeways. Human and animal movement along routeways will have played an important part in the introduction, not only of crop plants, but also weeds, some introduced with seed corn. Plaintain (*Plantago lanceolata*) is an obvious example; it is associated with trampled ground and often taken to indicate grazing animals, yet it is particularly likely to be associated with paths. Many weed species are known to be introductions and often become particularly aggressive colonisers in their new environment. People and animals moving around disperse seeds, in animal coats, clothing, mud on feet etc.

Herbivores also disperse seeds and Salisbury (1964) gives a list of those propagated from herbivore dung: cattle (42 species), horse (25 species), pig (6 species), goat (4 species). The significance of routeways in the spread of weedy species has been illustrated in recent times by successive waves of weedy invasions along the corridors created by the construction of canals, railways and motorways. These provide some indication of the much smaller scale and less dramatic introductions which must have been associated with the paths, trackways and roads in the past, especially since the introduction of weed species with crop plants. This is illustrated by the many Old World weeds introduced to the New World of the Americas and also by the case of New Zealand where the weed flora is largely of European origin; of 500 introduced species, 80% are from the Old World.

Neolithic monuments in Britain

Movement was evidently part of Neolithic life: seasonal movement with animals; the movement of rocks and other artifacts; and the emergence of ceremonial complexes which clearly served groups from wide catchments. It is inherently likely that ceremonial sites lay on significant routes. In a review of the distribution of Neolithic monument complexes in Scotland, Noble (2007) concludes that they are on natural routeways, and have an association with objects of non-local origin. For instance, he identified cross peninsula routes which avoid hazardous sea crossings.

Some monuments can be argued to mark, or respect, routes, stone avenues being obvious examples. Henges with opposed double entrances can also be convincingly argued to represent routes. Some other alignments such as cursus monuments are far more enigmatic although many people think they have something to do with movement through the landscape (Johnson 1999; Last 1999; Loveday 2006). Very often, however, we infer routes based on recurrent use of the same site, for instance in Mesolithic and then in successive Neolithic phases. In a largely wooded landscape it is reasonable to expect that recurrent activity will take place (as outlined in Chapter 3) at key points along paths, such as where the route overlooks tracts of landscape, approaches open water, or especially where paths intersect. Some intersection places may have achieved social significance as the places where groups moving in different directions met.

It is a reasonable, and generally argued proposition, that many of the paths on which Neolithic communities moved around the landscape will have been established earlier in the Mesolithic (Tilley 1994; Edmonds 1997). The creation of monuments of various types, but especially tombs, created fixed points in the landscape to which subsequent activities referenced. The making of monuments creates in Gosden's (1994) words 'patterns of recursiveness'. Barrett (1994) saw Neolithic sites such as megaliths, long mounds and causewayed enclosures, not so much as the centres of land territory, but at the beginning and end of paths, the intersection and meeting places in a wider seasonal cycle. Phenomenological approaches envisaged the monuments as drawing their significance in part from their relationship to specific topographic features, hills, mountains, prominent rock outcrops and the coast in relation to which they were positioned (Tilley 1994; Cummings 2002, Cummings and Whittle 2004; Cummings and Harris 2011). Certain outcrops and types of rock were attributed particular significance and this led to these being transported great distances, either as portable artefacts such as axes, or as huge stones which formed part of later Neolithic monuments. Perceptual and phenomenological perspectives have made a significant contribution to the study of prehistoric movement; however, the application of these approaches, for instance to Neolithic tomb location, has not always been sufficiently rigorous or evidence-based, as demonstrated by

Fleming (1999; 2005b). Up to now phenomenological approaches have made little use of palaeoenvironmental evidence (Chapman and Gearey 2000). The extent of tree cover will have been a key factor in the visibility of landscape features and thus landscape perception, for instance routes themselves would open up a more distant vista in an otherwise wooded landscape. Bradley (2012) has further proposed that, for people living in a closed woodland, this may engender a linear perception of space, whereas for those living in an open landscape, the perception of space may be more circular, such perceptions perhaps being reflected in the form of monuments.

In seeking to identify ancient routeways, a rigorous and questioning approach is essential in order to avoid the pitfalls which waylaid Alfred Watkins (1925) almost a century ago (p. 2). Former routes will be more readily recognised where they follow a relatively straight course, but there can be no assumption that routes were straight, they are just as likely to be curved if following topography. Identification of curvilinear routes connecting sites can be aided by GIS approaches using least cost paths (Conolly and Lake 2006). Instructive as such tools can be, we also need to keep in mind that routes may be governed by non-functionalist, non-utilitarian factors such as ritual observance, cosmology etc.

Megalithic tombs

Neolithic tombs are often seen as fixed points in a landscape in which earlier Neolithic communities and their herds remained mobile. It is striking that many of those tombs excavated under modern scientific conditions have produced evidence of long histories of environmental disturbance and episodes of occupation, by both Mesolithic and Neolithic communities, before the period of tomb construction. Where environmental evidence is available, it generally indicates construction in grassy clearings. Thus, they were already places with a history.

On a terrace above the River Usk, the tomb at Gwernvale, Wales focused on a natural group of boulders, round which were early and late Mesolithic artefacts and a hearth dated to 5880–5710 cal BC (Fig. 5.2; Britnell and Savory 1984). There was also an early Neolithic rectangular building with cereals and bones of domestic animals, all sealed by the tomb (Fig. 5.2b–c). The coincidence of activity over such an extended timescale could well reflect a long-lived routeway with the boulders providing a focal point. Indeed, this is the route that Parker-Pearson (2012, 289) advocates for the later dryland transport of the bluestones from Preseli to Stonehenge. There are a number of standing stones along the same route, which is a classic 'natural' routeway later taken by a Roman road and today by the A40T (Fig. 5.2a).

Below the Neolithic chambered cairn at Hazleton, Gloucestershire there was evidence of late Mesolithic activity, then a midden dated 3900–3800 cal BC (Saville 1990). Micromorphological and environmental evidence indicates cultivation, then reversion to scrub, before construction of the tomb around 3600 cal BC (Meadows *et al.* 2007). Below the chambered cairn at Ascott-under-Wychwood, Oxfordshire there were earlier and later Mesolithic artefacts and mollusc evidence for a phase of more open vegetation followed by closed woodland (Evans 1972; Benson and Whittle 2006). Clearance created grassland which was followed by occupation, a building and a midden in the fortieth and thirty-ninth centuries BC, then the tomb was constructed *c* 3760–3695 cal BC (Bayliss *et al.* 2007). Below a long mound at Raunds, Northamptonshire, constructed 3940–3780 cal BC, was evidence of repeated Mesolithic visits during the fifth millennium BC (Harding and Healy 2007, 47).

In each of these cases activity was periodic rather than continuous. As Thomas (J. 2013) says, 'it is hard to avoid the conclusion that particular places had significance maintained by social memory'.

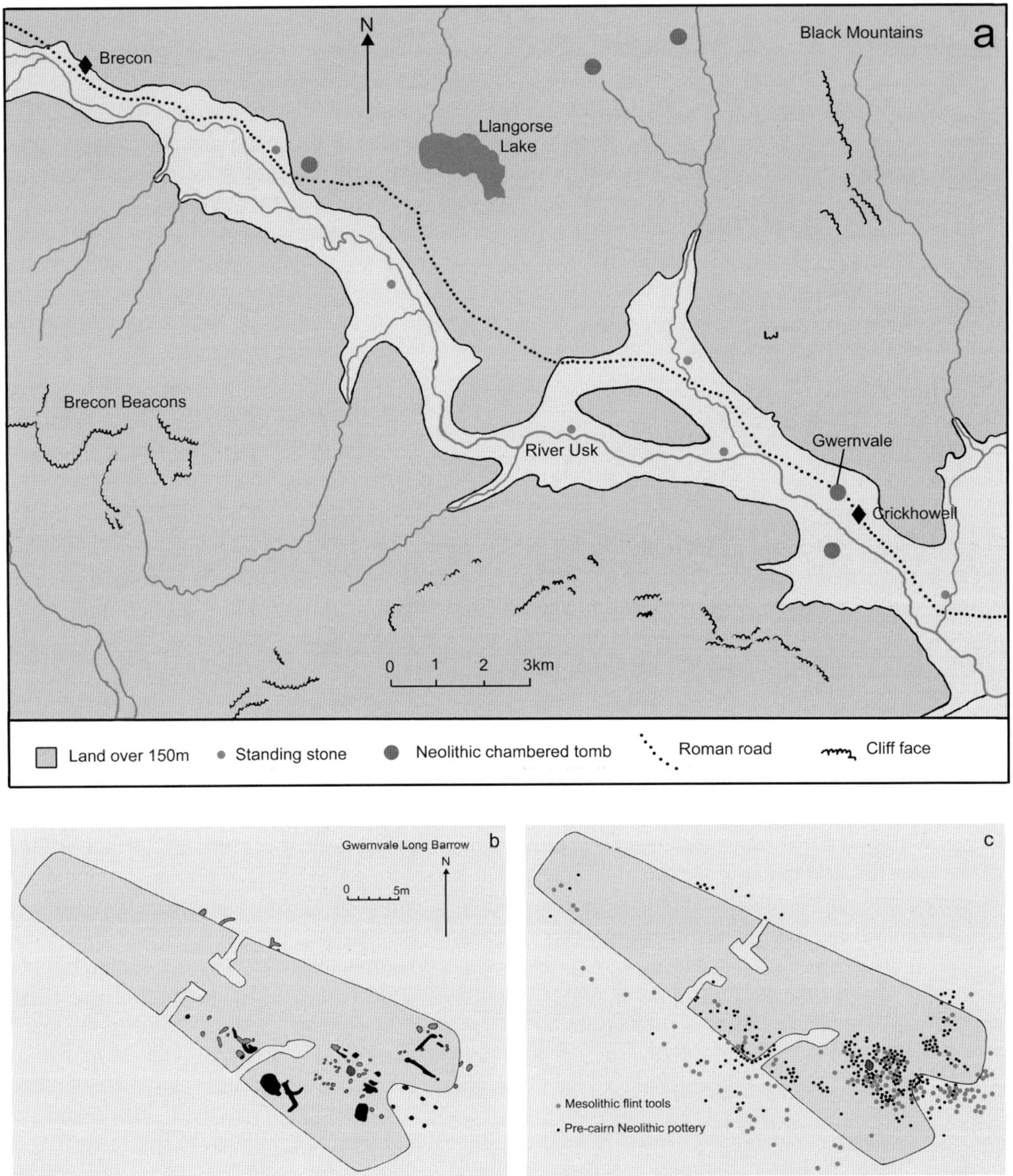

Figure 5.2 Neolithic chambered tomb at Gwernvale, Wales showing successive phases of activity: (a) Position of tomb in relation to other tombs and standing stones; (b) pre-tomb features and building; (c) pre-tomb Mesolithic and Neolithic artefact distributions (graphic: J. Foster; source Britnell and Savory 1984).

Trackways and especially their intersection points, and maybe the stories that accompanied the recurrent use of trackways, are likely ways in which this social memory was maintained.

Tree-throw pits and causewayed enclosures

Pits created by natural tree-throw are often revealed by excavations and can be recognised by their distinctive shape and fill: roughly circular with a banana shape of humic fill and the remainder of the fill redeposited subsoil weathered from the tree bole (Fig. 3.5). Such pits often contain palaeoenvironmental evidence of former woodland. More surprisingly there are numerous examples containing artefacts, such as flint, pottery and charcoal, pointing to activity in the clearing created by tree throw. Sometimes it is evident that the artefacts have been deliberately deposited. Evans *et al.* (1999) give an example at Barleycroft, Cambridgeshire where tree-throw features dated 3780–3650 cal BC contain artefacts and are cut by later pits *c* 3690–3530 cal BC, demonstrating recurrent visits. Deposition of artefacts has been interpreted as possibly relating to settlement closure and hiding of human presence, the fallen trees perhaps serving as settlement markers or foci for camps. Artefacts may sometimes have been cached (stored) in tree-throw holes pending scheduled return as part of seasonal mobility. Evans (1988) makes the very interesting suggestion that enclosures may have mimicked a concentric spatial principle presented by clearings and this resonates with the case developed in Chapter 10 at Bishopstone, Sussex (Fig. 10.3) where a probable Neolithic clearing seems to have been bounded by a ring of pits including at least one tree-throw (Bell 1977). It was not a formal enclosure, but might have developed into one, had its development not been arrested.

Causewayed enclosures were created from 3800 cal BC with a peak around 3700 cal BC (Whittle *et al.* 2011). They seem to have been communal gathering places, perhaps seasonal rather than permanent settlements. Surprisingly, those which have been subject to detailed palaeoenvironmental study appear to have been constructed in woodland or scrub, for example those at Windmill Hill (Whittle *et al.* 1999); Hambledon Hill (Bell *et al.* 2008); Knap Hill (Connah 1965); and the Sussex causewayed enclosures (Thomas 1982). The artefactual evidence certainly suggests that materials, pottery, axes, stones etc were brought to the site, sometimes from a significant distance, and may have been exchanged at the site.

Cursus monuments

Cursus monuments are among the most enigmatic sites in prehistory and some may wonder how they feature in a book on prehistoric tracks. The origin of their name is no less curious. It derives from Stukeley (1740) who thought they were for chariot races; that at least we can dismiss because they mostly date from 3650–3050 cal BC (Barclay and Bayliss 1999), some 2500 years before the first chariots in Britain. They can be defined as long enclosures, sometimes several kilometres long, defined by parallel ditches; they may possess banks and sometimes a central mound; at least one end is closed (Loveday 2006). Lengths vary from 170 m to 4 km and the conjoined Gussage and Pentridge cursuses in Dorset together cover 9.8 km. Related sites are defined by lines of pits or posts (Thomas 2006). Around 110 convincing and possible cursuses are known in Britain (Barclay and Harding 1999; Loveday 2006). Some examples have an entrance gap at the middle of one end, encouraging the idea that they may have created processional space (Loveday 2006, fig. 56), which is how Atkinson (1955) interpreted the Dorset cursus. Other examples have opposed gaps suggesting that they may have been crossed by paths (e.g. Holywood north; Dorchester; Drayton; and recent discoveries at the Stonehenge cursus). They are often located on river terraces and roughly parallel to rivers, terminating near streams. A number cluster around confluences, for instance in the Thames valley at Lechlade and Dorchester, and elsewhere they coincide with historically important river crossings, for instance that of the River Welland at Maxey and the River Severn at Welshpool (Loveday 2006, 134).

The ditches of cursuses tend to produce very few artefacts but once established they, or the significant places which they represent, tend to become the focus of later Neolithic and early Bronze Age monuments such as barrows and sometimes henges, as demonstrated by extensive excavations at Drayton (Barclay *et al.* 2003) and Dorchester-on-Thames, a site respected by other monuments for a thousand years (Whittle *et al.* 1992).

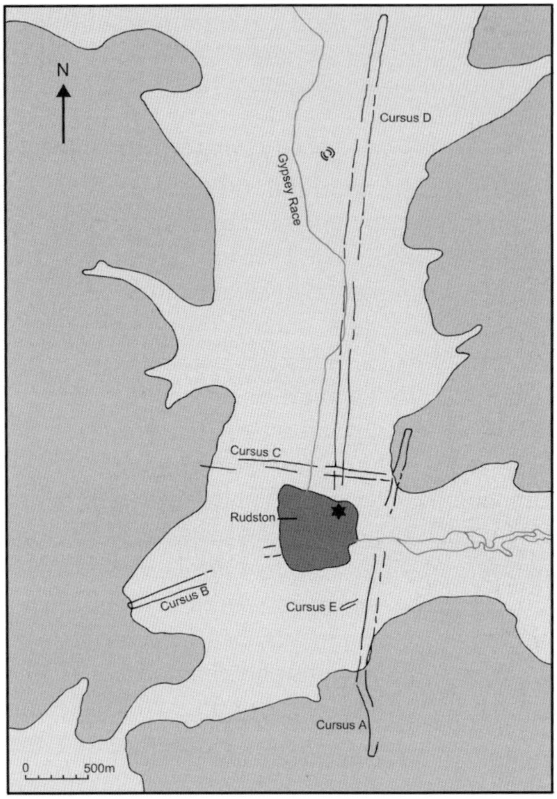

Figure 5.3 Rudston, Yorkshire: five cursus monuments in relation to topography and the river Gypsey Race. The tall standing stone is marked by a star (graphic: J. Foster; source Woodward 2000, fig. 5.1).

The very nature of these long linear monuments is suggestive of construction in an open grassland environment and hints perhaps at openness resulting from movement of animals along a path, or clearance by people along a path. Where detailed palaeoenvironmental analysis has taken place it supports the idea of an open environment: in the Thames valley (Barclay *et al.* 2003); round the Stonehenge Cursus (Richards 1990); and the Dorset Cursus (Barrett *et al.* 1991). However, in the Dorset case there is growing evidence that the chalklands of Cranborne Chase may never have been fully wooded in the Holocene (French *et al.* 2007). The 2 km long cursus, or bank barrow, Cleaven Dyke, Scotland was in a landscape of heath and grassland with birch and hazel woodland; turves were used in construction and there was evidence of grazing before the monument (Noble 2017), which would be consistent with elaboration of an existing path. Some cursuses are in areas with preceding Neolithic activity; at Drayton, for instance, there are tree-throw pits with artefacts and charcoal (3960–3360 cal BC) which pre-date the cursus (Barclay *et al.* 2003). A cursus at Stanwell (Lewis *et al.* 2006) was 3.6 km long along a break of slope between a Thames terrace and the floodplain of the River Colne. There was evidence of Mesolithic activity in the form of pits some 3000 years earlier and the excavators concluded that it was constructed along a pre-existing path of great antiquity. The Dorset Cursus incorporates earlier long barrows in its construction and is so designed that a barrow at its end is illuminated by the midwinter sunset when observed from a position along the cursus (Bradley 1993). Johnston (1999) interprets that cursus as representing a pre-existing path which linked earlier monuments and may have also related to cosmological alignments. He sees earlier processions as defining the route but argues that the route was put out of use by the act of construction. The

Stonehenge cursus (Fig. 5.6, below) meets dry valleys at either end which might have formed part of a pre-existing route and also seems to reference a longer alignment with long barrows to east and west and the later Woodhenge further east. Whether that axis relates to a real or metaphorical route, cosmology, or some other factor remains a matter of debate (Bowden *et al.* 2015).

Particularly striking in terms of the marks they made on the landscape are the four cursus monuments which converge/frame a turning point in the Gypsey Race river at Rudston, Yorkshire where there is a huge standing stone at the end of the main cursus (Fig. 5.3). Three of the cursuses run from the chalk uplands to the river valley and one crosses the valley between uplands. Woodward (2000) sees the cursus as providing a connection between the lowlands and the uplands where barrows, some perhaps contemporary, others Beaker and Bronze Age, are located.

Most of those who have written about cursus monuments recognise some connection with movement through the landscape such as paths. Last (1999) sees them as monumentalised paths, Tilley (1994, 149) as a pathway for rites of passage, Barrett (1994, 137) as formally demarcated paths, their relationship to other monuments suggesting they were parts of a longer route. Bradley and Chambers (1988) speculate they may have been paths connecting the living and the dead. Many provided a focus for subsequent burials and some were preceded by earlier burials, a mortuary enclosure in the case of Dorchester-on-Thames, and barrows in the case of the Dorset Cursus. However, there is little evidence of formal burial directly associated with the cursuses themselves, although fragmentary human remains sometimes occur.

Whilst many authors accept the idea that cursuses may reflect earlier paths, there is some ambiguity as to whether these preceding paths represent everyday paths, originating perhaps in the Mesolithic, as argued at Stanwell, or metaphorical paths for the ancestors, or the gods. Astronomical evidence from the Dorset cursus in particular does seem to point to actual physical movement by people because the cursus structures people's experience of landscape and encounters with specific monuments. Given wide recognition that cursus monuments may have originated in paths it remains unclear whether they served as active routes once they were demarcated as a cursus. For instance, Lewis *et al.* (2006, 52) argue for the pathway origins of the Stanwell cursus but evaluate a list of possible functions which does not include a route, which may have been considered too functionalist an interpretation. However, we should at least consider the possibility that they continued to mark routes in some way, perhaps the parts of routes which required a particular form or permission, ritual observance, or status. The clustering of later ritual monuments round many cursuses might support the idea that they lay on routes, as might the association of some with previously grazed areas, with river crossings and floodplain edges. It must also be acknowledged, however, that the absence of artefacts in the ditches, and the fact that in no case, known to the writer, is there an obviously later path leading out of them may be more consistent with the hypothesis of metaphorical paths, or the idea that they put paths out of use.

Henges and avenues

In the case of henges the relationship to movement and routes is often more tangible than with the cursus monuments. Henges are circular monuments, often with an external bank and internal ditch; some have circular arrangements of posts or stones in the interior (Burl 2000). Stone settings sometimes relate to a later phase in the history of the monuments. Henges were constructed in the later Neolithic with some modest examples before 2500 cal BC and large ones after that

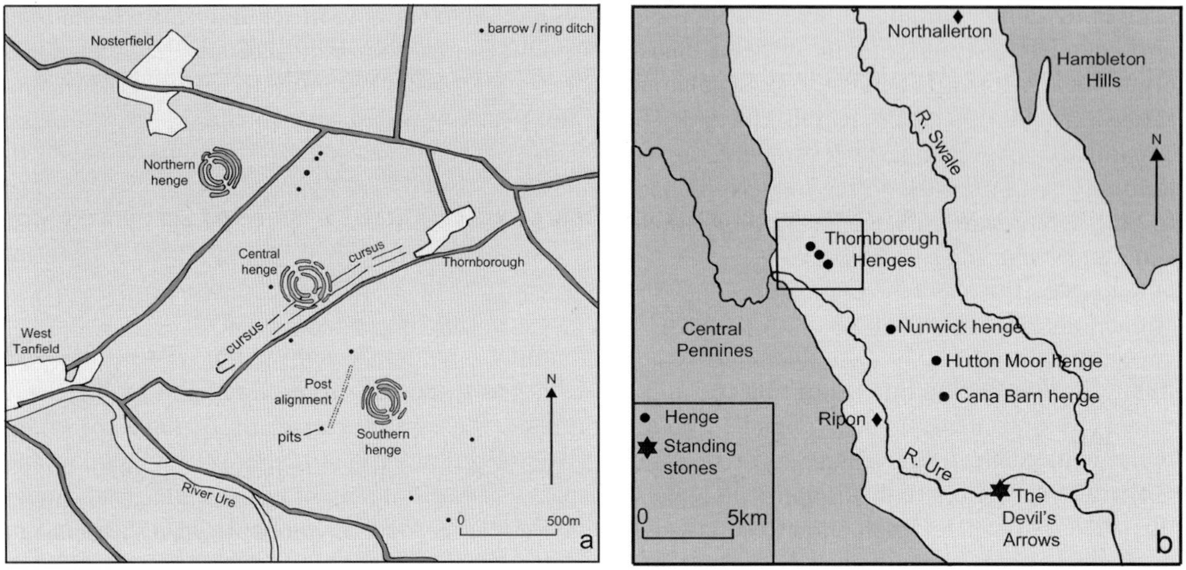

Figure 5.4 (a) Thornborough Henges, Yorkshire; (b) associated monuments indicating a routeway (graphic: J. Foster; source Harding 2013).

date (Bradley 2007). Some continued in use into the Bronze Age. A few henges are approached by avenues flanked by stones; some have two opposed entrances suggesting they were aligned on a route; some have four entrances, perhaps implying the intersection point of two routes; some are aligned with other monuments which may imply an axis of movement. On mainland Orkney, the double entrance henge, the Ring of Brodgar, is orientated with one other henge, many barrows and several standing stones along the axis of an isthmus between expanses of open water, so it is easy to conceive of this as a natural processional route.

The sequence at Thornborough Henges, Yorkshire (Fig. 5.4) begins with a cursus on a north-east to south-west alignment, which was succeeded by three henges. These lie 0.5 km apart with double entrances all aligned south-east to north-west, which is at right-angles to the cursus which underlies the central henge (Harding 2013). A double line of pits 350 m long stopped on the common axis between the two southern henges. More widely, the three henges are roughly aligned with three other henges situated 5–10 km to the south-east and also, 15 km to the south-east, a line of four or five giant standing stones, the Devil's Arrows, at a crossing place of the River Ure. The alignment may have related to the winter solstice perhaps legitimating the direction of travel which Harding likens to pilgrimage. Other henges, nine in all, may extend the hypothetical routeway along the lowland east of the Pennine upland to a remarkable 80 km (Harding 2012). That may represent a long-distance route which led eventually through Wensleydale across the Pennines to Cumbria. Artefactual evidence at Thornborough suggests it was not a place of permanent residence but somewhere visited from time to time.

The relationship between henges and routes will be further explored using the two best known henges at Avebury and Stonehenge where archaeological and palaeoenvironmental research over the last 40 years means that we have a spatial picture of environmental change which can be related to evidence for prehistoric routeways.

Case Study: Avebury Henge, Wiltshire

This site is defined by an enormous encircling bank and interior ditch with huge stones forming a circle inside the ditch; within this are two stone circles and a recently discovered rectangular setting (Fig. 5.5). The stones are sarsens, which are the silicified remnants of a Tertiary deposit which once covered the chalk. Four entrances define quadrants. The south and west entrances are approached by long avenues of standing stones. Land molluscs from below the bank demonstrate that the area had formerly been wooded, but by 3620–3350 cal BC, well before henge construction, the landscape was grazed grassland (Evans 1972; Evans *et al.* 1985). Valley sediment sequences *c* 350 m west of the henge show Mesolithic activity, early Neolithic clearance, grazing and some artefacts (Evans *et al.* 1993). Thus, the henge was made *c* 3000–2600 cal BC in a grazed clearing at least 800 m in diameter. The four henge entrances meeting at a crossroads provide a possible explanation for the location of the clearing and recurrent activity in the Mesolithic and early Neolithic. The site is at the intersection points of two 'natural routeways': it is at the beginning of the interfluve between the headwaters of the Kennet which flows east to the Thames and the River Avon which flows west into the Severn Estuary. This east–west route is crossed by a lesser northerly route along a valley.

Palaeoenvironmental studies on several sites in the immediate area show that it was wooded in the mid- Holocene; there was a progressive opening up of the landscape from the early Neolithic, so that by the mid-Neolithic the landscape was perhaps half open; it was largely cleared by the early Bronze Age. These changes have been documented in the form of a sequence of landscape reconstructions (Allen 2005; Gillings *et al.* 2008; Fig. 3.7). The significance of routes into Avebury is emphasised by the great avenues lined by large sarsen stones which are thought to have been constructed 2600–2200 cal BC (Gillings *et al.* 2008). The West Kennet avenue approaching the south entrance is the best preserved, 2.4 km long curving down from the Sanctuary on Overton Hill. The Sanctuary comprised successive circular timber circles, then two concentric stone circles. Significantly, or not, the end of the Avenue at the Sanctuary meets a hypothetical north–south route (the Ridgeway) which many have supposed is a prehistoric routeway, an hypothesis further discussed in Chapter 8. The east entrance to Avebury also leads up onto Overton Down via an old trackway called Greenstreet, to cross the Ridgeway 2.1 km away; it then continues along the downs to the east where it was a significant post-medieval route. Here it is frustratingly not possible to identify one clear prehistoric line among the many well-preserved early fields and associated tracks which have been mapped in this area, perhaps because there was not one prehistoric route but many (Fowler 2000). This east–west route into Avebury is of special significance since it was probably down this that the huge sarsens forming the henge were transported from the Fyfield Down area, where even today sarsens litter the downland surface in their thousands.

The north entrance to the henge faces up a valley which was dry in the Neolithic but now carries the headwaters of the Kennet up to the chalk escarpment and a tributary of the Thames beyond. The west entrance to the henge was approached by a second avenue 1.2 km long flanked by sarsen stones. Few now remain, though they were recorded in

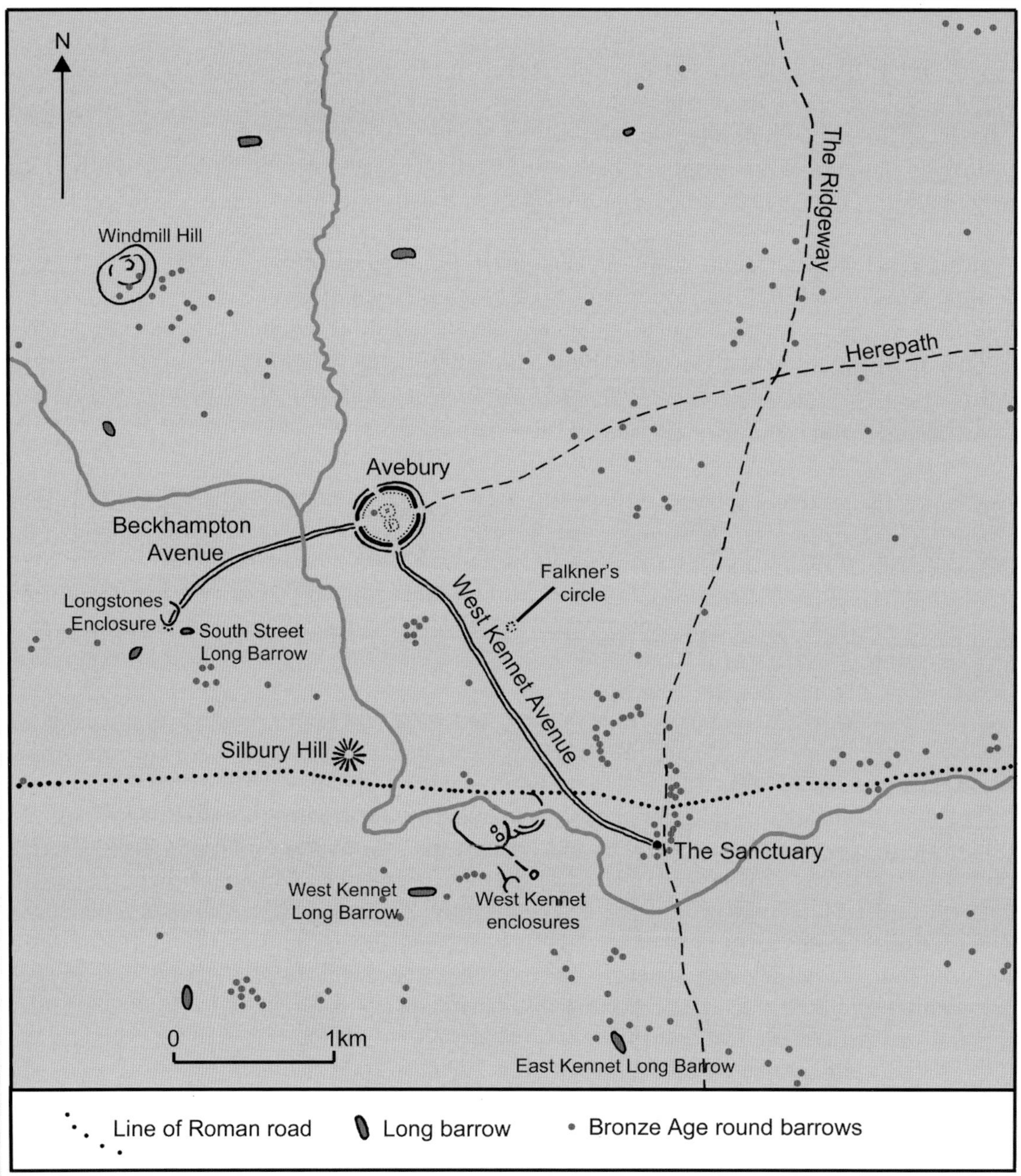

Figure 5.5 Avebury, Wiltshire: the henge in relation to the avenues and other key prehistoric sites (graphic: J. Foster).

historical sources and have recently been relocated by a combination of geophysics and excavation. That avenue led from Beckhampton where an earlier earthwork enclosure was levelled for its construction (Gillings *et al.* 2008). One puzzling point is that where the two avenues approach Avebury they are offset in relation to the entrances, leading some to argue that they were not processional routes but were set alongside processional routes (Gillings *et al.* 2008).

In the surrounding landscape there are many other major prehistoric monuments, including the early Neolithic chambered tomb at West Kennet, which, as environmental evidence shows, was made in a grassy clearing, and the long barrow at South Street, another place with a long history. Here barrow construction *c* 3660–3020 cal BC was preceded by a sequence of open woodland, clearance and ploughing followed by grassland (Evans 1972). Perhaps the most enigmatic of the neighbouring monuments is the huge mound of Silbury Hill, constructed *c* 2400–2300 cal BC, below which turves and exceptionally preserved biological evidence show it too was constructed on grazed grassland (Leary *et al.* 2013).

Case Study: Stonehenge, Wiltshire

Stonehenge is an iconic prehistoric site recognised worldwide, indeed it shares the status of a World Heritage site with Avebury (Fig. 5.6). Understanding of this landscape has been greatly increased over the last 40 years by a programme of publishing previous excavations (Cleal *et al.* 1995), and new fieldwork (Richards 1990; Parker Pearson 2012; Bowden *et al.* 2015). Associated palaeoenvironmental research has facilitated development of a spatial picture of the changing environment in time slices between 8100–1600 cal BC (Allen, M. 1997). As a result it has become clear that Stonehenge was a significant place long before erection of the iconic sarsen trilithon circle, or the arrival from Wales of exotic bluestones. In common with cases outlined earlier in this chapter, activity on the same site was over an extended timescale and was episodic rather than continuous, providing a clue that significant routeways lay behind recurrent use. That notion is persuasive because routeways became monumentalised in the Neolithic. The sequence starts in the early Mesolithic around 5000 years before the stone circle, when four or five large pits were dug 200 m north-west of the later stone circle and dated 8800–6590 cal BC. At least two of the pits held enormous freestanding posts; there is no evidence of a domestic site and a ritual or symbolic function for the pits has generally been favoured; possible analogies have been suggested with the totem poles of the American North-west Coast (Allen, M. 1995). Mollusc, charcoal and pollen evidence shows that they were in a cleared area within pine and hazel woodland. The recent discovery of a major Mesolithic site with dates between 7600 and 4700 cal BC at Blick Mead beside the River Avon 2.2 km east-south-east of Stonehenge strengthens the case for significant Mesolithic activity in the landscape (Jacques *et al.* 2018). Palaeoenvironmental evidence from the Avon Valley at Durrington Walls shows that here also there was vegetation disturbance in the later Mesolithic and Neolithic (French *et al.* 2012).

In the early Neolithic (*c* 3800–3400 cal BC) the basin, later to be occupied by Stonehenge, was encircled by 13 Neolithic long barrows, marking an ancestral grassy clearing with some secondary woodland development and, beyond the basin itself, primary oak woodland (Allen, M. 1997). The evidence of a sizeable grassland clearing of some duration is strengthened by the laying out of the greater cursus *c* 3600–3000 cal BC (Parker Pearson 2012) which might be suggestive of some sort of axial east-west movement across the plain.

Then, *c* 3000–2920 cal BC, the first phase Stonehenge I was constructed, a ditch with internal bank, and within this a circle of Aubrey Holes, which might originally have held bluestones (Cleal *et al.* 1995; Parker Pearson 2012). This monument became progressively elaborated, first as Stonehenge II with timber settings, then Stonehenge III *c* 2620–2020 cal BC, with increasingly sophisticated settings of stones including the giant sarsen trilithons and the settings of bluestone brought from Wales for which the site is so famous. During Stonehenge III, 2480–2200 cal BC, the great embanked avenue was constructed, forming a curving processional way from the River Avon to the henge entrance (Parker Pearson 2012, 219). Its end at the river is interesting in two respects. First, it is close to the main concentration of early Mesolithic activity at Blick Mead which might hint at a long established axis of movement linking river-side settlement to the ritual focus of putative totem poles. Secondly, and more tangibly, recent excavations at the end of the avenue beside the river Avon have revealed a further henge, perhaps the original site of a Bluestone circle, the West Amesbury Henge. This went out of use *c.* 2460–2190 cal BC, after which perhaps the stones were transported up the newly constructed avenue and erected at Stonehenge itself (Parker Pearson 2012).

The connection between the avenue and the River Avon very clearly demonstrates the significance of riverine movement in prehistory, a theme developed in Chapter 9. That significance is further highlighted by recent discoveries 6 km up the meandering course of the River Avon at Durrington Walls. Here recent excavations have revealed a sizeable settlement of many small rectangular houses covering perhaps 17 ha and dating between 2505 and 2465 cal BC. There were also great wood structures of concentric rings of huge posts. A large enclosure was defined at first by a ring of large posts, then later by an internal henge ditch and external bank made *c* 2480–2460 cal BC (Wainwright and Longworth 1971; Parker Pearson 2012). That enclosure originally had four entrances so perhaps, as suggested above at Avebury, the site's original location reflected the intersection of two routes. The routeway running north-west to south-east pre-dated the henge enclosure and was blank of houses, suggesting that route pre-dates the earthwork entrances. Excavations found clear evidence of a road leading from the south-east entrance down to the river Avon, confirming the significance of the river as a key transport route in the Neolithic. The idea of an earlier routeway also gains a measure of support from evidence of woodland openings in the Mesolithic and early Neolithic (French *et al.* 2012). It was argued, originally on the basis of analogy with monuments in Madagascar, that the river connected the world of the living represented by settlement and timber structures at Durrington Walls with the world of the dead, represented by stone structures downstream at Stonehenge (Parker Pearson and Ramilisonina 1998). Loveday (1998) makes the interesting point that one route through a crossing point of the Avon at Durrington Walls was still a significant route in the medieval period.

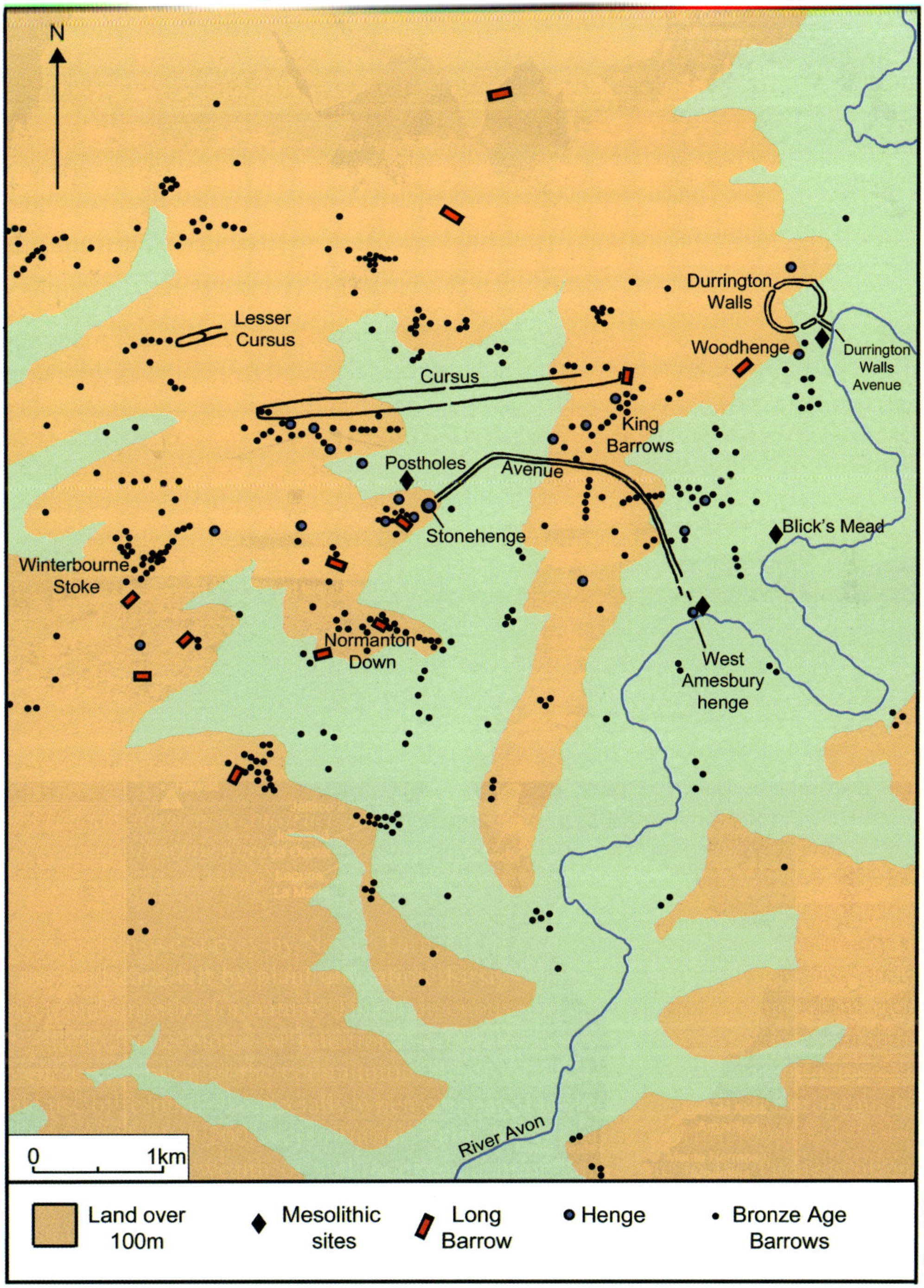

Figure 5.6 Stonehenge landscape, Wiltshire (graphic: J. Foster after figures in Bowden et al. *2015).*

The Stonehenge landscape remained a particular ceremonial focus beyond the Neolithic and through the early Bronze Age: flat axes and daggers were carved on the sarsen uprights *c* 1750–1500 cal BC (Field *et al.* 2015) and large numbers of round barrows were also constructed in a dry grazed grassland landscape. Some of the barrows are in very pronounced alignments. One of the clearest is parallel to and 120 m south of the cursus. The possibility that some barrow alignments reflect former routeways is more fully discussed in the next chapter. At Stonehenge, however, supporting evidence for this theory is lacking and the main pattern which has been recognised is a loosely concentric arrangement of barrows, perhaps demarcating the sacred space of the Stonehenge basin (Woodward and Woodward 1996).

Though the development of the Stonehenge landscape is now very well known thanks to some outstanding recent archaeological work, it must be acknowledged that the interpretation of that evidence in terms of routes is in many respects highly speculative. We cannot be sure of any connection between early Mesolithic and Neolithic activity, or that routes later monumentalised as the cursus or avenue reflected pre-existing axes of movement. However, taken together the cumulative evidence does at least provide food for thought and one hopes may stimulate further work on routes in the future. The culmination of the recent archaeological work at Stonehenge has been the opening of a new visitor centre at the end of 2013. That includes some fine dioramas reconstructing the changing environment and landscape from the Mesolithic to the Bronze Age. These reconstructions include various paths linking contemporary monuments. Such things must obviously have been an important aspect of the way it was structured and encountered. The problem is that their precise location and form, whether curving or straight, is inevitably largely a matter of speculation, although it would be interesting to see if any traces have been detected in the recent comprehensive geophysical coverage of the landscape. The dioramas show the paths as curving, wavy and bifurcating; one suspects that may be a reaction to Alfred Watkins' continuing legacy (p. 2) in that archaeologists dare not depict a straight path in prehistory.

Other henges and linear landscapes

The case for an associated routeway is particularly clear at Thornborough, with three pairs of aligned entrances, while at Avebury and Stonehenge routeways are identified by the avenues and by archaeological and palaeoenvironmental evidence. These sites are not alone; many other henges and related sites show patterns which suggest a relationship to routeways, highlighting the potential significance of this theme. At Stanton Drew, Somerset is a complex of three partially-preserved stone circles and two other stone settings, together forming at least two distinct alignments. The larger Great Circle was preceded by concentric circles of large posts similar to Durrington Walls (David *et al.* 2004). Two of the stone circles have traces of stone avenues leading to the River Chew, reminiscent of the situation at Stonehenge and Durrington and highlighting the role of riverine transport (Darvill 1986; Burl 2000). Other henges approached by stone avenues include Broomend of Crichie, Aberdeen and Callanish, Outer Hebrides (Burl 2000). In the Milfield Basin, Northumberland there is a group of three henges, and a routeway has been hypothesised passing through the opposed entrances of one and tangentially skirting the two others (Harding 1981).

Waddington (1998; 2005) makes a case that this is part of a longer processional way, which he argues, on the basis of the sequence of vistas and monuments presented, was approached from north to south. The northern part is marked by what he interpreted as a double-ditched avenue. South of this he suggests some pronounced changes of direction indicated by the alignment of entrances, pits and the disposition of monuments, thus demonstrating that in the identification of routeways from monument relationship we are not limited to straight lines. A prehistoric date for the double-ditched avenue has been questioned by Bradley (1993) in favour of an association with a later Anglian seventh century AD site at Milfield, the status of which he suggests was legitimised by association with the much earlier prehistoric monuments. If we accept the case for an earlier route linking the monuments, the double-ditched avenue, whatever its date, contributes to the case for legitimisation proposed by Bradley.

Around Dorchester, Dorset (Fig. 5.7) there is a major complex of Neolithic monuments, which includes a large henge at Mount Pleasant with a recently identified hollow way, perhaps Neolithic in date, leading from its east entrance to the river Frome (Barber 2014). This is one of a sequence of monuments along the Allington Ridge (Wainwright 1989; Smith *et al.* 1997). Activity on the ridge starts in the earlier Neolithic, *c* 3800 cal BC, with pits at Flagstones and some activity at Mount Pleasant. From this phase at Flagstones, and more widely on the ridge from the middle Neolithic, there was an open grassland environment. Then in the middle Neolithic, *c* 3300 cal BC, an enclosure was created at Flagstones and a long mound at Allington. Subsequently in the late Neolithic the Mount Pleasant henge was made and a site created at Maumbury Rings, where a great circle was defined by 45 shafts, some 10 m deep (Bradley 1975). In the early Bronze Age barrows were constructed along the same alignment, in the centre of the Flagstones enclosure and at either end of the Allington long mound, forming an alignment of at least seven barrows over 2 km, or 4 km if a more distant further group of four to the west are included. This linear monument complex evolved over an extended timescale, from the early Neolithic to the end of the early Bronze Age, some two millennia.

Another site where Neolithic monuments of contrasting character and date have a linear arrangement is at Fengate, East Anglia (Fig. 5.8a; Pryor 1991). The sites include an alignment of paired ditches, probably a long barrow, a rectangular ditched enclosure, a house/ritual structure and multiple burials. Pryor made the reasonable inference that the alignment reflects a Neolithic path which is on a slightly different alignment (25° difference) to a very prominent alignment of ditches and droveways laid out in the Bronze Age, as shown on Figure 6.10a. On a smaller scale the Neolithic ceremonial complex at Catholme in the west Midlands shows some similar traits (Fig. 5.8b; Buteux and Chapman 2009). It begins with a short cursus type monument of third millennium BC date. Then, aligned on a similar axis 200 m east, is a radial 'sunburst' of pits which evolved into a henge, and 200 m east of that a structure (Woodhenge) comprising multiple circles of posts constructed *c* 2500 cal BC. What gives the aligned relationship between all three monuments added significance is that in the late Bronze Age or early Iron Age pit alignments were laid out on both sides *c* 50 m outside the Neolithic monuments. Buteux and Chapman favour an interpretation of the linear pattern and the pit alignments at Catholme as boundaries, a hedge or wood edge; however, such longevity again over 2000 years, may alternatively be explained as a routeway, a possibility they acknowledge (Buteux and Chapman 2009, 107).

Loveday (1998, fig. 2.2) identifies 14 double entrance henges which are on the same alignment as nearby Roman roads at distances ranging from 0.5 km to 5 km (average 1.65 km). That implies

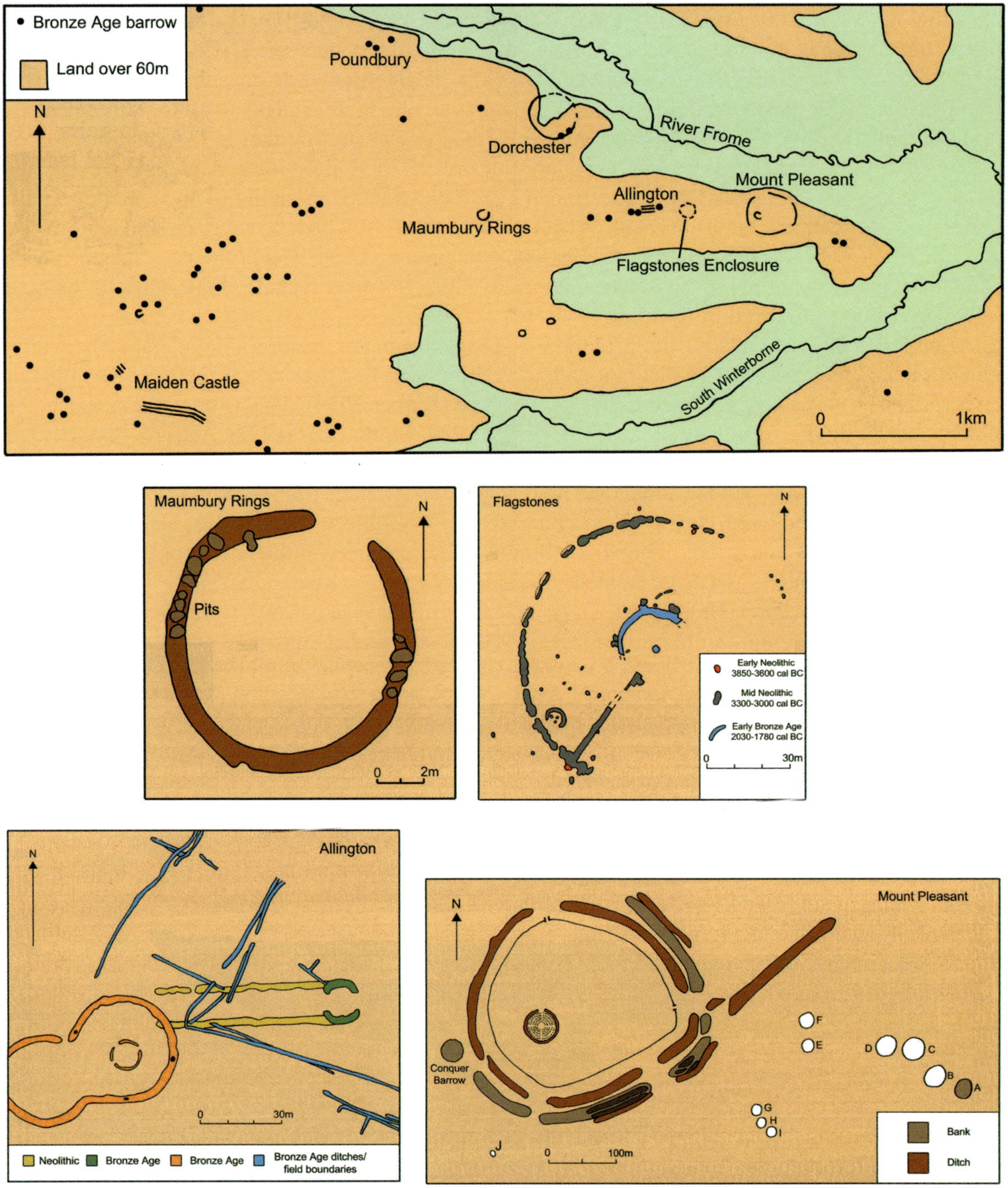

Figure 5.7 Dorchester area, Dorset, a linear pattern of Neolithic and Bronze Age monuments (graphic: J. Foster after Smith et al. *1997).*

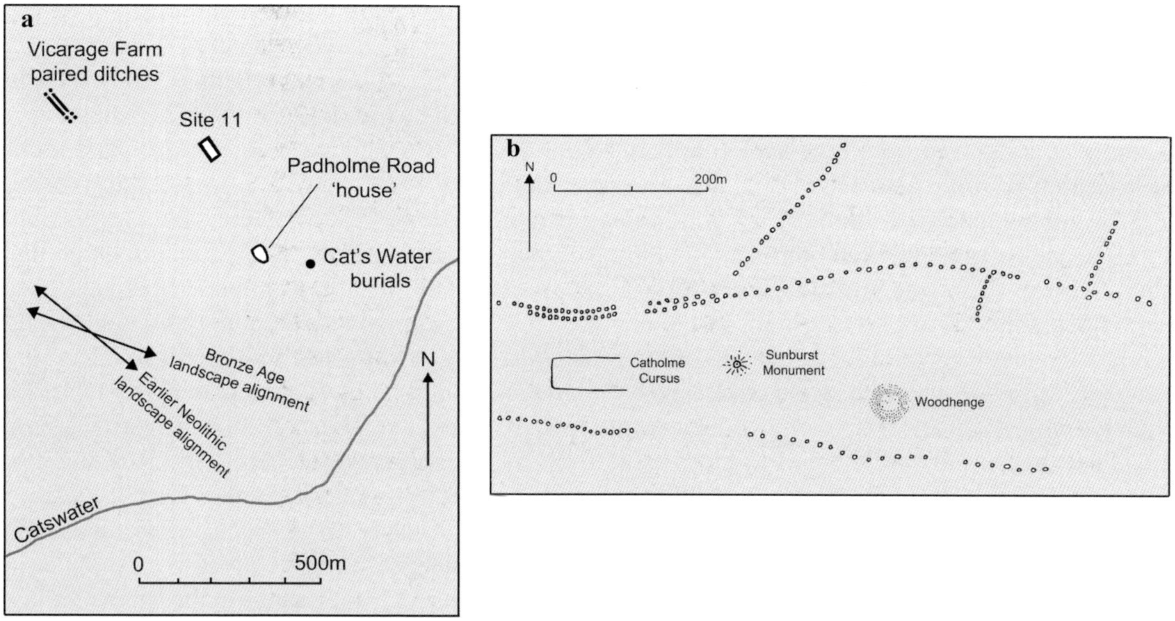

Figure 5.8 Monument alignments: (a) Fengate, Cambridgeshire: Neolithic sites in a linear relationship (graphic: J. Foster after Pryor 1991, fig. 28); (b) Catholme, Midlands: linear arrangement of prehistoric monuments and pit alignments (graphic: J. Foster after Buteux and Chapman 2009).

that double entrance henges were on, or close to, 'natural routes,' perhaps of wider utility rather than of a restricted ritual or ceremonial role.

Beside these somewhat speculative inferences of Neolithic routeways is one example of a hollow way at Oversley Farm, Cheshire (Garner 2007). It was a prominent feature, 7.5 m wide and 0.5 m deep, traced for 300 m; the basal fill had early Bronze Age artefacts and a radiocarbon date *c* 2100 cal BC, so it is reasonable to infer that erosion of the hollow way occurred in the Neolithic. The site produced Neolithic structures and it is inferred that the hollow way was in existence from the Neolithic to the Iron Age and determined the orientation of some of the later prehistoric features. It was interpreted as a drove road leading to a river fording place. One suspects that the early origins for such features may often go unrecognised.

Geological evidence for Neolithic mobility

Physical evidence for routes is by no means the only way of looking at prehistoric mobility and there is thankfully a growing battery of scientific techniques which can contribute to this question. The most long-established is the technique of artefact sourcing of stone axes. Especially far travelled are the Jadeitite axes from the Alps which reached Britain, including an example ritually deposited below a plank of the Sweet Track 3807–6 BC (Chapter 6). The old hard rocks in western Britain were also the source of polished axes which were widely distributed. Thin-sections enable the identification of petrological groups, many of which can be attributed to specific outcrops, and in some of these places quarry traces, debris and roughouts from axe production have been found, as for instance, Group VI Langdale axes from the Lake District; and Group VII Graig Lwyd and Group XXI Mynydd Rhiw axes, both from north Wales.

The movement of rocks formed a key part of the most significant henges, particularly Stonehenge. The bluestones of Stonehenge are of spotted dolerite which has now been geochemically traced to specific outcrops in the Preseli hills of south-west Wales, a distance of 225 km; other rocks of Welsh origin were also used (Williams-Thorpe *et al.* 2006). Up to 80 Welsh rocks were moved, most weighing around 4 tonnes. How this was achieved remains a matter of speculation; probably a sledge and rollers and large numbers of people were involved. Generally it has been assumed that a large part of the journey was by sea, although experiments and Madagascan analogy suggested to Parker Pearson (2012) overland transport was more likely. Those parts of a journey by land would surely have involved careful preparation of the route, given the points made in Chapter 3 concerning the likely hummocky nature of the landscape before it was smoothed by millennia of agriculture. An equally impressive feat was the movement of the sarsen stones, the largest at Stonehenge weighing about 35 tonnes; they are thought to have come from the Marlborough Downs some 29 km away. Huge sarsen blocks, 166 in all of up to 20 tonnes, were also transported the rather shorter distance *c* 3.5 km down Greenstreet to Avebury henge (Gillings and Pollard 2016). Transported rocks also feature prominently in other great Neolithic monuments, including some very impressive examples in megalithic tombs. The greatest of all megalithic tombs, Knowth and New Grange in the Boyne Valley, Ireland, employed boulders brought from an offshore island on the coast some 75 km away, and a huge quantity of small quartz boulders brought 60 km from the Wicklow Mountains (Hensey 2015).

Conclusions

We have considered mobility from the perspectives of skeletal and isotopic evidence and artefacts and geology; each implies greater mobility than is evident from the more enigmatic evidence of actual routeways. There are hints of these from the repeated and discontinuous activity preceding Neolithic tombs, the enigmatic cursus evidence and the more tangible evidence of henges and avenues. In lowland areas Neolithic routes may have been obliterated, or obscured, by subsequent arable agriculture, yet the question remains: why are Neolithic tracks so rare outside of ceremonial contexts and wetlands? A partial explanation may be that mobile pastoralists will not stick to rigidly defined routes along which grazing for animals would be rapidly depleted. Perhaps during this period we should think in terms of broad corridors of movement, which fits in with the evidence outlined for linear landscapes, for instance around Mount Pleasant. That concept gains strength in later chapters where extensive excavations of later prehistoric routeways have revealed Neolithic or early Bronze Age barrows, pits, postholes and activity areas clustered along roughly the same alignment, suggesting the possibility of much earlier origins. Examples illustrated by plans include the Yorkshire Wolds (Fig. 8.7); Peacehaven (Fig. 10.5), and Whitehorse Stone (Fig. 10.11b). In some more open areas defined routes may not have existed and very broad corridors of movement will be more difficult to define, as for instance in the ethnohistorical case of the movement of Lapps with reindeer depicted in Figure 3.2d.

6. Wetland trackways and communication

Introduction

The study of trackways in wetlands has, more than most aspects of archaeology, depended on the activities of a small band of pioneers working at different stages in the second half of the twentieth century: Hajo Hayen in Lower Saxony, Wil Casparie in the Netherlands, John and Bryony Coles in the Somerset Levels, UK and Barry Raftery in Ireland. A major expansion of wetland archaeology in the 1970s and 1980s is reflected in the publication of Coles and Coles (1989) and a succession of conferences, organised by the Wetland Archaeological Research Project which the Coles founded in 1986. Recently these pioneering studies have been followed by reviews of wetland archaeology by Van de Noort and O'Sullivan (2006), Menotti (2012), and Menotti and O'Sullivan (2013). One of the primary arguments made in these recent reviews is that wetland archaeology should not be treated as a separate class of evidence but needs to be contextualised by consideration of its relationship with dry ground archaeology. It is also argued that wetland archaeology has a tendency to be descriptive and has favoured environmentally deterministic rather than cultural/social explanations. They suggest that the result has been limited engagement with archaeological theory; this contrast has been especially apparent in Britain and Ireland where, since the late 1980s, social theory has been a much more dominant theme in prehistory than, for instance, in the German-speaking world (Menotti 2012). The challenge is to demonstrate what may be achieved by the effective linking of both wetland and dryland perspectives by integrating wetland sites in theoretically-informed wider syntheses, a task just as much the responsibility of dryland as wetland archaeologists.

It might be anticipated that relating patterns of wetland and dry ground prehistoric activity would be relatively straightforward. However, as the critiques of Van de Noort and O'Sullivan (2006) and Menotti (2012) demonstrate this is something which has infrequently been achieved in practice. One problem is that the two types of environment present such contrasting forms of evidence. Frequently the wetlands have abundant evidence for particular activities such as trackways or fishtraps, but there is very little known about settlement on the neighbouring dry ground. For instance, the wealth of trackways in the Mountdillon area of the Irish Midlands is seen in the context of very limited evidence for prehistoric dry ground sites (Raftery 1996). Where dry ground sites are present they may be just flint scatters disturbed by subsequent cultivation and containing little opportunity for dating or environmental analysis.

There are a surprisingly limited number of cases in which archaeologists have extrapolated from wetland tracks to make inferences about patterns of dryland movement or settlement locations. It has proved difficult to find continuations of wetland tracks on dry ground. Because of peat wastage the terminals of trackways are often poorly preserved and seldom investigated. Many wetland trackways are relatively ephemeral and short lived but often very well dated. Conversely some dry ground routes are clearly long lived but generally undated. The question is, however, have we really tried hard enough, and often enough, to make connections? The writer suspects not, furthermore suspecting that in certain cases, where there is clear evidence that wetland routes were in use for an extended period and continued onto dry ground, the significance and lessons of that relationship have been obscured by what may well be a false dichotomy between the functional properties of a routeway and its social and symbolic role demonstrated by the structured (e.g. ritual) deposition of artefacts. This case will be made with reference to the multiple post alignments/bridges of the late Bronze and Iron Age in Europe including the sites of Flag Fen, England and La Tène, Switzerland.

Wooden trackways in North-west Europe are mostly Neolithic to Iron Age, *c* 4700 cal BC–AD 500, a phenomenon of the peatlands in the Netherlands, north Germany, Denmark and the British Isles (Supplementary Appendix Table 6.1). By far the greatest number is in Ireland where there are an estimated 3500, of which 10% are dated (Brunning and McDermott 2013). The most substantial and longest trackways are in Lower Saxony and its border with the Netherlands where peat bogs once covered very extensive tracts between the rivers Elbe and Ems. In the eighteenth century AD one-third of north-west Germany was peat covered and some 350 trackways are known. Current knowledge suggests that long trackways are concentrated in North-west Europe, although shorter wooden routes approaching and around settlements are reported from various other parts of the world (Menotti 2012, 168). The distribution must reflect in part the distribution of peatlands where trackways were needed, the extent of peat cutting to reveal them, and also whether there has been archaeological monitoring leading to their discovery.

Wooden trackways display a considerable diversity of constructional techniques as Figure 6.1 shows. Some are simple discontinuous bundles of brushwood along the axis pegged in places at the edges (Fig. 6.1i), others are corduroy (roundwood) or planks laid at right angles to the axis and held in place with pegs (Fig. 6.1h). Some have a cradle-like construction of opposed diagonals to create a raised walkway of planks (Fig. 6.1a) or brushwood (Fig. 6.1f) over flooded areas. Some are made of woven hurdles (Fig. 6.1b), others are of planks forming a range of types from simple pedestrian walkways (Fig. 6.1c–e), to very carefully constructed wide trackways in which considerable work has been invested and with quite complex forms of construction with multiple pegs, lateral timbers below (Fig. 6.1j and k), and sometimes above, to help maintain a level surface for wheeled transport (Fig. 6.1g). In many cases considerable woodland resources and communal effort must have been committed. There is no simple evolutionary sequence; the Sweet Track, although a pedestrian walkway, is both one of the most sophisticated in the UK and also the earliest. In Saxony there are substantial and wide examples suitable for wheeled vehicles in the later Neolithic, although those with the most carefully constructed level surfaces tend to be later Bronze Age and Iron Age, especially the eighth–sixth centuries cal BC (Menotti 2012, 166). The most sophisticated tracks tend to be the longest and of greatest duration but this

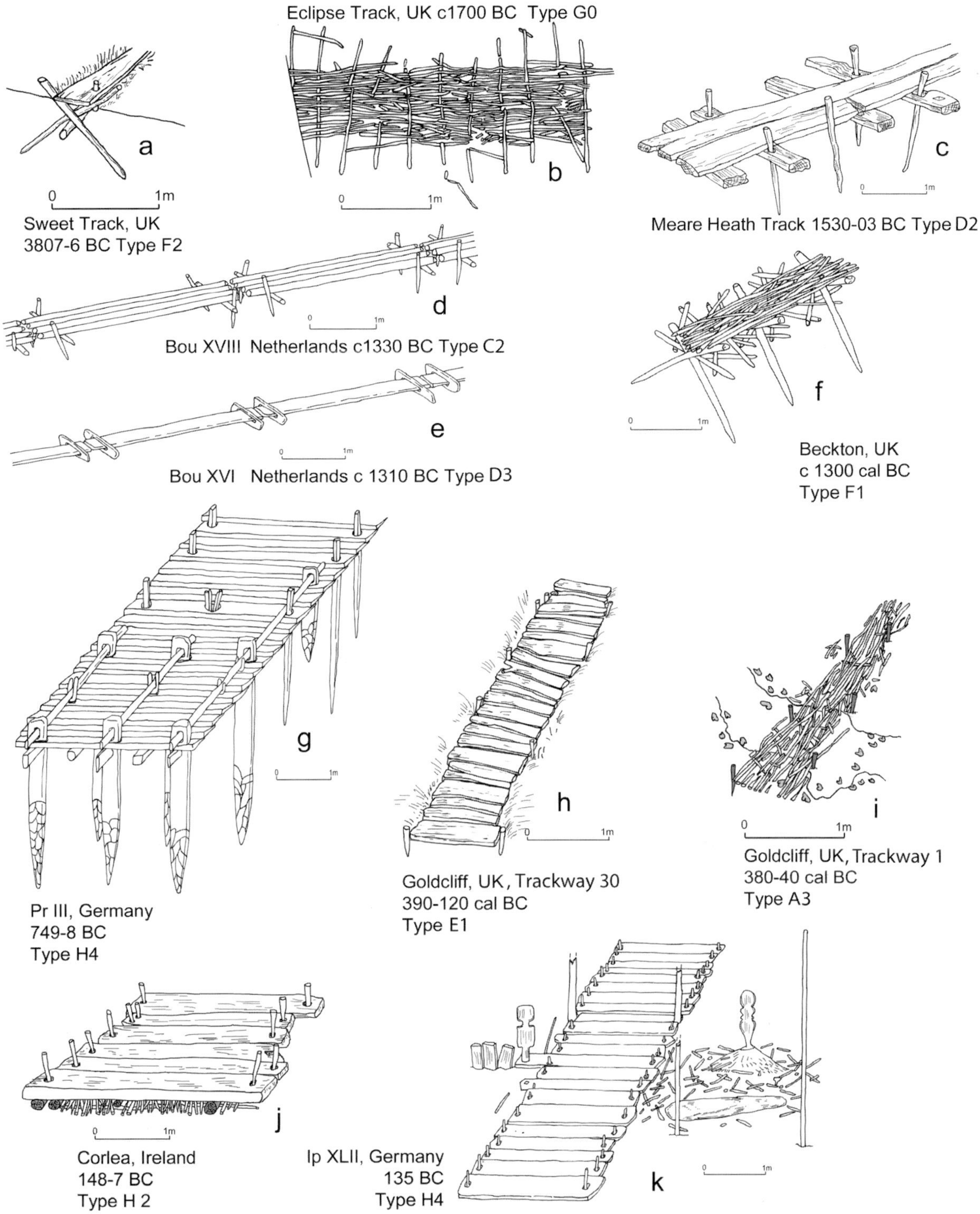

Figure 6.1 Examples of the diverse types of prehistoric wood trackways (graphic: J. Foster, multiple sources in text).

is by no means invariably the case: the Sweet Track had a life of less than a decade (Coles and Brunning 2009) and other major trackways such as the Corlea road in Ireland have provided very limited evidence of use (Raftery 1996). The simplest brushwood track may only have been used for one season. There are only a few for which a century or more of use is indicated. Limited periods of use create limitations in trying to infer wider patterns of communication.

Wood trackways are of particular importance because they are so readily datable using radiocarbon, or dendrochronology, sometimes to the precise year, with all the possibilities that creates for correlation with other sites and environmental changes. Trackways stratified within peats provide independent evidence for the environment at the time of construction in the form of pollen, plant macrofossils, testate amoebae, insects, peat humification or isotopic composition. In addition to wooden trackways there are also stone causeways in wetland and riverine contexts; examples have long been recognised in Denmark but short causeways are increasingly found in river valley contexts in Britain and Ireland where large scale excavations have exposed extensive tracts of floodplain. The more permanent character of these stone causeways means that some of them have proved of particular importance in establishing a connection between wetland and dry ground routes.

Wheeled vehicles

Wheeled transport would have been a huge advantage to past communities in moving bulk goods such as ore, wood and stone as well as in seasonal movements and military campaigns. The use of the wheel depends on two things; first, knowledge of the technology of wheeled transport and secondly on a suitable landscape, or surface, for wheeled vehicles. Vehicles could have been drawn by people but would have been vastly more useful when draft animals were available. The wheel is first attested at about the same time *c* 3600 cal BC in the Near East (late Uruk period), in North-west Europe (Funnel Beaker Culture; Bakker *et al.* 1999), the Ljubljana marshes, Slovenia (Gasser 2003) and the grass steppes of eastern Europe where many cart burials are known (Anthony 2007). By 3000 cal BC there are examples of wheels in north Germany, the north Alpine lake areas, Holland and Denmark (Fansa and Burmeister 2004) and apparent cart burials in Denmark (Johannsen and Laursen 2010). Of the 14 single piece disc wheels from peat bogs, all but one dates from a short period *c*. 2600–2500 cal BC (Fokkens 2005, 358). The wheel is represented in Egypt, India and China in the second millennium BC (Piggott 1983, 14). It was unknown in Africa south of the Sahara, South-east Asia and the Americas before historical contact. Wheeled vehicles are widely represented in the Bronze Age rock art of Europe in Norway and Sweden (Coles 2005), Val Camonica and Monte Bego in Italy, and on stelae in Iberia. Wheels are present on model carts of the Bronze Age of which the best known is the Trundholm sun waggon from a bog in Denmark dated *c* 1400 BC. Waggon graves are numerous in Urnfield and Hallstatt contexts (thirteenth–fifth centuries BC) in central Europe north of the Alps (Pare 1992). In the British Isles evidence for wheels is more limited than in continental North-west Europe, and is apparently somewhat later. In Ireland there is part of a wheel from the Edercloon 45 trackway (1206–970 cal BC; Sands 2013) and an example from Blair Drummond, Scotland 1035 cal BC. There are also later Bronze Age examples from Flag Fen, Peterborough and Cottenham, Cambridgeshire (Pryor 2001) and two recent finds from a waterlogged settlement at Must Farm *c* 900–800 cal BC (Knight *et al.* 2016).

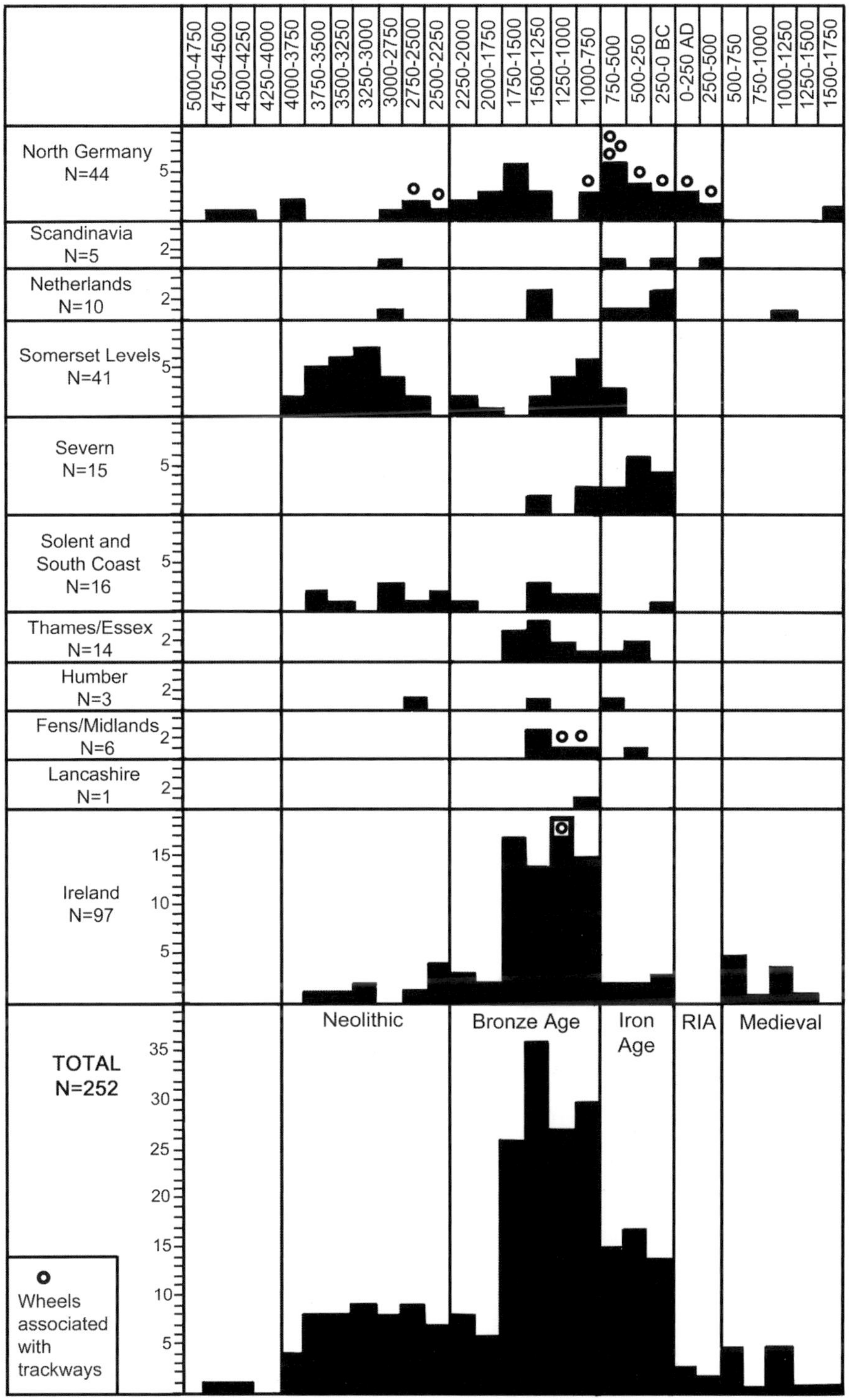

Figure 6.2 The dates of trackways in areas of North-west Europe (see Supplementary Appendix 6.1 for full details of sites included) (graphic: J. Foster).

The widespread evidence for various kinds of wheeled vehicles in Europe from *c* 3600 cal BC contrasts markedly with the very limited surviving evidence for roads on which they might have been used. In Germany and the Netherlands there is a clear association of wheel and other vehicle finds with wooden trackways (Fig. 6.2; Both and Fansa 2011a), some of which are manifestly constructed to provide a level surface for wheeled transport. Such trackways often combine lateral planks overlying transverse runners and with a second layer of transverse runners above, morticed into vertical posts to maintain a level surface; a particularly sophisticated example is the IIIPr track in Lower Saxony (Fig. 6.1g). Other trackways are thought to be for wheeled vehicles by virtue of their size, solidity and width, such as Corlea 1, Ireland (Fig. 6.9). Beyond the north European wetlands, evidence for suitable routes is much more limited and tends to be found within excavated settlements. Wheeled vehicles could of course have been used in areas of open country such as the grass steppes of Eastern Europe, or the grazed grasslands established in western Europe as a result of agricultural activity (Behre 1988; 2008). However, in prehistory people moving around would have encountered much more uneven surfaces with hollows and gullies, boggy areas, the legacy of glaciation, solifluction, generations of tree throw, erosion, etc, which would have restricted the use of wheeled vehicles in many areas without the preparation of defined routeways. Where they have suitable surfaces, trackways in wetlands do at least provide us with some indication of the areas where wheeled vehicles may have been in use.

Trackways dates

Figure 6.2 illustrates the date distribution of trackways which are securely dated by radiocarbon or dendrochronology in North-west Europe. Full details of sites and sources are given in Supplementary Appendix 6.1. It has been compiled mainly using the data from Both and Fansa (2011a) for Germany; Casparie (1987) and van der Sanden (2002) for the Netherlands; Brunning (2007) for the UK; and for Ireland, Raftery (1996), supplemented by Moloney *et al.* (1995), Brindley and Lanting (1998), Gowan *et al.* (2005) and a data-base of dates compiled by Chapple (2016). Figure 6.2 is compiled in 250 year time slices; in the case of radiocarbon dates the mid-point of the calibrated range is taken. It should be emphasised that these are recorded trackways which will be a biased sample of those which once existed. Most have been found through peat digging. Later tracks are likely to have been cut away before archaeological recording began, and this may partly account for a paucity of tracks in the last two millennia. Some older trackways may still remain buried in peat. In order to avoid confusion in areas of Germany and the Netherlands where there are multiple trackways, in the following discussion they are numbered using a scheme developed by Hayen (1963), a Roman numeral followed by two letters to designate the district.

Mesolithic trackways?

Movement through the landscape by hunter-gatherers has been discussed in Chapter 3 and the traces of their footprints in Chapter 4. This chapter concerns the physical construction and marking of routeways and here the evidence for Mesolithic trackways is perhaps surprisingly inconclusive. The most famous and intensively investigated British Mesolithic site at Star Carr, Yorkshire saw periodic lake edge occupation over some 800 years between 9300 and 8500 cal BC (Mellars and

Dark 1988; Milner *et al.* 2018). There were pit houses on dry ground and along the lake edge three distinct platforms of split timber made over about 175 years, each between 11 m and 17 m long. Two were roughly parallel with the shore and one at 45° to the shore. They were not interpreted as trackways, nor were they activity areas, but they seem to have been created to facilitate access to the lake shore where small discrete episodes of artefact deposition occurred between the timbers.

Lake edge platforms and settlement mounds are further attested on Irish sites such as Derragh on Lough Kinale (O'Sullivan 2007; Fredengren 2010). A possible Mesolithic track has been reported from Lullymore Bog, Ireland but its dating in the early sixth millennium BC cannot be regarded as secure (Brindley and Lanting 1998). Possible trackways of sand with some wood have been identified in the Federsee, southern Germany, leading to a small island with Mesolithic finds (Brindley and Lanting 1998). Mesolithic communities certainly had the capability to create trackways: the platforms noted, the evidence for wooden houses and fishtraps and the quality of woodworking attested particularly from sites in Scandinavia makes that clear. However, given the abundance of Mesolithic discoveries in wetland contexts, particularly in Denmark and parts of northern Germany, it is striking that there do not seem to be any fully convincing examples of wooden trackways, which should perhaps lead us to question whether hunter-gatherers chose to mark their favoured routes in this way, just as, with a few exceptions (Chapter 3), they did not create the monuments which are such a distinctive feature of early farmers. It is tempting to see this in terms of the distinction which Ingold (1980) draws between the fluid and cooperative territoriality of hunter-gatherers as opposed to the tenure of land by agriculturalists. If so, the trackways of the latter might have as much to do with ownership as access.

Neolithic trackways in mainland Europe

Germany

The earliest of the fully convincing wooden trackways is found in north Germany in the peatlands west of Lake Dummer at Campemour (XXXIPr) dated by dendrochronology to 4629–4545 BC; the evidence suggests it was in use for a century (Both and Fansa 2011a). This site is some 130 km south of the North Sea coast, just beyond the loess lands occupied by the first Neolithic Linearbandkeramik farmers. The trackway coincides with initial clearance and occurs at about the time of the earliest cereal cultivation and domestic livestock in the area (Hartz *et al.* 2002). It is thus assumed to be Neolithic, although to its north hunter-gatherer communities persisted for up to 700 years later, so it may be unsafe to make assumptions about the economy of the people who built the Campemour trackway. That first trackway is quite substantial with a foundation of lateral runners and a surface of transverse corduroy roundwood 2.5–5 m wide. Two later Neolithic trackways are of particular note in having evidence of vehicles associated with them. A Meerhusener Moor trackway (XVLe) dated to *c* 2490–2160 cal BC, was a substantial affair, 3.3 m wide and 2.5 km long; associated with fragments of six wheels, and other cart fragments, and there were many broken hooves of cattle between the track timbers demonstrating their somewhat hazardous employment as draft animals (Hayen 1987; Both and Fansa 2011a). The Grosse Moor north of Lake Dummer is crossed by a succession of prehistoric trackways including two Neolithic examples. One of these, VII (Pr), dated 2650 cal BC was of substantial corduroy construction, 2 km long, 3.3 m wide; part of an axle was also found here. It is very clear from these finds that in Germany wheeled vehicles were being used on these wooden tracks, presumably to transport bulk goods, from at least the later Neolithic, and probably from 3000 cal BC, when wheels are first found in peat bogs in the area. In the south German Federsee bog a wooden pathway was found within the Neolithic settlement of Seekirch Stockwiesen, dating to 3030–2890 cal BC, where disc wheels were also found (Schlichtherle and Strobel 1999).

Netherlands and Scandinavia

In the Netherlands there is just one Neolithic trackway (XXI Bou) dated to 2610 cal BC; it is of transversely laid roundwood, 3 m wide and 1 km long. Casparie (1987) argues it was intended for wheeled transport and indeed a broken disc wheel was associated. It extends for 1 km into the bog (see Fig. 6.6) then ends, indicating that access to the bog rather than crossing it was the objective. By the time of its construction the dryland woods were already subject to localised clearance (Casparie 1982). A Neolithic track has also been found at Tibirke in the North of Zealand, Denmark dating to *c* 3000–2500 cal BC; it comprises a vertical stake-built causeway 3 m wide, 150 m long, which Schou Jørgensen (1988a) describes as suitable for waggons. It runs across what at the time was a tidal estuary to a settlement on an island. A wooden causeway was also associated with a large wood platform of Pitted Ware mid-Neolithic date at Alvastra, Sweden. This has been interpreted as a gathering place for votive activities as suggested by finds including human skulls, smashed battle axes and miniature axes but also numerous hearths and some evidence for agricultural activities (N. Hinders, pers. comm.).

Neolithic trackways in the British Isles

Somerset Levels

The Sweet Track is the best known trackway in Britain (Coles and Coles 1986) and was once the earliest known trackway in the world; now pre-dated by German examples. A recently discovered possible trackway at Belmarsh in London appears to be as early, maybe slightly earlier (3960–3710 cal BC) than the Sweet Track (Hart 2010). The Sweet Track (Fig. 6.3) has the great benefit of precise dating by dendrochronology to 3807/6 BC and was preceded by the earlier, and fragmentary, Post Track dated 3838 BC (Hillam *et al.* 1990). Repairs to the Sweet Track were made 4–6 years after construction and it is thought that the trackway only lasted about a decade (Coles and Brunning 2009). The dates for the main Sweet Track are highly significant because the latest Bayesian modelling of radiocarbon dates suggests that the earliest Neolithic in western Britain is *c* 3800 cal BC (Whittle *et al.* 2011). It is striking that trackways are present from the very beginning of the Neolithic in both Germany at Campemoor and in Britain. The Sweet Track also stands out by virtue of its length (1.8 km) and the sophistication of its construction, which exceeds nearly all of the later British trackways. Construction involved a rail on the bog surface, stakes driven in at 45° from either side, a carefully fashioned plank with mortices laid in the V-formed and a peg through the mortice to hold the plank in place (Fig. 6.1a) to create a raised single file pedestrian walkway.

Tree ring studies showed that the wood used for most of the trackway came from 400 year old mature oak woodland to the north, whilst that for the southern 200 m was from secondary woodland 100–150 years old (Morgan 1988); the wood grew following clearance around 3950 BC, which might date or predate the arrival of farming in this area. Pollen evidence shows that the trackway was constructed at the time of the elm decline (see p. 115), but there was evidence of woodland disturbance

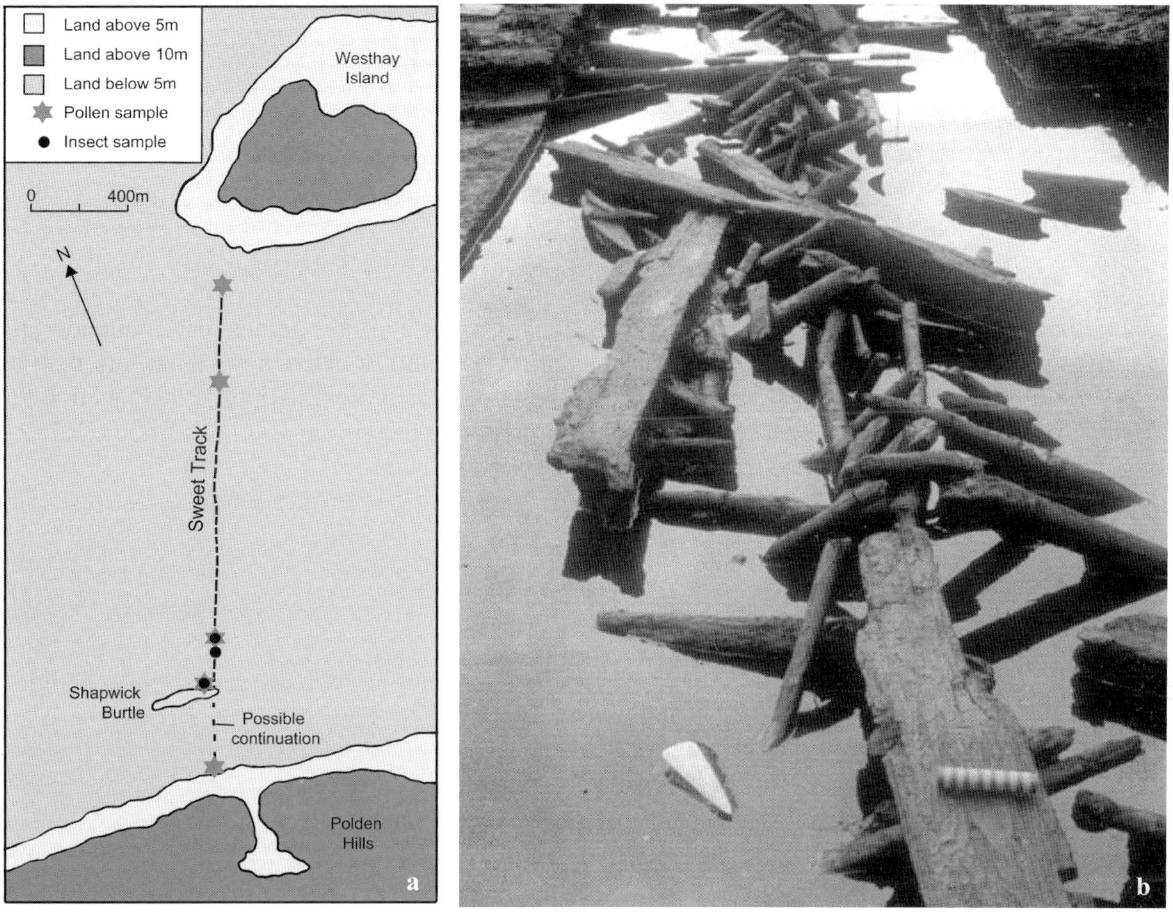

Figure 6.3 The Sweet Track, Somerset, UK: (a) the line of the trackway in relation to topography (graphic: J. Foster after Coles and Orme 1984); (b) the excavated trackway (© John Coles/Somerset Levels Project).

some decades before the trackway construction and some clearance herbs just after construction. Dung beetles suggest that parts of the wetland may have been grazed (Girling 1984; Thomas, J. 2013).

The Sweet Track is also exceptional in terms of the amount of associated material culture. This includes deliberate ritual deposition of many of the distinctive artefacts which we associate with the first farmers: 10–15 fine ware pots, leaf arrowheads and axes, as well as wooden tools. An axe of jadeitite had come from 1000 km away in the high Alps (Pétrequin *et al.* 2008) and was deposited below a trackway plank without ever being used; it was of a type produced between 6000 and 4000 cal BC and may have entered circulation before farming came to Britain (Parker Pearson 2012, 21). A flint axe beside the track apparently came from mines established at the very beginning of the Neolithic, 180 km away on the South Downs. It appears that geological materials were moving around before the evidence for the first gathering places in the form of causewayed enclosures *c* 3700 cal BC (Whittle *et al.* 2011). Such finds seem to underline the role of the trackway, not just in local communication, but in far more distant patterns of communication. Interpretation of the Sweet

Track has tended to follow the fashions in archaeology. At first seen in largely functional terms, more recent writers have highlighted the social symbolism represented by the deposited artefacts without particularly focusing on the social significance of the act of communication itself (Bond 2004; van de Noort and O'Sullivan 2006). Bond (2004) styles it 'a place mediating culture and spirituality'. It might equally be interpreted as mediating people and nature or mediating between communities? Interestingly the Somerset Levels is not characterised by any great concentration of dryland early Neolithic activity. This could fit a model of relatively mobile communities for whom the creation and marking of connections, both local and long distance, was of special social significance.

In total there are 25 Neolithic trackways and platforms from the Somerset Levels and, as Figure 6.2 shows, this is a greater concentration than in the other key areas of trackway discoveries in Germany and Ireland. The main period of trackway construction 3600–3000 cal BC is marked by increasing clearance, demonstrating a connection between use of the wetland and agricultural activity on the islands and bedrock ridges within the Levels (Coles and Coles 1998). The trackways include substantial laid roundwood tracks such as the Abbot's Way, 2.2 km long and dated *c* 2500 cal BC, and the woven hurdle trackways at Walton Heath and Rowlands, dated *c* 2800 cal BC, and made of coppiced hazel wood demonstrating woodland management.

Other British sites

Elsewhere in the British Isles there are a number of wooden structures which are apparently trackways in the intertidal zone at Wooton-Quarr, Isle of Wight (Tomalin *et al.* 2012). These are notable in running at right-angles to the shore and thus apparently providing communication across the transition from freshwater wetland to saltmarsh. Neolithic trackways are also reported from the Thames Estuary with the already noted very early example at Belmarsh and a brushwood and roundwood trackway at Ebbsfleet (Bates and Stafford 2013). A corduroy track on Hatfield Moors, Yorkshire is dated 2950–2500 cal BC (Chapman and Gearey 2006). It ended in a small platform out in the bog and certain features of its plan suggested to the excavators that it may have had a ceremonial role. Eight Neolithic trackways are reported from the peatlands of Ireland, but this is a small number compared to those of the Bronze Age. Of particular note is a 400 m long track at Cloonbony, Shannon, dated 2500 cal BC; at one end it forks into two lines, which has been interpreted as suggesting a route leading to two distinct settlements (Raftery 1990).

Bronze Age and Iron Age trackways in northern Europe

Germany

In Lower Saxony there are 15 trackways dated to the Bronze Age, with numbers peaking at the end of the early Bronze Age (Fig. 6.2; Both and Fansa 2011a). This is largely accounted for by a great concentration of trackways on Ipwegermoor, just north of Oldenburg, where there are over 30 trackways, 11 dated and with the main concentration between 1770 and 1350 cal BC. These are essentially pedestrian walkways of brushwood, corduroy and planks (Hayen 1963; Both and Fansa 2011b). To the north at Jethauser Moor (XXXVI Ip) a much more substantial trackway is dendrochronologically dated to 1358 BC. It comprises broad oak planks notched and fastened between rows of posts and carefully constructed lateral timbers to maintain a level surface. Hayen (1987) is of the view that such trackways were for faster, more sophisticated vehicles with spoked wheels and a pivoting front axle to facilitate steering. A late Bronze Age trackway (X Pr) dated 790 cal BC in the Grosse Moor is associated with a wheel, though not spoked (Both and Fansa 2011b, 183).

In the Iron Age and Roman Iron Age there are 18 well-dated trackways in Germany. Ten of these are more than 2 km long and one (XII Ip) is 6.4 km long. It is dendrochronologically dated to 754 BC. It is a particularly sophisticated trackway of substantial planks held level with lateral timbers and its clear suitability for wheeled vehicles is underlined by a wheel find. There are six of these particularly sophisticated trackways, three dated to the eighth century BC. A later example is on Wittemoor (XLII Ip), dendrochronologically dated 135 BC. This trackway seems to have been particularly associated with the transport of bog iron from a dryland settlement to a navigable river, but was abandoned soon after construction. It is especially well known for two large carved wooden objects which are interpreted as highly stylised male and female human figures; they were placed in an area where there was also evidence for fires, apparently marking a particularly treacherous part of the route (Fig. 6.1k; Coles and Coles 1989). That trackway encapsulates two key themes: the use of trackways as a means of practical communication, and as places of special ritual and symbolic importance. The same point is made by the wheels. Altogether seven of the dated trackways have associated wheels or axles and they are mainly associated with the more sophisticated, or substantially constructed, trackways which were clearly intended for wheeled transport.

In the southern German peat bog of Federsee (Schlichtherle and Strobel 1999; Heumüller 2000), a major track at Bad Buchau comprised three successive plank-built walkways, the first made in 1505 BC, the others made over a period of 170 years. They connected the dryland to a gravel ridge in the bog. Another track, dendrochronologically dated 1767–1481 BC, led from the entrance of the palisaded settlement at Sielung Forschier and a further example is dated dendrochronologically to the early Iron Age, 721–621 BC.

Scandinavia

This area has much less evidence of trackways. In Norway a single prehistoric example at Båsmyr, in Vestfold dated from 500–800 BC crosses a shallow coastal sound at high water (Smedstad 2001). From about AD 900 there are more trackways near the coast of central Norway and many of these are on 'natural' communication routes. Short paved causeways across wetlands are also found in southern Sweden in the Bronze Age and Iron Age (Westerdahl 2006). Denmark has an early Bronze Age track at Speghhoje (Schou Jørgensen 1988b) and a wooden trackway at Tibirke (Kunwald 1984; Schou Jørgensen 1988a; 1988b) overlying an earlier Neolithic track; this is in turn overlain by a paved road of 200 BC, thus demonstrating the longevity of routes in key topographic locations. The emergence of stone paved causeways in Denmark from the late Bronze Age is of special interest, the more so because some are associated with settlements, fields etc. At Borremose, Jutland a substantial stone causeway leads across a bog to a small island with a defended Iron Age settlement of rectangular houses comprising living accommodation and byres for animals (Fig. 6.4; Martens 1990). The settlement has every appearance of an agricultural village but the discovery of four bog bodies, one strangled with a rope, highlights the recurrent association of everyday activities and ritual behaviour in wetland contexts. The point is further highlighted just 4 km north of Borremose; in a small bog at Gundestrup was a great bronze cauldron, which is one of the greatest art works of the European Iron Age; it is decorated

Figure 6.4 Borremose, Denmark, paved road across wetland leading to enclosed settlement (photo: M. Bell).

with strange animals, warriors and scenes of sacrifice, and wheels (Klindt-Jensen 1979).

At Broskov, in southern Zeeland, Denmark successive stone trackways, some 60 m long, cross the narrowest point in a bog valley (Fig. 6.5; Kunwald 1962). The first is dated to the Danish Iron Age, *c* AD 300, and comprises flat stones with larger stones forming an edge. This was succeeded by a later trackway of smaller stones in the twelfth century. Thus, the route had a duration of around 1000 years. What makes this site of special note is that hollow ways converge from the neighbouring slopes on the trackway at its ends. To the north is a series of three distinct hollow ways, 4–14 m wide, some sunken up to 4 m. The hollow ways partly delimit fields, represented by lynchets of the type which Nielsen (2010) has shown formed from the late Bronze Age to the Roman Iron Age, in landscapes first intensively cleared in the middle Bronze Age (Rasmussen 2010). Also of note is a prominent stone with many cup marks at the southern end of the road (Fig. 6.5); such stones are shown by Nielsen (2010) to be particularly found in wet places at the edges of fields, highlighting the symbolic, as well as functional, aspects of these routes.

Netherlands, Bourtranger Moor

Key themes which have emerged in Denmark in terms of ritual associations and dryland connections can be further explored in the context of evidence from the Netherlands. The vast Bourtranger Moor on the border between the Netherlands and Germany is some 13 km across and here most of the known Dutch trackways are located (Fig. 6.6). This area is one of the few where direct connections between dryland and wetland archaeology can be made, likewise between categories of the wetland evidence, such as trackways and ritual deposition, which, even in this area, have generally being discussed separately. The tracks lead across the moor from the Hondsrug sand ridge, which in prehistory was surrounded to east, west and south by bog and riverine sediments; here a seemingly complete prehistoric settlement pattern has been reconstructed. There are clear indications from alignments of barrows etc for a main communication route down the axis of the Hondsrug ridge (discussed in Chapter 7). Discrete areas of 'Celtic fields' have been mapped from the air (Brongers 1976, pls 23 and 25); there are farmsteads, each associated with one, or sometimes two, successive urnfield cemeteries enabling the approximate territories of each grouping to be defined (Louwe Kooijmans 2005).

Figure 6.5 Broskov, Denmark: stone roadway of several phases between AD 300 and 1000, with a cup marked stone at one end. Of particular significance are the multiple hollow ways converging on the road which are associated with the lynchets of a field system (graphic: J. Foster, source National Museum, Denmark; colour photos, M. Bell; black and white photo courtesy of George Kunwald and National Museum of Denmark)

The vegetation history of Bourtranger Moor has been established by a series of pollen diagrams linked to trackway investigations, coupled with analysis of peat stratigraphy and the wood used for trackways (Casparie 1982; 1984; 1986). By the time trackways were constructed in the middle Bronze Age, *c* 1300 cal BC, the dryland of the Hondsrug was substantially deforested, settled and used for agriculture as the distribution of prehistoric sites and fields shows (Fig. 6.6). Woodland was quite limited and the construction of trackways would have represented a very considerable investment, both of woodland resources and social capital. This highlights the importance which would have attached to trackway construction. Six are known, covering 2000 years, and several seem to have been short lived. XV Bou may have been dismantled soon after construction because of a bog burst. The trackways lead east from the sand ridge, one, already noted (XXI Bou), is Neolithic, the others are Bronze and Iron Age in date, thus broadly contemporary with the sand ridge landscape. It is tempting to see each trackway as providing access to the bog for individual settlement territories, thus implying the extension of territories out into the bog. However, the trackways are by no means all contemporary and most of them seem to have had a short life before they were engulfed by peat. Casparie (1987) argues that several of the trackways went part way onto the moor because they provided access to areas rich in bog iron. Interestingly one of these (XVII Bou) is middle Bronze Age in date (*c* 1410 cal BC) which may indicate precocious local iron working, as seemingly confirmed by the discovery of an iron pin on the trackway surface (Casparie 2005).

There is another side to this bog, however, which belies any purely functionalist explanation for the trackways, whether in terms of communication, or bog iron winning. This is the evidence for ritual deposition in the wetland. On the peat at Barger-Oosterveld is a cult site, a ring of boulders surrounding a four post structure with horned lintels, dating to the middle Bronze Age (van den Broeke 2005). In the surrounding bog, although not closely associated with the cult site, were a wide range of objects which had apparently been ritually deposited in the peat (Fig. 6.6b). These include bog bodies, some close to a trackway (XV Bou) dated 690 cal BC. Other deposited items include bronze weapons, wooden artefacts, bracelets, beads, two wheels and pottery. The Valtherbrug trackway (I Bou) is particularly substantially constructed of transverse planks and roundwood on a substructure of longitudinal roundwood; its construction required 60,000 logs from 10,000 trees (Casparie 1987). It crosses the moor for a distance of 12 km and is 2–3 m wide. There are a wide range of dates between 390 BC and AD 112 suggesting the track may comprise more than a single structure. It is thought to be associated with the ritual deposition of bog bodies, parts of carts and querns in the bog (van der Sanden 2002). It seems that the wetlands were of special significance for ritual purposes, and the trackways may therefore have provided processional routes for these purposes, although it should be noted that many of the deposited artefacts suggest small-scale domestic acts rather than major ceremonies, a theme to which we will return later in the case of Flag Fen (p. 154).

Also notable among the Dutch trackways are two at Smilde (I and II Sm), near Assen. The wooden trackways of *c* 220 cal BC cross two narrow bog depressions (280 and 170 m wide) but they also form part of what Casparie (1987) identifies as an 11 km long routeway along sand ridges. At the end of the wooden trackway it continued as a sand route and the mottled sand surface shown in Casparie's photographs (1987, fig. 24) looks as if it may be created by cattle footprints.

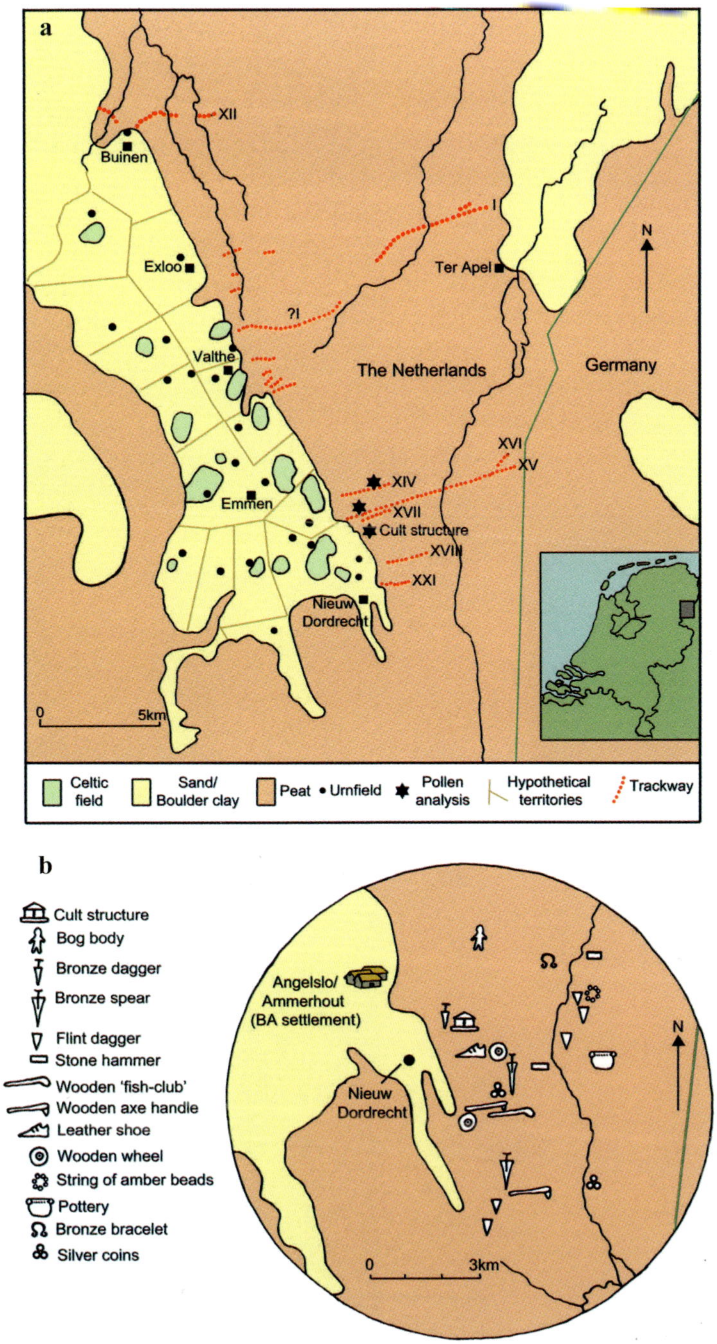

Figure 6.6 The Emmen area of the Netherlands showing wetland and dryland activity in later prehistory: (a) Bourtranger Moor trackways (after Casparie 1987), 'Celtic fields' urnfields and hypothetical territories (after Louwe Kooijmans 2005); (b) ritual deposition of artefacts in the south-east part of this area (after van den Broeke 2005) (graphic: J. Foster).

Bronze Age and Iron Age trackways in the British Isles

Case Study: Somerset Levels

In the Somerset Levels there are 13 trackways of Bronze Age date, three in the early Bronze Age and ten in the middle and late Bronze Age from 1400–800 cal BC. These principally traverse the area between the Polden Hills and the bedrock and sandy islands, especially Meare and Westhay, to the north (Fig. 6.7). Coles and Coles (1986) made the highly significant observation that Bronze Age trackways north of the Polden Hills converged on points which they suggested represented the locations of Bronze Age wetland edge settlements spaced at intervals of *c* 1.2 km. The most sophisticated and substantial is the 2.5 km long Meare Heath Track, made of oak planks laid on transverse sleepers held in place by vertical stakes through mortices (Fig. 6.1c; Coles and Orme 1978a; Coles and Coles 1986); this is dendro-chronologically dated 1530–1503 BC (Brunning 2013, 68). It is one of a group of trackways

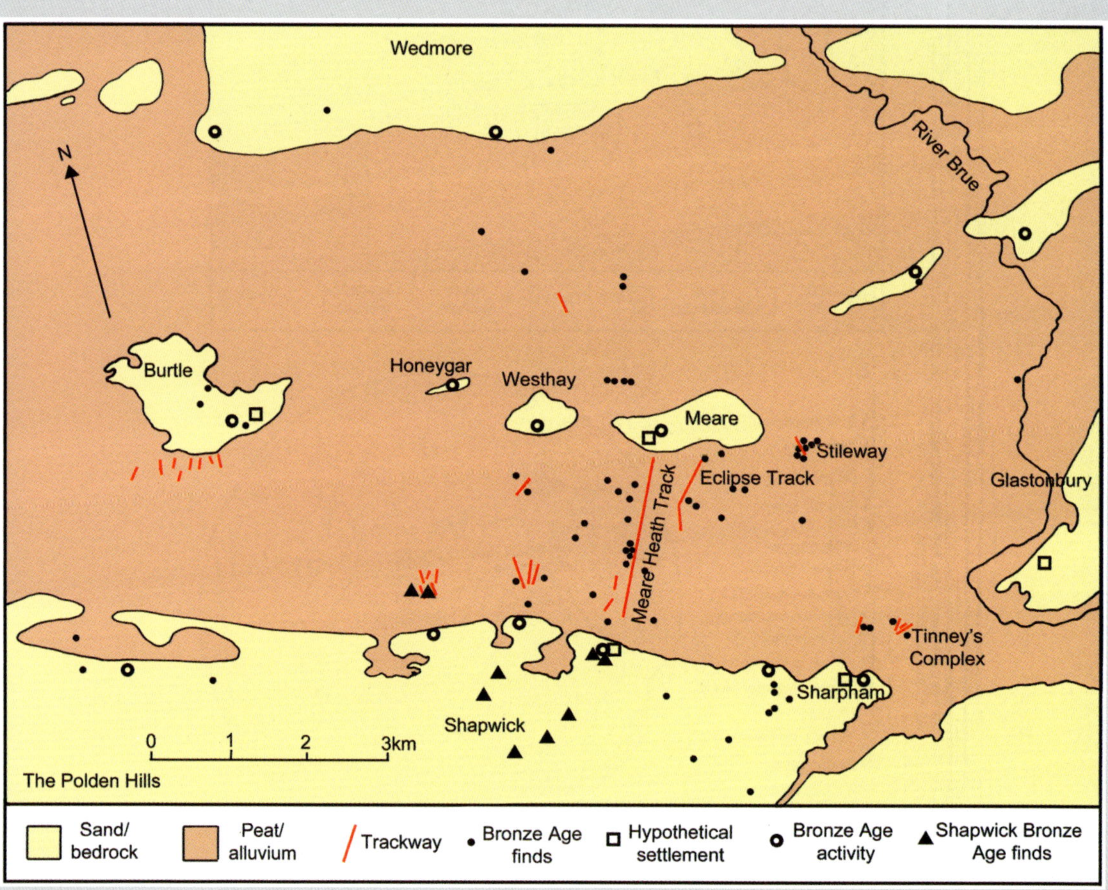

Figure 6.7 The Somerset Levels, UK, showing the Bronze Age trackways and other finds (graphic: J. Foster after Coles and Coles 1986, fig. 35).

converging on the same point and spanning most of the Bronze Age from *c* 1790–830 cal BC. It is difficult to escape the conclusion that there must have been a dryland track upslope. The Shapwick project revealed traces of Bronze Age settlement along the north edge of the Polden ridge and it was further hypothesised by Aston and Gerrard (2013) that coaxial field systems (see p. 186) at Shapwick may have their origins in the later Bronze Age. The longevity of one boundary was demonstrated by a later Bronze Age ditch on the same alignment as Roman and later boundaries and a modern hedge. If elements of the field layout originated in the Bronze Age then the same may be true of routeways. This is not to imply continuity of the Meare Heath track route, which would surely have left greater traces in the wetland. What it could mean is that existing routes on dryland imposed a structure on wetland during those episodes when particular routes were passable, or socially important. This is no more than a hypothesis sketched to illustrate the type of field investigation necessary in order to establish links, or the lack of links, between dryland and wetland routes. Techniques for dating routeways within field systems are explored in Chapters 8 and 10.

The Tinney's Trackway complex is another of these radiating groupings, in this case of multiple successive, mainly brushwood, trackways, perhaps at times replaced almost annually (Coles and Coles 1986). One was reinforced with oak planks dendrochronologically dated 1703–1478 BC (Brunning 2013, 68), others are later, *c* 1400–1000 cal BC. They radiate from a point north of the Sharpham Peninsula where it is probable that a settlement was located: pollen shows quite intensive clearance and cultivation, with dung beetles indicating grazing animals. Drawing on environmental evidence from the wetland it has also been possible to infer locations of settlements on the dry islands to the north (Coles 1989), including Meare island where there was clearance at the time of the Meare Heath Track and this continued well after (Beckett 1978). The Somerset Levels is an area where, drawing on a combination of trackway orientations, spatial evidence from multiple palaeoenvironmental studies, and retrogressive map analysis as part of the multi-period Shapwick Survey, we may be beginning to approach an understanding of a complete landscape, wetland with dryland, and how communication routes through it *may* have developed through time.

Case Study: Severn Estuary

Just to the north of the Somerset Levels, prehistoric peats and associated archaeological sites are extensively exposed in the intertidal zone of the Severn Estuary. Mesolithic and later footprint-tracks from the same area were discussed in Chapter 4. Twenty-nine years of monitoring sites revealed by erosion has led to the identification of two settlement sites of rectangular buildings at Redwick and Goldcliff, the first middle Bronze Age, *c* 1500–1000 cal BC, the second Iron Age *c* 400–100 cal BC (Bell *et al.* 2000; Bell 2013). The occupation evidence is at the interface between peat and overlying estuarine sediments showing that on both sites activity occurred in the early stages of marine transgressions. These sites are interpreted as seasonal settlements, mainly used in spring and summer for animal husbandry on saltmarsh. One of the strands of evidence supporting this hypothesis is the animal footprints (Fig. 4.14). Peat surfaces which were inundated by the transgressions are covered by a pattern of curvilinear depressions, just a few metres across, which drained the saltmarsh. Many of these channels have traces of short, very simple trackways; a few bundles of brushwood were laid across the channel and held in place at the edges with pegs; generally only a few pegs and a little brushwood remained. Rather more substantial tracks crossed a few larger palaeochannels. The Bronze Age trackways are generally parallel to the shoreline suggesting that this was the primary axis of movement during spring tidal periods of maximum tidal inundation. This suggests movement from settlements to river and stream channels where evidence has been found for fishtraps. Similar patterns of trackways parallel to the shore and crossing channels have been noted by Van de Noort (2004) in the Humber Estuary, e.g. at Melton during the Bronze Age.

At the Iron Age seasonal settlement at Goldcliff there is rather more to go on in detecting patterns of local movement. Eighteen trackways (shown in Fig. 6.8d) have been exposed by erosion on the foreshore. All are pedestrian walkways but they are of a wide range of types: some simple brushwood; some brushwood with the edges held in place by pegged round wood

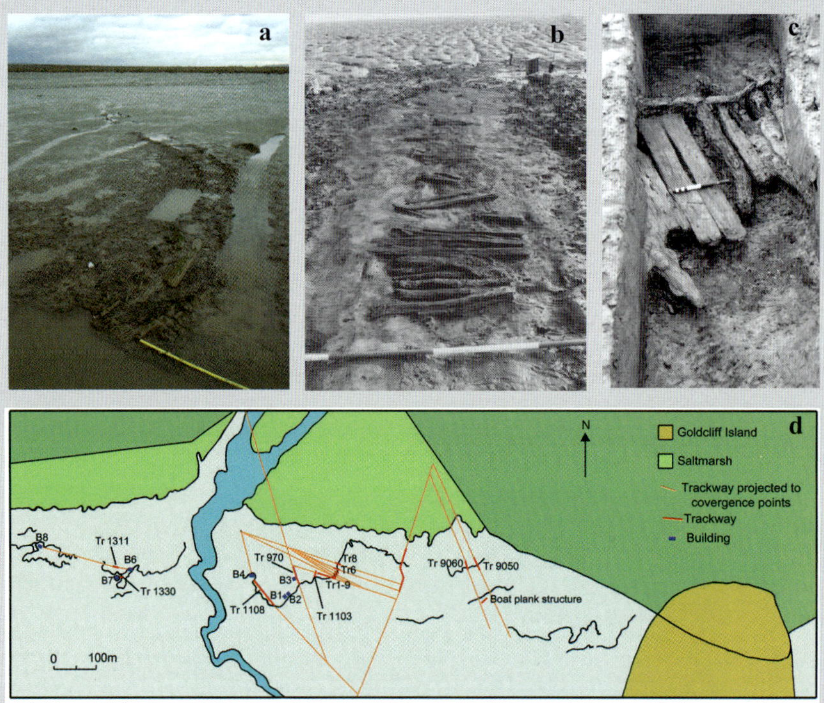

Figure 6.8 Goldcliff, Wales: Iron Age trackways: (a) brushwood track 1103, dated 390–70 cal BC; (b–c) corduroy track 1130 dated 390–120 cal BC; (d) The orientation of trackways showing the relationship to buildings and convergence points which may indicate lost or buried activity areas (photos: M. Bell; graphic: J. Foster).

(Fig. 6.8a), some V-shaped formed of opposed diagonal stakes with brushwood and planks between; some roundwood and planks with pegs (Fig. 6.8b–c). The differences may partly reflect the different environments in which they were made: reedswamp, fen woodland and raised bog. One trackway (1108) is dendrochronologically dated 336–318 BC; other radiocarbon dated examples are between 500 cal BC and cal AD 100. Trackway 1130 (Fig. 6.8b) could be traced for 100 m across the foreshore. Some trackways converged; Trackway 1108 led directly to excavated Building 4, while discontinuous brushwood Trackway 1311 led between Buildings 6 and 8. Since some of the tracks lead to, and between, buildings, it can be argued that they link activity areas and thus provide clues to patterns of habitual movement. On Figure 6.8d the lines of the trackways have been projected. This is a speculative exercise, as not all trackways will be straight as Figure 6.9a shows, but many are, and the exercise has possible predictive value. Several trackways converge on three locations to the north, perhaps the locations of still buried buildings or activity areas. Three trackways converge to the south, perhaps the location of some activity now lost to erosion. What is striking is that many of the trackways are not aligned east-west to the bedrock island of Goldcliff, where incidentally no evidence of Iron Age activity has been found. The orientation of communications is more north-south and that suggests perhaps the importance of channels for boat communication and other estuarine resources such as fishtraps.

Thames Estuary

From the Thames Estuary east of London there is also growing evidence for trackways, mostly found in development led excavations. They span the period from the late third millennium to the late second millennium cal BC with a peak *c* 1500 cal BC (Stafford *et al.* 2012). One example at Hayes Storage Depot was a 4 m wide track of gravel, flint and sand with evidence of animal trample (Bates and Stafford 2013). The trackways were made at a time when the palaeoenvironmental evidence points to intensified land use on the neighbouring dry ground with creation of extensive coaxial field systems, some with droveways leading towards in the river (Yates 2007). The East London trackways are comparably situated, running from the edge of gravel river terraces onto the floodplain wetland and are thought to be associated with seasonal grazing of animals on the floodplain (Meddens and Sidell 1995; Meddens 1996).

Later Prehistoric trackways in Ireland

In the Midlands of Ireland, where peat bogs cover such a large area that, even today, they are extensively milled as fuel for power stations, trackway finds are more numerous than anywhere else. Archaeologists did not take much notice of them until the discovery in 1984 of the most impressive example at Corlea led to the instigation by Professor Barry Raftery of a programme of wetland research which has transformed understanding. Some 97 of the trackways have now been scientifically dated (Fig. 6.2). A few, as noted above, are Neolithic but the vast majority (73%), are Bronze Age; there is a very marked dropping off in the Iron Age (except in the Derryville area; Cross *et al.* 2001) and none from cal AD 0–500 (Raftery 1996, pl. 51). Even allowing for the probability that earlier trackways may still be buried, and later examples are more likely to have been milled away, this suggests a particular focus on bog-related activities during the middle and late Bronze Age (Fig. 6.2). The trackways are of a great diversity of types: most are of simple brushwood, often short lengths designed to get across a particularly wet patch, others are much more substantial, providing a defined path across extensive bogs. Some areas, such as the Mountdillon Bogs, Derryville and Edercloon, have produced great concentrations of trackways. At Edercloon 45 individual structures, including 26 trackways and several platforms, have been recorded. Some tracks were repeatedly rebuilt and used throughout the Bronze and Iron Ages. They were not orientated to cross the bog but apparently provided access to the wetland for activities (Brunning and McDermott 2013). Of the Edercloon tracks no 5 is dated 1206–970 BC and was associated with a range of wooden artefacts apparently deliberately deposited in the foundations (Sands 2013). The objects included a disc wheel, which was unfinished and deposited as part of a trackway wholly unsuitable for wheeled transport (Brunning and McDermott 2013).

Corlea 6 was a 750 m long construction of split timber on roundwood laterals and brushwood, dated by dendrochronology to 2259±9 BC. It is of particular note because the timbers were worked with a metal axe, providing a secure date for early metal use in Ireland. This highlights the significance of trackways in wider material cultural studies, as well perhaps as hinting at the high status association of the more substantial structures from the earliest Bronze Age (Raftery 1996). Some other more substantial trackways are of the late Bronze Age and comprise mortised oak planks and split timber as at Derryoghil 1, a 650 m track dendrochronologically dated 938±9 BC, and Annaghcorrib, Galway, a 1.5 km route from dryland to the River Suck dendrochronologically dated 892±9 BC. Very extensive excavation of bog and bog margins was occasioned by the construction of a mine at Lisheen (Gowan *et al.* 2005) leading to the discovery of 34 trackways and two stone causeways dating between the early Bronze Age and Iron Age. Only two of the trackways crossed the bog completely: Killoran 18 of stone and wood dated dendrochronologically to 1400±9 BC, and Cooleeny 31, a substantial and well-constructed route destroyed by a bog burst in 600–595 cal BC, perhaps as a result of drainage being blocked by track construction (Casparie 2001). This track is thought to have connected high status regional sites. The other tracks were short lengths in what was described as a vernacular tradition. They were mostly located in areas of settlement and *fulachtafiadh* (cooking places) on the bog edge. Repeated use of the same sites was considered to be because they lay at the end of dryland routes, some of which were continued into the bog by short tracks. The middle Bronze Age period of most intensive trackway construction equates with a period, between 1550 and 1350 cal BC, when greater activity is attested by pollen analysis. However, the Bronze Age clearances are small, evidence for agriculture limited and the economy was mainly pastoral. Some trackway construction also took place at times when woodland was regenerating and there is little other evidence of activity (Caseldine *et al.* 1996).

Of particular significance among the Irish trackways is Corlea 1 (Fig. 6.9), the discovery of which in 1984 sparked the subsequent campaigns of wetland research. It is a road so impressive and out of scale with the other Irish examples, that it has been dubbed by its excavator 'the colossus of roads' (Raftery 1996). Corlea 1 runs for 1 km from uplands to a small island and it is clearly part of a longer routeway since the contemporary Derraghan 1 trackway, extends west of the island. Altogether they form a route of 4.7 km (Fig. 6.9b). Corlea 1 is dendrochronologically dated 147/148

Figure 6.9 Corlea, Ireland, trackway 1: (a) the excavated trackway (photo: courtesy the late Prof Barry Raftery); (b) the Corlea/Derraghan trackways in relation to topography, lakes and the Celtic ritual centres at Uisneach and Cruachain (graphic: J. Foster after Raftery 1990, fig. 57).

BC. It is built of large oak planks and split logs on lateral runners pegged in place through mortices (Fig. 6.1j). The scale of the trackway made Raftery confident that it was built for wheeled transport, although careful examination produced no evidence of wear, leading some to question, on the basis of areas of disrupted timber, whether it had ever been completed and used, though in places repairs were made as a pedestrian walkway (Raftery 1999). The heavy timbers mean that its life would have been limited as Casparie and Moloney (1996) have estimated that it would have sunk into the bog within a decade. Construction of the Corlea trackway is very close in time to that of other major constructions in Ireland (Raftery 1996, 420), such as a dendrochronological date for part of a defensive earthwork, The Dorsey (150 BC), and other radiocarbon dated linear earthworks are of similar date. A huge building at the Navan Fort is dendrochronologically dated a little later to 95 BC. It appears therefore that Corlea coincides with a period of great communal works in Ireland. Its surrounding area contains few Iron Age sites and finds suggesting, as does the scale of the trackway, that it has more to do with regional communication than local exploitation. Raftery (1990) has made the fascinating suggestion that the trackway may have been part of a longer route connecting two ceremonial foci, Ulsneach 22 km to the south-east and Cruachain 35 km to the north-west, via the major historical fording point at Lanesborough at the head of Lough Ree (Fig. 6.9b).

Bridges, post alignments and associated ritual deposits

This brings us to a group of sites of later Bronze Age and Iron Age date which are arguably central to the theme of movement but for which current interpretations generally emphasise associated ritual deposits of artefacts, particularly weapons, in wetland contexts (Bradley 2016). The question is whether these two aspects of communication and ritual are really so separate. The issue is especially highlighted by the site of La Tène, Switzerland where the artefact assemblage was so spectacular that the site gave its name to the later Iron Age culture in Europe. Interpretation is made more difficult by the fact that the main finds were in the second half of the nineteenth and first two decades of the twentieth century (Servais 2007). Finds came to light when drainage works and lake level changes revealed at least two separate alignments of large timber piles crossing a former river channel at the north end of Lake Neuchatel. Between the two was a substantial deposit representing many episodes of metalwork deposition, mostly weapons (swords, shields, spears, tools) with human skeletons. Dendrochronological dating indicates construction of one alignment 278±6 BC and with a continuance of activity in the area to 65 BC (De Navarro 1972). The most economical interpretation is that it represents a crossing point utilised by successive bridges where ritual deposition took place. Pryor (2001, 434), in the context of discussion about Flag Fen in the UK, highlights the ritual deposition role. He questions whether the alignments were functional bridges rather than ritual demarcations. But are these two roles incompatible, or have we privileged one aspect of social communication, ritual deposition, over another, the physical act of movement, in ways which may have made us play down some of the best evidence for axes of movement in later prehistory?

In the UK the adjacent Flag Fen and Fengate sites pose similar questions. The sites were the subject of an outstanding campaign of excavation and research led by Francis Pryor over a period of 40 years (Pryor 1998a; 2001). This site complex is central to the theme of this book, particularly in terms of the relationship between fields and trackways as developed in

Chapter 8. At Flag Fen the connection between dry and wetland sites is very clear (Fig. 6.10a) but there remain questions of interpretation. The dry ground of Fengate comprises a series of prehistoric settlement areas set within a system of coaxial fields established *c* 1800 cal BC and abandoned about 1000 cal BC. Through the fields run five parallel drove roads which lead down to a wetland embayment. One of those droveways leads to an alignment of multiple posts (Flag Fen) which runs for 1.2 km across the wetland with a timber platform part way along the alignment. At the east end of the alignment at Northey there is crop mark evidence of other, more fragmentary, fields. The hypothesis is that the Fengate fields served a largely pastoral function, with animals being driven down the droveways to pasture on the wetland during the summer. Cattle footprints were found at the end of the drove by the post alignment (Pryor 1998a, fig. 46). The post alignment was made at a time of marine transgression and was at about the tidal limit. It is dendrochronologically dated between 1300–924 BC. Near its west end the ritual deposition of 275 metal artefacts took place, including weaponry, many broken, axes, as well as smaller items brooches, pins and rings. Finds included part of a wheel and axle (Pryor 2001). The artefacts are largely late Bronze Age but there are also items deposited in the Iron Age (Pryor and Bamforth 2010), including some well after the post alignments would have decayed. Careful excavation and recording of the artefacts indicated that they were not sacrificed in large-scale ceremonies but in small, sometimes domestic scale, acts of deposition (Pryor 2001). The post alignment comprises five rows of posts and five levels of horizontal wood. Interpretation of the structure has ranged from a symbolic barrier against marine transgression, to successive post alignments mainly related to ritual acts of deposition. The final report (Pryor 2001) makes it clear that during at least some phases of its development a trackway formed part of the structure, initially of laid roundwood and later planks or other wood laid along the axis. The fact that a clear droveway leads to the alignment surely suggests that it is primarily a trackway/bridge where the ritual deposition of artefacts also took place. It is thus perhaps our clearest physical demonstration of direct wetland and dry ground relationships in later prehistory. The significance of this place as a crossing place is further emphasised by the fact that a similar route was taken by the only Roman road to cross the Fenland, the Fen Causeway (Pryor 2001; Malim 2015).

Close to Flag Fen (2.4 km south; Fig. 6.10b), and very close in date to the construction of that post alignment, is another alignment recently excavated at Must Farm (Gibson *et al.* 2016; Knight *et al.* 2016; 2019). It was a simpler structure, a raised 2 m wide walkway supported on large squared paired posts and incorporating some large planks (Fig. 6.11). It is dated 1290–1250 cal BC and crossed a palaeochannel. Earlier footprint-tracks indicate that the causeway may have formalised an existing path. The causeway is interpreted as part of a routeway across the Fens, part of what, with Flag Fen, seems to be an emerging network linking dryland areas within the wetland. As at Flag Fen there was deposition of metalwork nearby in the river channel, mainly weapons (spearheads, a dirk, rapier) and pins. After the causeway had gone out of use the routeway to which it was connected may have determined the location of a settlement on the same site. This comprised five roundhouses dated *c* 900–800 cal BC destroyed by fire very soon after construction; the complete organic and inorganic artefact inventory of this Bronze Age community, including two wheels, was preserved in the waterlogged sediments of the channel.

Other possible Fenland trackways or causeways are associated with approaches to the Isle of Ely, one to its south-east at Stuntney, the other to its south at Barway dated 530 cal BC and Bronze Age artefacts are reported near both (Godwin 1978; Hall and Coles 1994, 84). Excavated

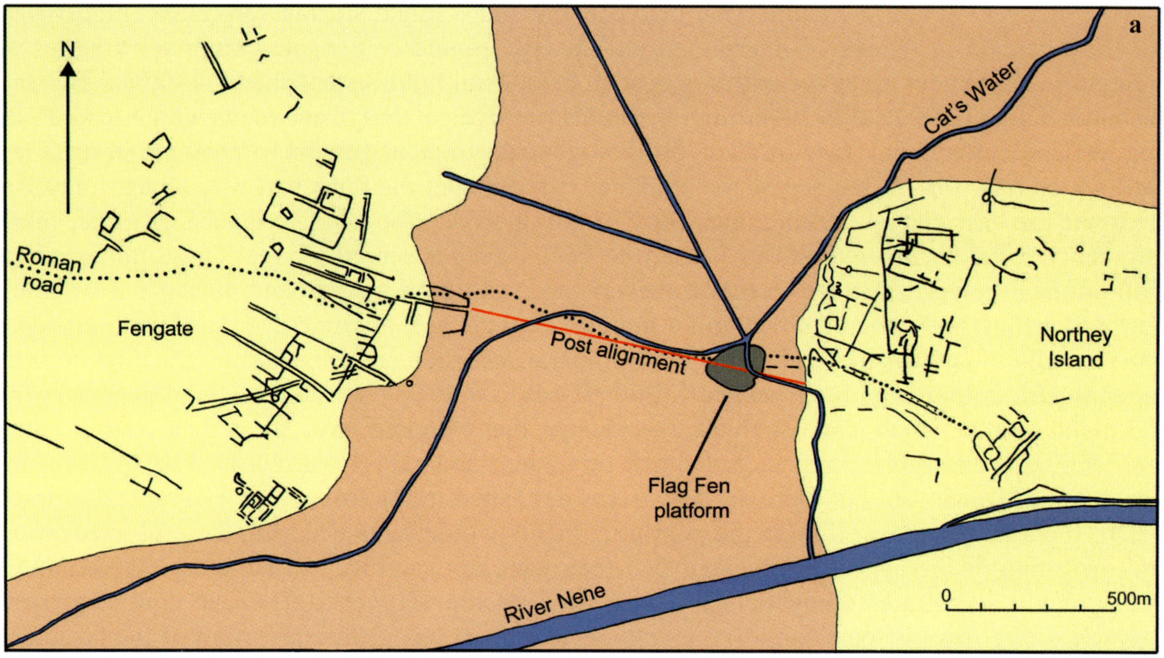

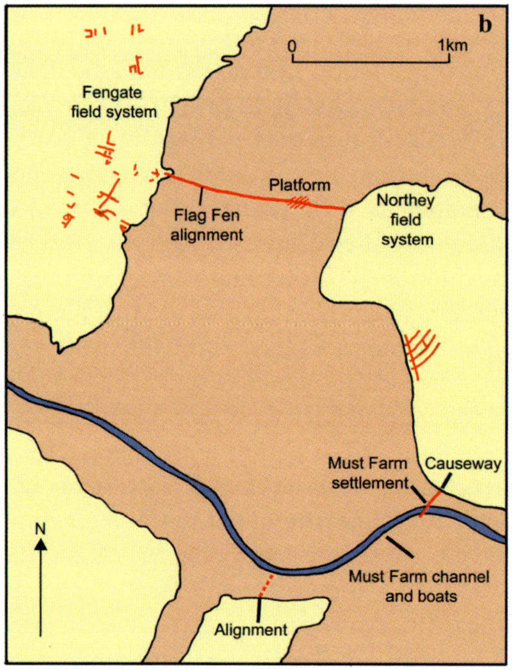

Figure 6.10 The Fengate, Flag Fen and Must Farm landscapes, UK: (a) Fengate and Flag Fen showing the main alignments of Bronze Age droveways and fields on dryland at Fengate and Northey and the post alignment and platform across the wetland between (source Pryor 2001); (b) the wider landscape of Flag Fen and Must Farm showing post alignments (source: Knight and Murrell 2012) (graphic: J. Foster).

Figure 6.11 Must Farm alignment formed of two rows of large posts crossing a palaeochannel and underlying the excavated Bronze Age palisade and settlement (photo: M. Bell, courtesy Dr M. Knight, Cambridge Archaeology Unit).

droveways on dry islands are also thought to approach causeways and areas of metalwork deposition (Evans 2002). This Ely evidence has echoes of Flag Fen but the similarities with the latter are particularly strong in the case of Fiskerton, in the Witham Valley of Lincolnshire where, since the eighteenth century, metalwork and 19 log boats have been found during river canalisation, the finds covering a long period from the Bronze Age to medieval period (Field and Parker Pearson 2003). The main concentrations of metalwork correspond to river crossing places still in use in historic times. At Fiskerton a double row of 195 large posts was erected, with limestone metalling between; dendrochronology showed that this took place in ten episodes over a period of 150 years between 457 and after 321 BC. The structure crossed what at the time was a reed swamp. It was associated with deposition of metalwork: swords, spears, and axes as well as pottery, amber, jet objects and fragments of human bone. Some 250 years later in the Roman period this was the site of a causeway, that too being associated with artefact deposition: bronze bowls, a bracelet and pottery. There are also medieval causeways in the area (van de Noort and O'Sullivan 2006). The published monograph concludes that it is uncertain as to whether the site is entirely ceremonial, or served as a crossing point or jetty (Field and Parker Pearson 2003). Despite its being 467 years later than the latest date of the Flag Fen alignment, the similarities are striking even to the extent of the same route being followed by Roman causeways in both cases; this raises some interesting issues about the extent to which Roman roads formalised existing routes, a point developed by examples in Chapter 8.

Causeways or bridge-type structures also occur more widely. On the south coast of England at Shinewater on the Willingdon Levels, Sussex, a substantial trackway was recorded, with three rows of vertical posts and horizontal beams and rails (Greatorex 2003; Jennings *et al.* 2003). The inclusion of sand and fine gravel along the alignment might be part of the construction but might perhaps relate to trample by animals. It crossed a freshwater fen with saltmarsh to seaward. Along the trackway was a wooden platform on which hearths and evidence of cereal processing may indicate occupation. Here again ritual deposition is indicated by axes, a bracelet, a sickle, amber beads and human bone. Shinewater, which is dated 900–800 cal BC, may represent a route from the Bronze Age agricultural landscape of the chalk in the south-west to the seasonal wetland pastures of the Willingdon and Pevensey Levels. Also on the south coast, inland of the Solent in the Test Valley, excavations have found lines of posts and planks representing three bridges dating between 1700 and 1400 cal BC; one is associated with a fragment of boat cleat and there is also a middle Bronze Age rapier from the site (Fitzpatrick *et al.* 1996; Van de Noort 2011).

In the Severn Estuary/Somerset Levels region there are three further sites which invite comparison with those under discussion. Of these the most fully investigated is at Caldicot, south Wales where palaeochannels contain a series of Bronze Age structures and stone scatters (Nayling and Caseldine 1997). Most notable is a double line of large posts, dendrochronologically dated 990–989 BC, crossing a river channel at about the limit of tidal influence. Artefact deposition is indicated by two sword chapes, a tin strip, pottery and wooden objects. In the Somerset Levels are two similar structures built at much the same date. At Greylake there are earlier reports of alignments of posts at a natural crossing point between two projecting areas of dry ground. This may be associated with recent finds of oak posts, dated dendrochronologically 963–952 BC, found with bones of humans and animals and a broken axe (Brunning 1997; 2013). The structure crossed a freshwater fen. Horizontal planks and some cut roundwood was found but the small excavation did not produce clear evidence of a superstructure and Brunning argued that it may have related to ritual deposition and demarcation rather than communications. Also on the Somerset Levels at Harter's Hill (Fig. 6.12) two parallel rows comprising at least 34 large posts, associated with plank fragments and cut roundwood, run from the hill slope out into Queen's Sedgemoor (Brunning 2013). The earliest pile is dendrochronologically dated to 1076–5 BC and the structure was in use for at least 30 years. Conditions were getting wetter at the time and previous alder woodland had been replaced by sedge swamp and shallow open water. The previous discovery of a bronze sword from the same field is likely to be associated. Although the sword and the substantial posts of the alignment shows similarities with the others discussed in this section it must be acknowledged that unlike the others it is not positioned at a natural crossing point but heads more or less to the middle of the Moor.

Figure 6.12 Harter's Hill alignment, Somerset, hypothetical reconstruction of the alignment by Peter Lorimer (image: Somerset County Council).

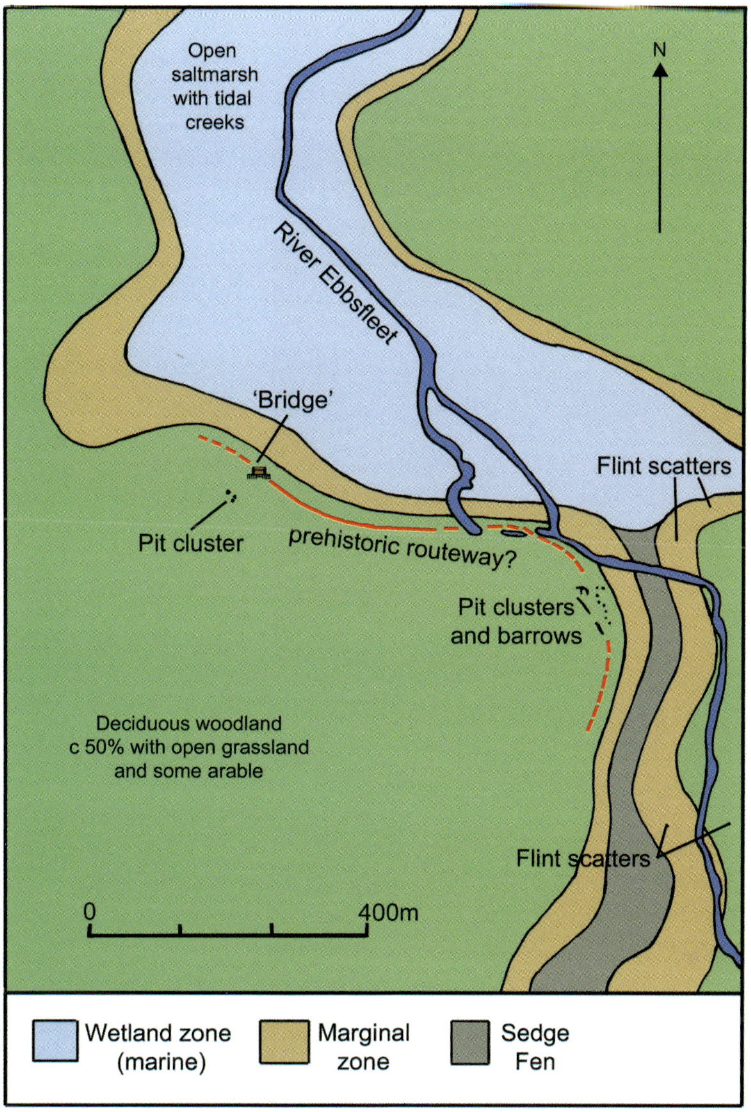

Figure 6.13 The Ebbsfleet Valley, Kent, a hypothetical Bronze Age routeway along the saltmarsh edge represented by a possible bridge, short lengths of brushwood track, pit clusters and barrows (graphic: J. Foster, after Bates and Stafford 2013, fig. 86).

In the Thames valley there are a number of finds of post alignments which appear to be bridges or jetties (Stafford *et al.* 2012). Beside the modern Vauxhall Bridge, London 25 oak posts occur in two parallel rows dated to 1750–1285 cal BC and were at about the tidal limit at the time (Cohen and Wragg 2017). Pottery and spear heads have come from the area. Another probable bridge at Freemasons Road is dated *c* 1600–1750 cal BC. At Ebbsfleet in the Lower Thames (Fig. 6.13) a double post alignment may represent a walkway or bridge dated 1410–1220 cal BC; this is particularly significant because a continuation of this route seems to be indicated by small brushwood

Figure 6.14 Yarnton, Thames Valley, UK Iron Age stone causeway (photo: courtesy of Oxford Archaeology).

tracks, clusters of pits and barrows. This hypothetical route is not across the wetland but along the edge of the Ebbsfleet Valley which at the time was saltmarsh (Bates and Stafford 2013, fig. 86). Upstream on the Thames at Eton Rowing Lake a succession of six later Bronze Age and Iron Age bridges were constructed across river channels. They were made with roundwood uprights and mortised planks (Lambrick and Robinson 2009). They form part of a Bronze Age landscape of pasture, settlements and fields which has been extensively excavated. Gravel and limestone causeways across the flood plain have been found at Thrupp, Abingdon dated to the middle Iron Age and at Yarnton where stone for the causeway (Fig. 6.14) was transported from 4 km away. Metalwork finds suggest this had been a significant crossing point from the late Bronze Age and wheel ruts are reported from the causeway. Lambrick and Robinson (2009) observe that there is little evidence as to how the bridges, trackways, causeways etc found in riverine contexts in the Thames valley link to wider dry ground routes. It is clearly a task for the future to extrapolate beyond the clues provided in riverine contexts.

Bronze and Iron Age metalwork was found in the Thames during dredging, including many finds from the Eton area and further downstream in London. It seems probable that some of this deposition took place from the bridges/post alignments which are increasingly being found along the Thames. The possibility should also be considered that what appear, to this writer at least, as bridges or causeways may also have served as territorial boundaries or barriers to upriver movement, or they may have combined multiple functions. Acts of ritual deposition, such as bog bodies, metalwork and other artefacts occur in Ireland on what were later recorded as minor kingship boundaries (Kelly 2006). Ritual deposition is not necessarily taking place in remote locations but in wetlands which, in some cases, the environmental or field archaeological evidence shows were close to agricultural land and settlement. This again highlights the importance of trying to join up the landscape, the tracks, fields, environmental evidence, settlement and ritual places and to avoid treating these as separate categories.

Conclusions

Wooden trackways are present from the very beginning of the Neolithic in north Germany and the Somerset Levels (Fig. 6.2), but only in the latter are Neolithic examples particularly well represented. It is during the middle and late Bronze Age that trackways are frequent in each of the main areas, implying that some characteristic of Bronze Age life made trackway construction especially prevalent. Indeed Kristiansen and Larsson (2005) draw on quite different evidence from material culture and wheels to contend that mobility was a particularly distinctive trait of the Bronze Age. Numbers reduce in the Iron Age especially in Ireland and Somerset but not in Lower Saxony. Overall there are far fewer in the last two millennia. However, early trackways are likely to remain buried in some areas and later ones will have been removed by peat digging before archaeologists came on the scene.

The factors that occasioned trackway construction have been much debated. Often trackways have been interpreted as maintaining communications in the face of wetter conditions. In other cases, it has been suggested that they reflect drier episodes when routes, previously impassable, became accessible. In Ireland evidence of testate amoebae and peat humification, testifying to temporal changes in bog surface wetness, have been compared with radiocarbon and dendrochronological dates for wetland structures, including trackways (Plunkett *et al.* 2013). Wetland activity was shown to be clearly episodic, with six pronounced episodes of wetland activity identified in the last 4500 years, separated by five lulls in activity lasting between 100 and 400 years. However, no correlation was found between activity and lull phases and the changing patterns of bog surface wetness. Instead, the episodes of wetland activity show greater correspondence to phases of clearance and activity on dryland during which the landscape was being more intensively used. In Ireland the concentration of tracks in the middle and late Bronze Age coincides with increased clearance and agricultural activity and for Raftery (1996) in the Mountdillon bogs this was the main factor prompting trackway construction. Similarly, in Somerset there is a good correspondence between the temporal concentrations of trackway construction and palynological evidence for clearance on dryland in the middle Neolithic and middle to late Bronze Age (Coles and Coles 1998, fig. 4). However, several trackways, such as those at Meare Heath, are also associated with periods of increased wetness, notably *c* 1500 cal BC (Brunning and McDermott 2013; Brunning 2013). The Bourtanger Moor tracks in the Netherlands can also be related to pollen and settlement pattern evidence for activity on the neighbouring Hondsrug dryland.

There can be no assumption, however, that trackway construction had the same driver in each case or region. In north-west Germany detailed comparison of dendrochronologically dated trackways and the deaths of trees growing on bog surfaces indicates that trackway construction frequently corresponded to a rise in bog surface wetness (Achterberg *et al.* 2015). Not only may wetland environmental changes have affected patterns of human activity, but people also influenced the wetland environment itself. The very act of trackway use created an artificial linear depression and Casparie (1986; 2001) has argued that at Derryville, Ireland and Bourtranger, Netherlands, the disruption of drainage by the creation of a substantial track may have triggered bog bursts, fundamentally changing bog hydrology and putting trackways out of use.

As a generalisation it seems that social factors were more significant drivers of trackway construction than has often been assumed. Many trackways seem to have been built to facilitate fairly local movement from dryland onto the edge of bogs, thus providing access for activities at the wetland edge, e.g. the Derryville examples, or activities on the bog surface at Edercloon. For a few in The Netherlands and North Germany they appear to have had a role in the extraction of bog iron ore. Only in a minority of cases are trackways presently understood as part of wider communication systems. Corlea 1 is clearly an example; similarly two trackways crossed Derryville bog and of these Cooleeny 31 is interpreted as a high status regional route. In the Netherlands only two tracks crossed from one side to the other of the vast Bourtranger Moor. Also fragments of much slighter tracks elsewhere at Smilde have been identified as part of a longer route linking sand ridges. In north Germany there is a greater concentration of substantially constructed trackways and they have been seen by Hayen (1987) as part of regional communication networks. Where bogs are especially extensive, as in Lower Saxony, they often formed natural barriers between cultural groupings (Hayen 1987). We have identified other examples where communication routes persisted over long timescales. However, these tend to be the exceptions. It is striking that many trackways, including some of the more substantial and carefully constructed examples such as Corlea I and the Wittemore track (XLII Ip), seem to have been abandoned not long after construction. This might be explained in environmental terms, but the possibility should also be considered that they reflect social relations such as alliances, trade connections, or periods of religious observance etc which were of a short-lived nature.

Wheels and other parts of vehicles are associated with trackways in Saxony and the Netherlands from the late Neolithic and those areas have more examples of substantial trackways suitable for wheeled vehicles. In Lower Saxony sophisticated trackways have upper lateral timbers to maintain a level surface suitable for vehicles; these appear in the middle Bronze Age and are especially marked in the eighth century BC. In Britain and Ireland only the Corlea 1 trackways seems clearly intended for vehicles. Perhaps the more substantial double post alignments/bridges and especially the stone causeways that occur in lowland England mainly in the late Bronze Age were intended for vehicles. However, aside from the wheel at Flag Fen, where the post alignment does not seem obviously suitable for wheeled transport, and wheel ruts on an Iron Age causeway at Yarnton, there is no direct evidence.

Trackway construction and use cannot be seen in purely functional terms because there is so much evidence for ritual deposition and symbolism associated with them. This is highlighted by the wooden outlines of human figures at the Wittemoor XLII Ip trackway, wooden human figures from Irish trackways (O'Sullivan 2007, 186), and the many other objects associated with some trackways such as Sweet Track and Edercloon 5. In the case of the British double post alignments, particularly of late Bronze Age date, the ritual deposition of artefacts is very clearly evident at Flag Fen and seen also to a lesser extent at most of the other sites. It has been argued here that the ritual deposition evidence has tended to eclipse evidence that these structures also served as communication routes and may therefore have obscured their full social significance. For the purposes of this study of prehistoric routeways, wetland tracks possess the huge advantage of being well dated and associated with a wealth of environmental evidence. In all but a few cases the real challenge remains to relate the wonderfully detailed local picture they provide to the wider patterns of communication.

7. Barrow alignments as clues to Bronze Age routes

Introduction

In Chapter 6 we have looked at some relatively clearly-defined evidence for wetland tracks but only in a few instances could those be related to dryland routes. In Chapter 5 we reviewed rather more tentative evidence for dryland routes used by the first farmers, concluding that quite often these related to bands of movement rather than closely demarcated paths. The clues that were suggested as significant in relation to these Neolithic dryland routes included recurrent patterns of activity on the same site separated by centuries, sometimes millennia, as for instance below megalithic tombs. A second key feature was the construction of monuments on the same alignment, again often over extended timescales. Alignments are not easy to interpret because, whilst they could indicate route-ways, they have also often been interpreted in terms of cosmology (e.g. astronomical alignment), not that one needs necessarily to exclude the other. Recurrent patterns of activity and significant alignments may also provide clues to patterns of movement in later prehistoric periods such as the Bronze Age. Not infrequently patterns that emerge in the Bronze Age also incorporate some Neolithic monuments, suggesting that the alignments may be of earlier origin. In the early Bronze Age we have a much larger number of monuments, particularly barrows, making alignments clearer and we can also begin to detect relationships to defined routeways which become formalised and demarcated in the agricultural landscapes of later periods, particularly the middle and late Bronze Age and Iron Age, which are considered in Chapters 8 and 10. However, the case is not simple. The evidence for barrow alignments on routeways is far stronger, and accepted as orthodox in Denmark, north Germany and the Netherlands, but less so in the British Isles, although it will be argued that even in Britain there are convincing cases and probably more which are deserving of detailed investigation.

Denmark

In Denmark there is far more evidence for ancient dry ground trackways than in any of the other areas considered. This is partly because a national inventory of these sites has been created for heritage protection purposes. Figure 7.1a shows the 2300 sites on that inventory with evidence of ancient tracks including hollow ways (Bang 2013). Three major long distance routes can be

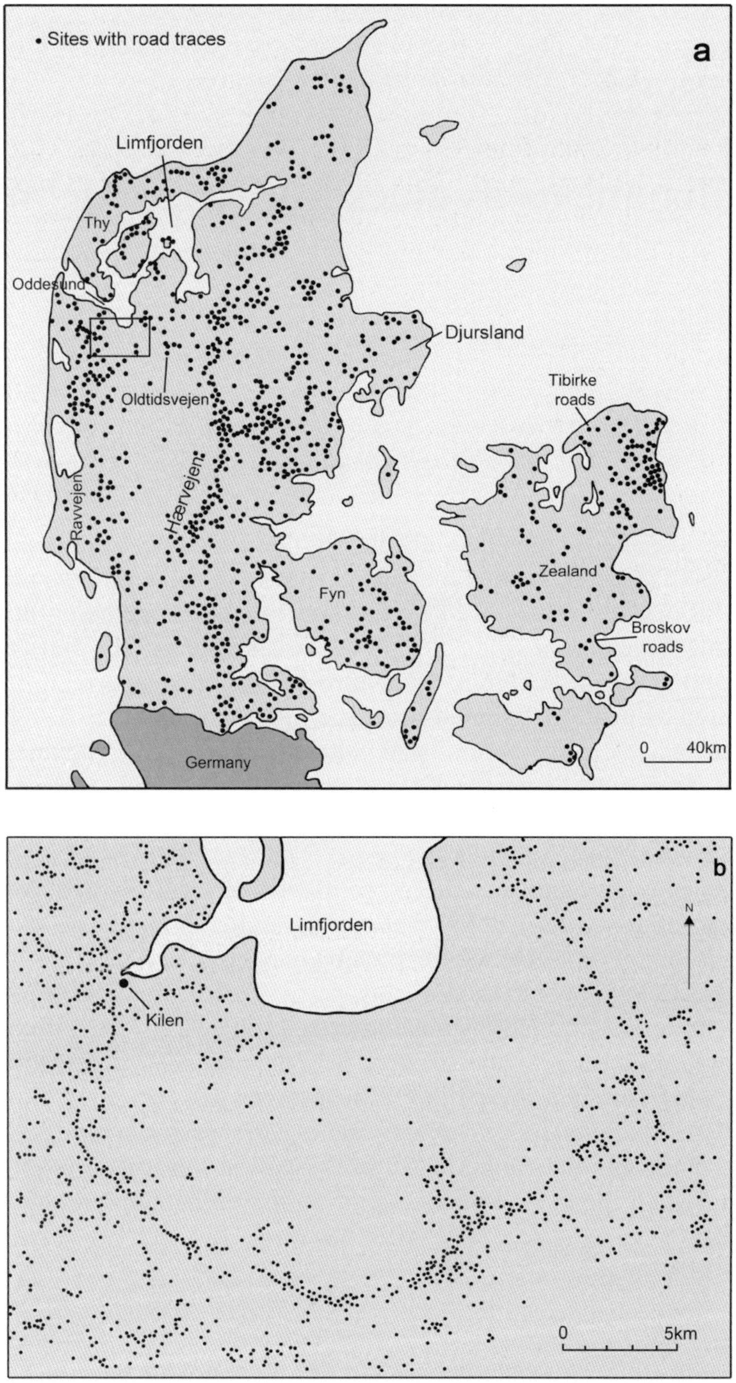

Figure 7.1 Denmark, early routeways: (a) prehistoric and historic road traces in the Danish National Inventory; (b) barrows marking the Oldtidsvejen prehistoric road just south of the Limfjorden (graphic: J. Foster after Bang 2013, figs 1 and 2).

identified. These are in a sense 'natural' topographically-determined routes, related to watersheds and crossing places over water. Each is supported by the evidence of trackway traces, alignments of barrows and other evidence. Linear arrangements of barrows have, since the work of Müller (1904), and subsequently by Bakker (1991), been recognised as routeways. The most prominent of these routes runs north to south from Viborg down the watershed in the centre of Jutland and is known as the Haervejen, or 'army road'. This is considered to be, not so much a closely demarcated line, as a broad communication corridor, the course varying seasonally according to weather conditions (Schou Jørgensen 1988b). The evidence is strengthened by the discovery at historical fording places of Bronze Age ritual deposits, wheel ruts below a few of the barrows and alignments of cooking pits corresponding to that of barrows. The evidence is assembled by Lovschal (2013) who contends that the barrow alignments should be seen, not just in functional terms, but in terms of the cues and prompts to genealogy and ideology created by the patterns of monuments, physical movement being related to paths of thought and recollections of social significance. This resonates with the ideas of Bradley (2000) who, in discussing acts of ritual deposition in natural places such as bogs and rivers over extended timescales, suggests that this relates to some memory of past acts. Such an explanation becomes more tenable if they were linked to stories attached to particular routes through landscape. The second major route in Denmark, the Oldtidsvejen is an east-west route just south of the Limfjorden which is picked out by the curvilinear grouping of barrows shown in Figure 7.1b. Another north-south route, the Ravvejen, is marked by a barrow alignment down the west coast of Jutland.

Case Study: Kilen, a Bronze Age cross roads in Jutland

Travelling from the north coast of Thy, where Bronze Age settlements and fields are preserved below blown sand and where amber was collected for transport south (Bech and Mikkelsen 1999), travellers would have taken a 'natural' route via a narrow strip of land across the Oddessund. At the head of another inlet of the sea at Kilen was a natural crossing point where two routes converged. The north–south route known as Ravvejen crossed with the east–west route, the Oldtidsvejen, both marked by prominent alignments of barrows (Figs 7.1–2). A particularly large barrow (Fig. 7.2b) occurs where the west–east route begins its descent to the Bredkaer valley. Here four alternative routes spread like the fingers of a hand down into the valley. Each is deeply incised into highly erodible Pleistocene sands and gravels. In their present form they are hollow ways formed by the passage of people, animals and carts; natural erosion processes have also contributed to their morphology (Fig. 7.2c). Numerous prominent hollow ways (Fig. 7.2d) mark alternative routeways from the plateau into the valley on both sides and these have been mapped in the surrounding area using LiDAR (Bang 2013, fig. 4). The alternatives may reflect the changing extent of the inlet and where the Bredkaer Baek was most easily crossed. On the south side of the valley two prominent hollow ways ascend to the plateau. On either side of these routes are spurs on which, within 100–200 m of the routes, are two Neolithic causewayed enclosures (Molbjerg I and Molbjerg II; Klassen 2014). From here the main route is marked by a prominent barrow alignment which includes two Neolithic long barrows and many Bronze Age round barrows (Fig. 7.2e), marking a clear line

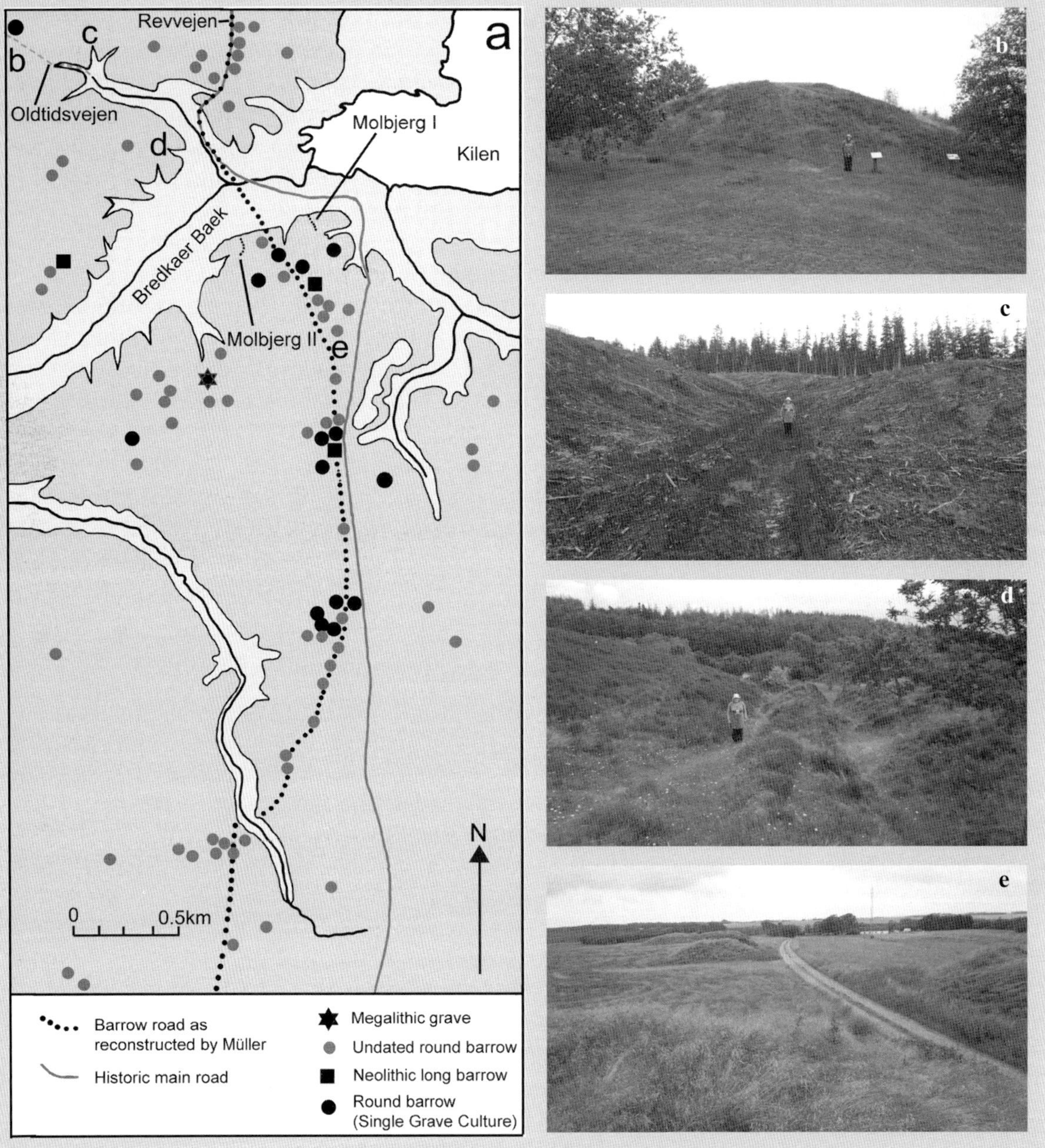

Figure 7.2 Barrow alignments in the Fousing/Kilen area in west Jutland, Denmark (location on Fig. 7.1b). The barrows are Neolithic and Bronze Age and the location of Neolithic enclosures at Molbjerg are marked. Photos b–e are marked on (a) and described in the text (graphic: J. Foster modified after Klassen 2014; photos: M. Bell).

which was originally identified as a prehistoric routeway by Muller (1904); part of this route is still followed by a historic road. The evidence in this area is particularly clear because it represents the crossing point of two important routes, clearly marked by barrow alignments and by multiple deeply-incised hollow ways. The multiplicity of hollow ways and alternative barrow alignments at Kilen is something often seen in prehistory, the result perhaps of changing local conditions as some routes became impassable. The hollow ways and barrow alignments in this area are marked by display boards informing walkers of their relationship to prehistoric routes.

Klassen (2014) drawing on evidence from Kilen and other Danish sites has observed regular relationships between routeways marked by barrows, river crossings and Neolithic enclosures. Based on these relationships he used GIS to build predictive models of routeways and Neolithic enclosure locations in the Djursland peninsula, east Jutland, testing these predictions by air photography, fieldwalking and geophysics. This led to the discovery of several new Neolithic enclosures, thus validating the predictive model and the inferred routeways. Fieldwork appears, however, to have been limited to the individual enclosure sites and did not include actual physical evidence for the inferred routeways such as the hollow ways which Bang (2013) recognised at Kilen. A further complicating factor in the Djursland peninsula is that marine inlets were very extensive in the Neolithic and much communication might be expected to be by water. Klassen (2014) has identified several other cases where routeways can be inferred from alignments of Neolithic graves and barrows, at Vroue Hede, Jutland and at Haslov, Sweden, where a TRB (Funnel Beaker Culture) road of 0.7 km is delimited by palisades and standing stones adjoining tombs along the edge of a wetland (Klassen 2014; fig. 19).

Germany

In the Uelzen area of north Germany between Hannover and Hamburg abundant *hunebed* (megalithic tombs) of the Trichterbecker Culture (TRB) (*c* 3500–2800 cal BC) also show evidence of pronounced alignments, many of which correspond to existing routes, often on higher terraces and watersheds, as distinct from the mainly riverine distribution of contemporary settlements (Bakker 2008, fig. 23.2). It is notable that two settlements in this area are on *hunebed* alignments at probable fording points in river valleys. At Uelzen evidence may point to a complementary system of dryland tracks on watersheds and higher terraces and a transport system which may have been mainly river based. South of Kiel at Flinkbek in Schleswig Holstein, a major alignment of barrows extends over 4 km along the ridge of a seemingly 'natural' route which may have extended to the head of the Kiel Fjord to the north. It was flanked by wetlands to the east and the valley of the River Eider to the west (Bakker *et al.* 1999; Klassen 2014). The barrows span some two millennia from the early Neolithic to the Bronze Age. Cultivation preceded some of the barrows, as ploughmarks and pollen evidence for agricultural activity showed (Diers *et al.* 2014). Below one Funnel Beaker Culture barrow dated *c* 3420–3380 cal BC were wheel ruts, providing some of the earliest evidence for the use of wheeled transport in Europe (Fansa and Burmeister 2004; Mischka 2011).

Netherlands

Case Study: Veluwe and Hondsrug barrow roads

On the sandy soils of the central Netherlands, in the Veluwe area between Epe and Vaassen, there are very distinct alignments of barrows which have been interpreted as marking one axial and two divergent side roads (Fig. 7.3; Bakker 1976; 2008). The main alignment runs for 6 km and comprises 35 barrows; the earliest date to the third millennium BC, and there are many Bronze Age and a few Iron Age burials. Further alignments of barrows to the west of the mapped area may hint at the existence of other routes (Fokkens 2005, fig. 16.7). Much of the heathland is now farmland, or forested, with the barrows in small clearings. Some of the barrows are marked by geocaches so that those walking along the prehistoric route can lift a handle raising a cylinder in which are models and information about the prehistoric barrows and the routeway they marked (Fig. 7.3c). The alignments are thought to have created landscape structures representing, perhaps, lineages, or dynastic successions of those buried, animated by movement along the axis (Bourgeois 2013). Many of the barrows have been excavated and were constructed of turves; pollen analysis demonstrates these were cut from heathland. This led Doorenbosch (2013) to infer that linear corridors of heathland existed from the Neolithic since pollen analysis showed that heath had developed well before barrow construction. Some woodland remained, but away from the alignments and settlement seems not to have been near. Doorenbosch calculated the numbers of animals, the burning and deturfing activities which would have been required to maintain these heathland corridors. She appears to envisage acts of deliberate clearance to maintain linear heathland tracts. A more economical explanation might be that the linear heathy tracts reflect regular patterns of animal and human movement, perhaps to seasonal wetland pastures. Continuing burials may indicate that these routes remained to some extent active into the Iron Age when 'Celtic fields' were laid out beside parts of the alignments, but without disturbance to the barrows (Brongers 1974). Cart tracks have been identified alongside some barrows but predating the fields (Bourgeois 2013). The field patterns do not incorporate identifiable trackways running alongside the alignments, so perhaps the inferred patterns of movement had lost at least some of their significance by the Iron Age.

In north Holland the Hondsrug is a ridge of Pleistocene glacial origin which stretches for 70 km south-east of Groningen to Emmen forming a natural routeway between peat bogs and riverine wetlands. Its significance, in terms of the relationship between wetland and dryland routeways, was discussed in Chapter 6. In the northern part of the ridge Jager (1984) has mapped a pronounced linear pattern of megaliths, barrows and urnfields which, from the date of the monuments, he argues indicates a route in use from the period of Funnel Beakers to the late Iron Age. Excavations along this line have uncovered trackways and wheel ruts beside, rather than below, the barrows; on one site a hollow way was sectioned. The tracks were abandoned by the early medieval period when axes of communication changed and earlier routes were buried by plaggen (man-made manured) soils of medieval date. In the southern part of the Hondsrug around Emmen the distribution of Bronze Age and later settlements, fields and urnfields has been illustrated in Figure 6.6 in relation to the wooden trackways

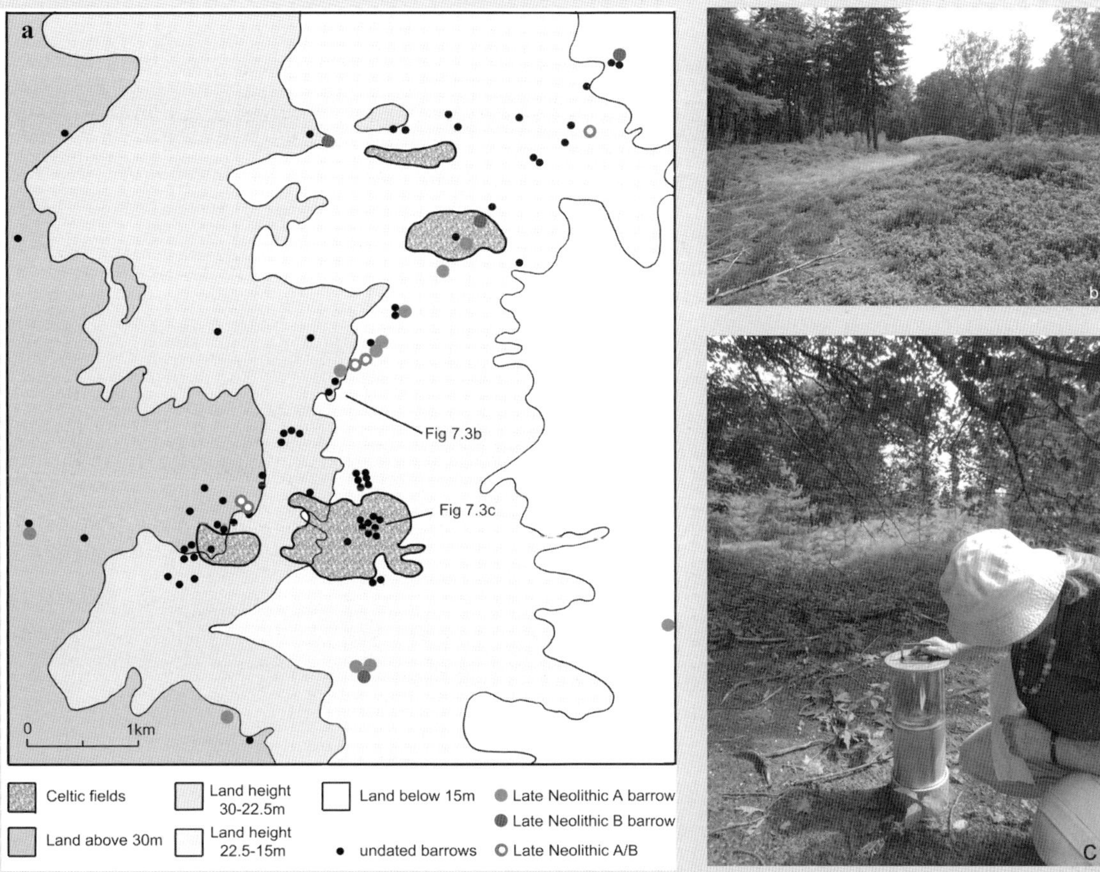

Figure 7.3 Barrow alignments in Veluwe, Netherlands: (a) plan of barrows; hatched areas are 'Celtic fields' (graphic by J. Foster after Bakker 2008); (b) barrows in former heathland, now wooded; (c) geocache beside barrow; present day walkers along the route learn about its past. (photos: M. Bell).

which run across peat east of the Hondsrug. Bakker (1976, fig. 10) maps the distribution of *hunebed* megalithic tombs in the same area; they are mainly down the centre of the sandy ridge, and track ruts are again found on the same axis. This alignment corresponds approximately to the hypothetical territorial boundaries down the axis of the sand ridge shown on Figure 6.6. This suggests firstly the possibility of a routeway going back to the Neolithic and secondly that it provided the boundaries between hypothetical settlement territories in the Bronze Age. Bakker (1976, 79) also notes that cart tracks in some Bronze Age urnfields are often aligned on the grave field and sometimes as at Sleen (4 km west of the area mapped in Fig. 6.6), with a reserved space for a road.

North European connections

Bakker (1976) argues that despite the obstacles to communication created by extensive coastal wetlands, some extending 150 km inland, there are considerable similarities between the TRB in Denmark, north Germany and the Netherlands. These he explains by evidence for the movement of lithic raw materials, for instance axes, from the Limfjord in northern Jutland, to north-west Germany where the distributions fan out from crossing points on the Elbe (Bakker 1976, fig. 13). Later in the Bronze Age the routes south from Denmark are considered to have played a part in movement of amber which was gathered in settlements such as Bjerre in northern Jutland and traded, or exchanged, as far as southern Europe. In the TRB period, artefact distributions highlight the importance both of north-south routes from Denmark and east-west routes inland of the coastal wetlands in northern Germany and the Netherlands. In the Iron Age from *c* 500 cal BC there seems to have been an interdependence between the *terp* (artificial settlement mound) sites on coastal saltmarsh in the Netherlands and north Germany, where the farmers were engaged in cattle husbandry, fishing and salt production, and the settlements inland on sandy soils with evidence of Celtic fields which produced crops, wood and iron (Brongers and Woltering 1973). Such interdependence assumes an effective system of communication routes of the type demonstrated along the Hondsrug ridge and probably by the network of wooden trackways traversing the wetlands of Lower Saxony discussed in Chapter 6.

England and Wales

In mainland North-west Europe we have seen some convincing evidence for the positioning of barrows on trackways. However, the impression should not be given that all barrow alignments indicate trackways. In Britain (site locations on Fig. 8.1) alignments of barrows are common, but usually without any convincing evidence that they had anything to do with tracks. Some follow topography, or may relate to cosmology, or lineages of those interred. However, even here there is evidence that sometimes barrow alignments do relate to routeways. One such is at Bishopstone, East Sussex where a line of Bronze Age barrows was aligned on a track of Iron Age and Romano-British date; this is outlined in more detail as a case study in Chapter 10. In the highland zone of Wales, Bronze Age monuments have been observed to cluster along mountain trackways (Lynch 2000). One such example is the extinct volcano at Penmaenmawr, a dramatic situation with panoramic views of the coast of north Wales (Fig. 7.4). Here a curving trackway, in places a hollow way, is marked by Bronze Age cairns. The track passes a stone circle, the early Bronze Age Druids' Circle (Fig. 7.4b-c), which is flattened on the trackway side apparently respecting the earlier trackway line (Griffiths 1960). 200 m north of the trackway is the rock outcrop of Graig Lwyd which was a Neolithic axe factory site. The axes from here, most likely given special significance not just because of the properties of the stone but the dramatic upland from which they originated, can be shown by petrological analysis to have been distributed widely in southern Britain (Clough and Cummins 1988). A standing stone also marks a continuation of the track at Penmaenmawr. Roughly parallel and 3 km south another routeway is marked by three stone circles, standing stones and cairns and a much later Roman road on the same line demonstrates the longevity of that route. Williams (1988) observed that standing stones sometimes mark the positions of ancient tracks.

Lest we get carried away with the notion of an inevitable association between barrows and routes, there are many more examples where arrangements of barrows seem to have a rather different landscape role which might for instance reflect cosmological alignments, genealogies, or territorial divisions. In Ireland, Cooney (2009) identifies linear arrangements of barrows of Bronze Age to Iron Age date interpreting them in terms of lineages of the people buried, although such a sequence might also imply an axis of movement. Bakker (1991, 518) from the position of his considerable experience of prehistoric trackways in mainland Europe, says that he has never

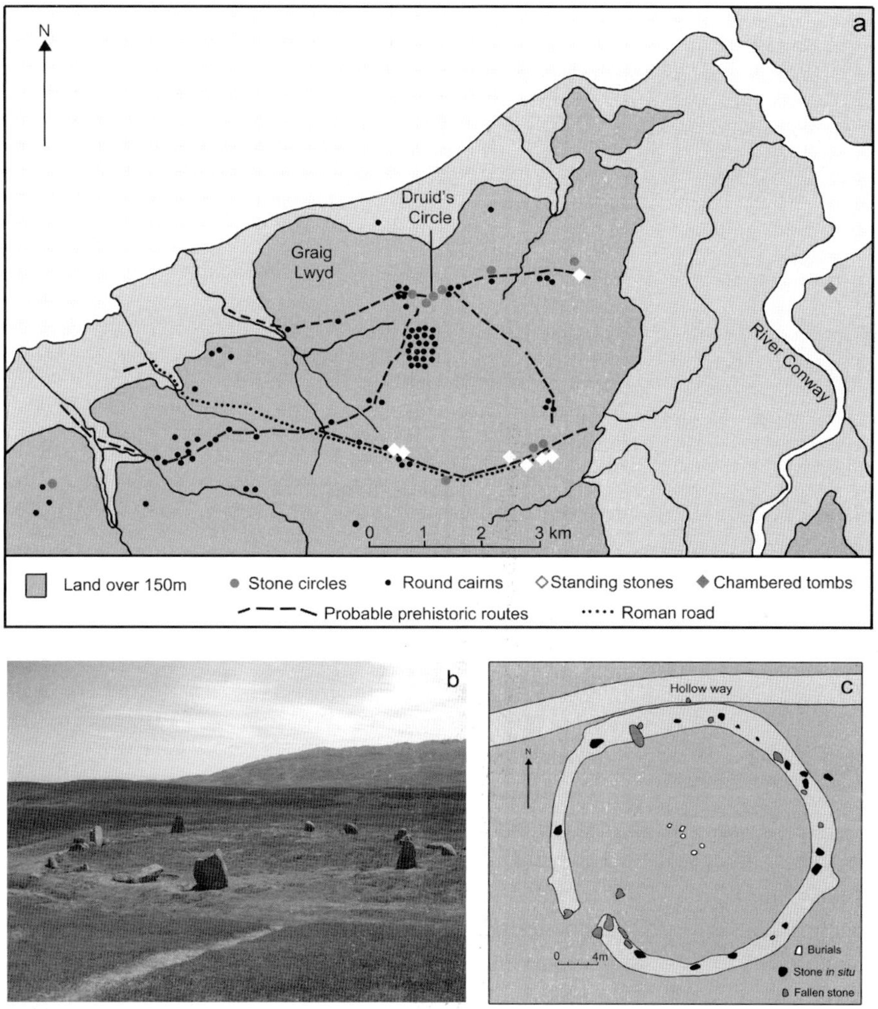

Figure 7.4 Penmaenmawr, north Wales, UK: (a) possible upland routes indicated by alignments of prehistoric monuments; the southernmost is followed by a later Roman road; (b) the Druid's Circle; (c) plan of the Druid's Circle which appears distorted respecting an earlier hollow way (graphics: J. Foster after Griffiths 1960; photo: M. Bell).

understood why the barrow alignments around Stonehenge were not interpreted as following routes. There are indeed very prominent barrow alignments, most obviously 2 km west of Stonehenge at Winterbourne Stoke cross roads comprising a Neolithic long barrow and 14 early Bronze Age round barrows (see Fig. 5.6). Other prominent alignments are on Normanton Down 1 km south of Stonehenge and on King Barrow ridge 1 km to the east of Stonehenge. These, and the other Stonehenge barrow alignments, do not have any clear association with routes, they do not lead to Stonehenge, or other monuments, which might be expected if they perpetuated routes of the Neolithic, or Beaker period. Rather they are, as it were, tangents, or curvilinear arrangements, framing and enclosing the basin within which Stonehenge lies. It has been argued that barrows may be so positioned that they are skylined from routes rather than being on routes themselves (Watson 2001). The King Barrow alignment cuts across the Stonehenge Avenue which is the one really clear prehistoric route in this landscape (Fig. 5.6).

Perhaps we should not exclude completely that they could represent routes tangential, or concentric, to Stonehenge by channelling access around, rather than directly to, Stonehenge (Woodward and Woodward 1996). That may not be not too fanciful if there was one correct way to approach the great monument as the avenue implies. However, in response to Bakker's point one must acknowledge that there are none of the distinctive clues to the former existence of tracks associated with the Stonehenge barrows that are convincing in the case of some of the continental alignments, or the Yorkshire Iron Age barrows discussed in the next chapter. It is, however, rather intriguing that there is a line of eleven barrows 130 m to the south of the Neolithic Stonehenge cursus. Cursus monuments and their very enigmatic relationship to movement through the landscape were discussed in Chapter 5. On Salisbury Plain east of Stonehenge some barrow groups of the early Bronze Age were linked by later Bronze Age linear ditches suggesting that the barrows may have lain on territorial boundaries (Bradley *et al.* 1994), an interpretation which is also proposed for curvilinear barrow groups in the South Dorset Ridgeway (Woodward and Woodward 1996).

On the North York Moors barrows define the watersheds and are thought to reflect seasonal grazing lands linked by deep hollow tracks to the valleys where there is evidence of Bronze Age settlement (Spratt 1982; Woodward 2000). That introduces us to another aspect of movement in the landscape relating to seasonal movement and transhumance. In the Brenig Valley of upland north Wales there is a complex of Bronze Age round barrows (Lynch 1993). Some barrows can be shown by pollen analysis to have been constructed, not just of local turves but of turves with more evidence of cereals and weeds of arable land, which are thought to have been brought from lowland agricultural landscapes. This establishes a direct connection between the seasonal pasture on the upland, where barrows were made, and the lowland of the parent settlements where farming took place. The great late Neolithic mound at Silbury Hill in Wiltshire likewise had turves at its core which came from several contrasting soil topographic and vegetation locations. Although in that case they were available relatively close to the hill, they do nonetheless provide biological evidence for the movements of people and materials in the landscape (Leary *et al.* 2013).

Conclusions

In Chapter 5 and this chapter we have seen there is clear evidence of distinct alignments of monuments in Neolithic and Bronze Age landscapes. In continental Europe the evidence mainly relates to barrows and the alignments seem generally to be of Neolithic origin and continue into the Bronze Age. In Britain the alignments are most clearly defined in the Neolithic in relation to ritual monuments but they sometimes include barrows. In the British Bronze Age barrow alignments are frequent but not as long or with the numbers of barrows seen in Denmark and the Netherlands. In Britain there is no consensus as to their significance. In Britain and to a less pronounced extent in continental Europe many ideas have been proposed for the origins of these alignments, they may relate to routeways, cosmology, sequencing of monuments may be related to religious practice or genealogies of those buried, or some combination of these factors. In the Netherlands, north Germany and Denmark it is widely accepted that the barrow alignments often mark routeways, although it is not argued that all, or even most, barrows are on routeways. There is persuasive evidence to support this hypothesis, including the evidence for linear grazed landscapes preceding barrows on the Veluwe where there are also aligned pre-Iron Age cart tracks. At Flintbek, Germany there are wheel marks associated with the barrow alignment. In Britain we have noted cases where barrows do seem to represent routeways, but there is not such clear evidence as in continental Europe. However, we will return to this question in Chapter 8 which looks at the very much more abundant evidence for routeways in middle Bronze Age to Iron Age agricultural landscapes with fields in Britain. Where trackways are defined by flanking ditches or field banks it is possible to identify many more examples of earlier barrows on the same alignments which suggests that some of those routes are of earlier Bronze Age or in some cases Neolithic date.

8. Trackways in later prehistoric agricultural landscapes

Introduction

We have seen in earlier chapters how hints and fragments of prehistoric routeways can be reconstructed from various sources, such as recurrent use of the same locations, alignments of monuments and palaeoenvironmental traces. Only in wetlands are routes clearly archaeologically visible and datable and only in rare cases can those be related to what was happening on dry ground. The situation is very different in later prehistory when we have extensive agricultural landscapes with droveways and fields, which makes it easier to identify routeways. As regards longer distance routes it has generally been assumed that prehistoric travel was mainly along the hill crests, the so-called ridgeways (Fox 1923). A particular challenge is to establish whether ridgeways do actually represent early routes, and in many cases, as we will see below, the evidence remains quite ambiguous.

An additional problem is that in dryland situations we have very little evidence for purposeful construction involving deposition of stone, sand, gravel etc to form a hard surface which would have created stratified sequences with potential for dating. Often early trackways are negative features formed by the erosion of bedrock by traffic. When deeply incised these are called hollow ways, or sunken ways. They were created by the erosive effects of travel along the route by people, animals, wheeled vehicles etc. Hollow ways, once established, also become accentuated by the erosive effects of rainfall runoff, and incision may often cut away any evidence of earlier routes along the same line. Over 50 years ago Bowen (1961, 61) identified the urgent need to establish how hollow ways can be dated and what light they throw on the possible existence of ridgeways; that challenge has, until now, gone largely unheeded.

In Britain, Neolithic and early Bronze Age landscapes are characterised by funerary and ritual monuments, particularly barrows. A more domestic landscape of settlements, fields, trackways and linear boundary ditches is widespread from the middle Bronze Age. After this there is more direct evidence for trackways associated with settlements and fields, sometimes as surviving earthworks where there has been little or no later ploughing. Where later ploughing has occurred these patterns can sometimes be recovered by air photography, geophysics or open area excavation. Recent large-scale, developer-funded excavations, often related to infrastructure projects and aggregate extraction, reveal great tracts of prehistoric landscape in which spatial and stratigraphic relationships between datable features, such as ditches and banks, provide evidence of the existence and date of routeways. Problems of dating may also be overcome by careful retrogressive landscape analysis, the observation of morphology, spatial relationships between features and strategic excavation. Examples will be outlined to demonstrate that the study of early routes is not as intractable as the sparse literature suggests. This chapter considers how tracks in prehistoric agricultural landscapes can be recognised and dated and develops this by examination of examples from various parts of the British Isles. Chapter 10 develops this picture with more detailed consideration of routeways in and around the Weald in south-east England. Figure 8.1 shows the locations of the trackway sites discussed in Chapter 7 and this chapter.

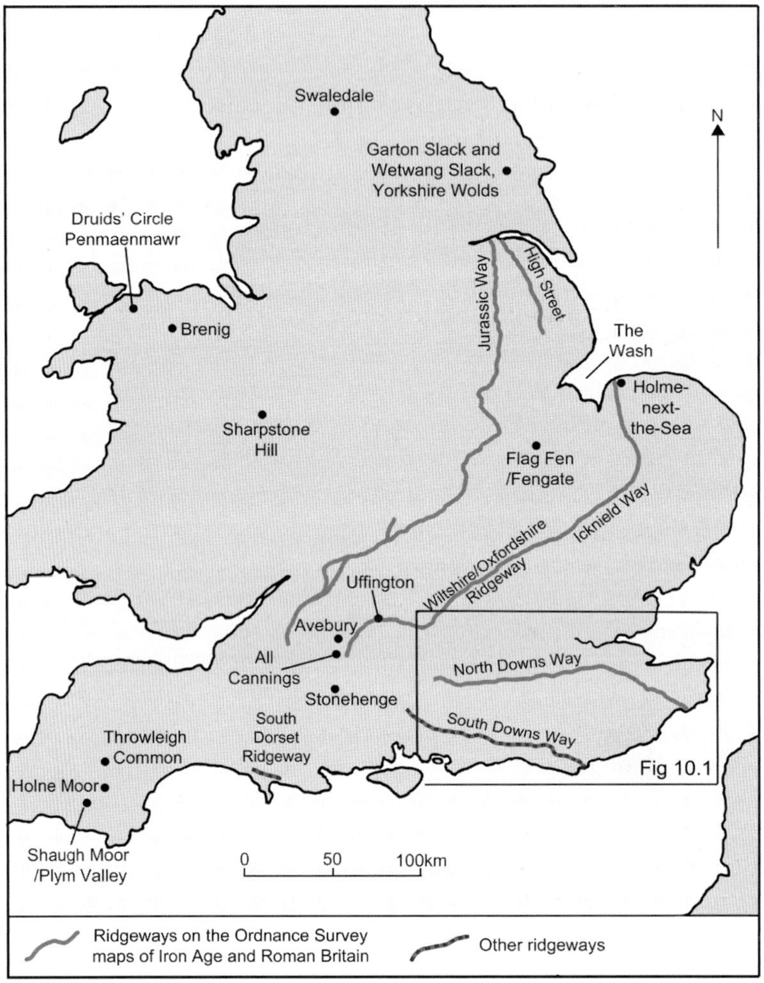

Figure 8.1 Map showing the distribution of sites in England discussed in Chapters 7 and 8. It also shows the supposed ridgeway routes, distinguishing those on the Ordnance Survey maps of the Iron Age (1962) and Roman Britain (1956) and other ridgeways. South-east England is shown in more detail in Figure 10.1 (graphic: J. Foster).

Recognising tracks in agricultural landscapes

Before considering individual case examples it is helpful to consider some of the ways in which trackways in agricultural landscapes may be recognised and to define some key terms. This is a basic first step towards a typology of trackways in agricultural landscapes without stone walls. Those with stone walls are clearer and more generally understood. The types are shown on Figure 8.2, which illustrates archaeological examples in isometric plan and below this in section. Table 8.1 provides reference to actual examples illustrated in this chapter, Chapter 10 and in other publications.

Tracks are often delimited by two parallel ditches, or in upland areas stone walls, marking a drove way, the ditches designed for drainage and both ditches and walls to keep animals on the track away from fields (Fig. 8.2a). Occasionally a metalled surface may be present, but usually this has been eroded by later cultivation; it is more likely to survive in environments of alluvial or colluvial deposition. Trackways may also be defined by linear strips free of other archaeological features; in the hypothetical example on Figure 8.2b definition is by the edge of quarry pits on the right and buildings aligned at right angles on the left. Trackways may also be represented by relatively slight linear hollows, the result of small-scale erosion (Fig. 8.2c). Sometimes a hollow track may be buried by later sediment (Fig. 8.2d1) and sometimes traces of earlier buried routes may survive at the sides (Fig. 8.2d2). On steep slopes and on more easily erodible bedrock much more dramatic hollow ways are created, sometimes many metres deep (Fig. 8.2e). Hollow ways are often enhanced by the erosive effects of rainfall demonstrated by observations during storms and meandering erosion gullies on their floor. Where there were ploughed fields alongside a trackway, cultivation often gave rise to erosion and deposition processes within a field which create linear features called lynchets (Bowen 1961) along the trackway edge. A positive lynchet is an accumulation of eroded soil called colluvium at the downslope edge of a field, in this case abutting a track. Positive lynchets can be easily recognised because there is a more level terrace-like strip along the top of the bank interrupting the general slope of the topography. A negative lynchet is formed by the erosion of soil on the upslope part of a field. This occurs where, as soils thinned, the plough cuts into subsoil, or bedrock, again producing a more level terrace strip, in this case running along the downslope edge of a trackway. Where a trackway is marked by a positive lynchet upslope and a negative lynchet downslope, it is known as a double-lynchet trackway (Fig. 8.2f). Far more frequently than is generally appreciated, hollow ways are enhanced by positive and/or negative lynchets as shown on Figure 8.2g. Tracks may also utilise level surfaces created by earlier field terraces (Fig. 8.2h). Figure 8.2i illustrates a more complex example where a droveway defined by parallel ditches has been succeeded by a larger linear bank and ditch representing a land allotment boundary.

Figure 8.3 further illustrates relationships between routeways, fields, landscape topography and colluvial sediments. In addition to lynchets bounding individual fields, the writer's observations on the ground have identified linear colluvial deposits which are associated with specific topographic contexts where cultivation perennially ceased and sediment accumulated, because, for instance, land became too steep or wet. It is not unusual to encounter such contexts where little or no trace of associated fields has survived subsequent agriculture. Two such contexts were originally identified by the writer at plateau edges and alluvium/floodplain edges (where sometimes they can be mistaken for, or mask, river terraces; Bell 1981, fig. 5.1). In the present context a third type should be added, Trackway edge colluvium: where soil has accumulated on the uphill side of a linear boundary created by a track. Where such deposits have the previously noted level step on their

Table 8.1. Hypothetical types of later prehistoric trackways in Figure 8.2 linked to illustrated examples in Chapters 8 and 10 and other publications.

	Type of track	*Site example*	*Fig. no*	*Published example*
a	Droveway between ditches or walls	Heathrow T5	8.8	Lewis *et al.* 2010
		Dartmoor	8.6	Fleming 1988
		Peacehaven	10.5	Hart 2015
b	Track marked by strip without features (eg in settlement)	Danebury		Cunliffe & Poole 1991, fig 4.1
c	Terrace Way	Wilmington		Curwen 1954, pl. xxxii
		Ports Way	10.6–7	Curwen & Curwen 1926
d	Hollow way (shallow, d1; buried d2)			
e	Hollow way (deeply incised)	Saddlescombe	10.7c	Boardman 2013
		All Cannings	8.5	Boardman 2014
		Rother Valley	10.10	Boardman 2014
		Shepton Beauchamp & East Coker		Boardman 2014
f	Double lynchet trackway	Thundersbarrow	10.8	Curwen 1933
		Plumpton Plain	10.2a	Holleyman & Curwen 1935
		Bullock Down	10.4	Drewett 1982
		Ports Way	10.6–7	Curwen & Curwen 1914; 1926
		Itford Bottom	10.2b	Bell 1983
		East Dean Woods	10.9	Manley 2016
g	Hollow way flanked by double lynchets	Lyminge	10.13	Bell *et al.* forthcoming
h	Terraceway on lynchet	Saddlescombe	10.7d	Curwen & Curwen 1914
1	Droveway succeeded by linear ditch boundary	Yorkshire Wolds	8.7	Stoertz 1997; Fenton-Thomas 2005
		Danebury		Palmer 1984

surface, extend lineally well beyond individual fields, and the track is not itself on either a positive or negative lynchet, it may reasonably be inferred that the trackway is contemporary with, or earlier than, the colluvial terrace, rather than something slotted into a landscape of existing fields.

Observation suggests that in the case of hollow ways cross section sometimes provides clues to their origins. A V-shaped hollow is formed by people or animals in single file, perhaps shepherd's tracks; a wider U-shaped form perhaps represents larger herd animals, maybe cows or pigs; and U-shaped forms with a flattish base and sometimes wheel ruts are those produced by the passage of vehicles. Many of course served multiple purposes.

Dating tracks in agricultural landscapes

But how *might* we date these tracks? As noted in Chapter 1, attempts to date trackways have been rather few and too often there have been an assumption they are undatable. On the contrary, provided we are sensitive to the problems of interpretation and the precise meaning of the dates obtained, there are dating techniques we can apply. Retrogressive landscape analysis using old maps

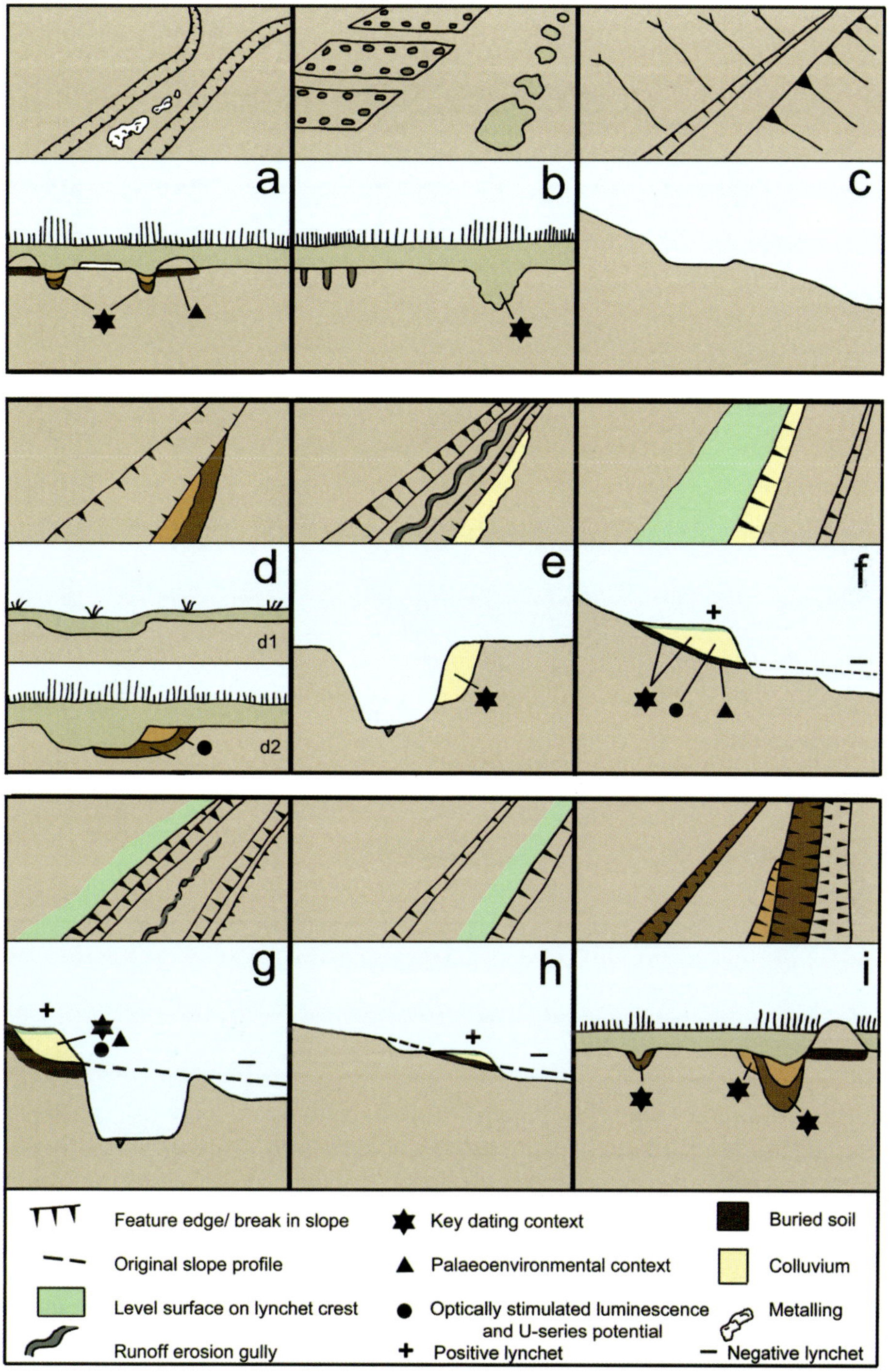

Figure 8.2 Sketches of types of evidence for trackways in agricultural landscapes. For details of actual illustrated examples see Table 8.1 (graphic J. Foster).

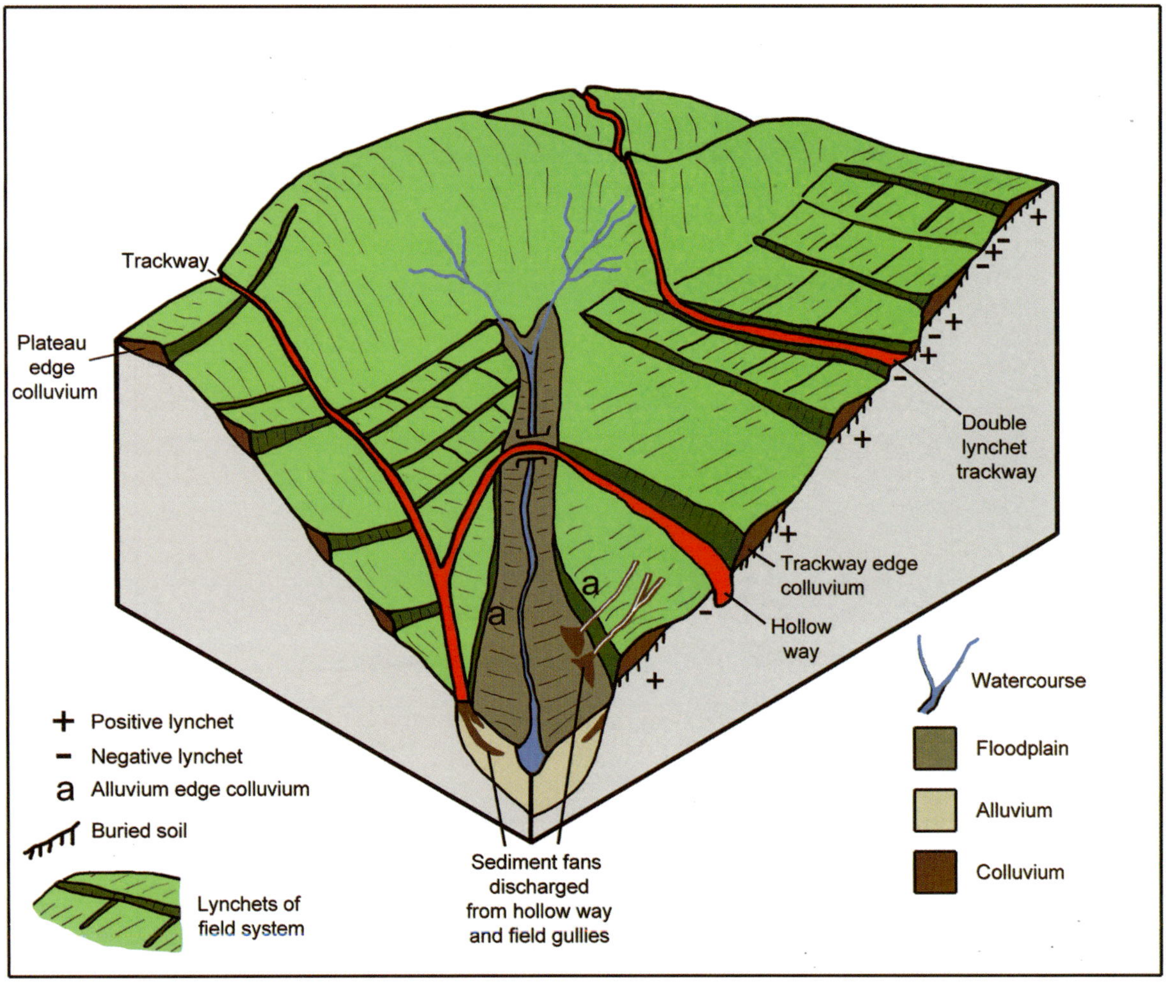

Figure 8.3 Isometric diagram showing some typical relationships between colluvial deposits, lynchets, trackways and hollow ways (graphic: J. Foster).

unaffected by recent changes can provide important evidence for the relative dating of trackways. Relationships between routes may show, as in Figure 8.4a, that one route (A) is continuous whilst others dog-leg (B) and are likely to be later, provided there is no evident topographic explanation. Figure 8.4b shows an early oval woodland enclosure at Rowfold, West Sussex round which a route-way and a long linear landholding curve suggesting both are later features (Chatwin and Gardiner 2005). Figure 8.4c shows a more complex case in Hertfordshire in which east-west coaxial tracks from higher ground to the River Lea are cut across by and predate Roman Ermin Street. Cheshunt Park is later than both and within the park a pre-Roman ditch has been excavated on the same alignment as the coaxials, which may therefore be prehistoric (Bryant *et al.* 2005). Continuity of land boundaries may be suggested where, for instance, excavated Romano-British boundaries align with surviving boundaries (Rippon *et al.* 2015). There are many cases where the spatial relationships

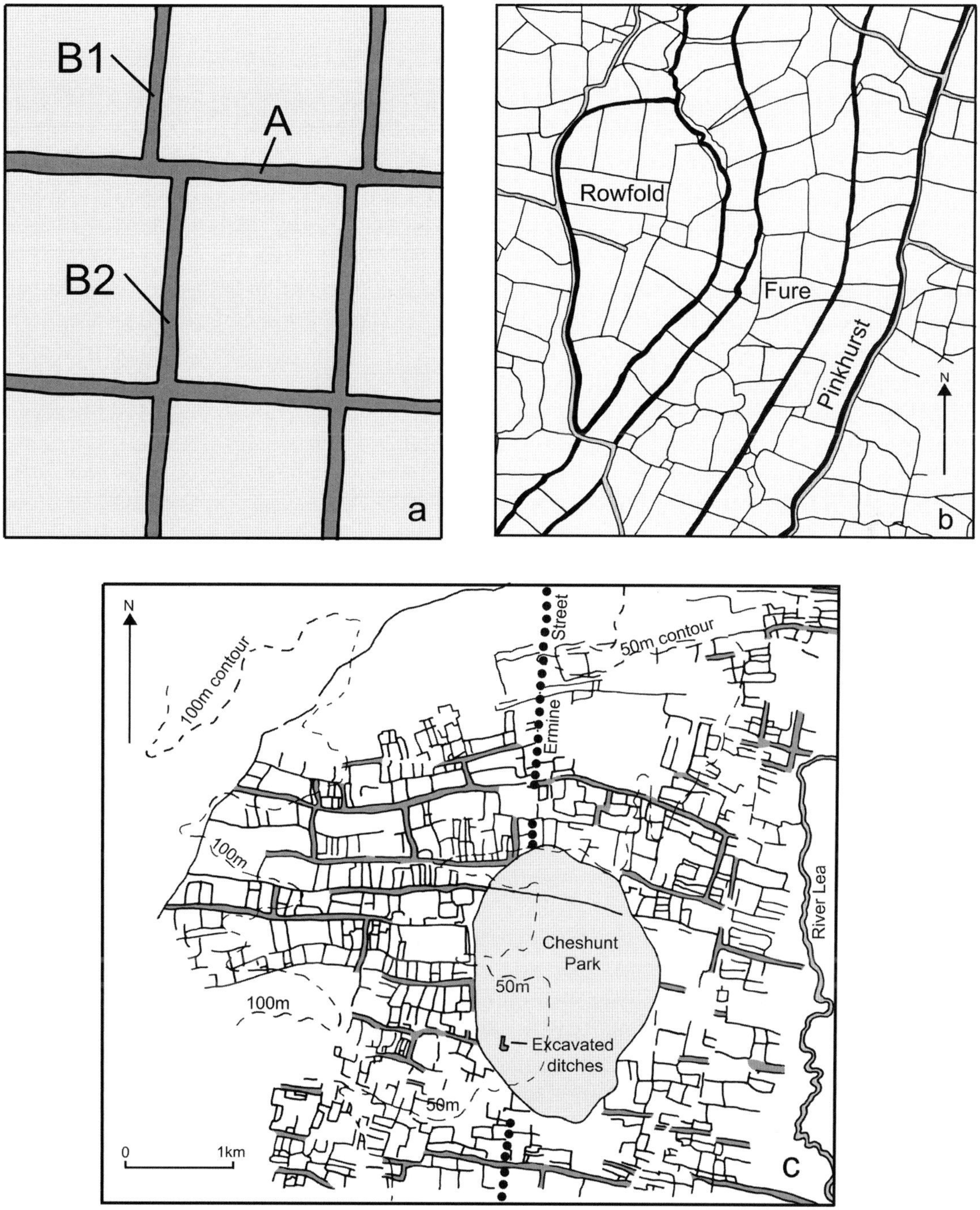

Figure 8.4 Retrogressive landscape analysis as a guide to relative dating: (a) hypothetical case in which doglegs indicate A pre-dates B; (b) oval enclosure at Rowfield, West Sussex which pre-dates north–south routes and long linear landholdings (after Chatwin and Gardiner 2005, fig. 4); (c) East–west coaxial routes pre-date Roman Ermine Street at Cheshunt Park, Hertfordshire (after Bryant et al. *2005, fig. 3) (graphics: J. Foster).*

between trackways and sites of various types may indicate date but the strength and significance of those relationships requires critical evaluation.

Also applicable to dating trackways are a range of methods that have been previously applied to the dating of fields. First two concepts need to be introduced: *terminus post quem* (TPQ or date after which), meaning in this context that the trackway is later than the date obtained. Secondly, *terminus ante quem* (TAQ or 'date before which'), meaning in this context the trackway is earlier than the date obtained. Perhaps the most straightforward means of dating is when a trackway is integral to a settlement or a field system which has been dated, either typologically, or by excavation. For instance, a trackway may be integral with a hut circle, a hillfort, or a field system or, as in the case of Figure 8.2b, be delimited by datable features such as quarry pits. Artefacts such as pottery and metalwork can also assist with dating. In these contexts one has to be aware that some artefacts might be residual, or reworked. However, colluvial deposits often contain large numbers of artefacts, many probably derived from manuring, and if each of these objects is plotted on a section then a good indication of the periods when the colluvium accumulated can be obtained; this is especially so in areas where the sequence of pottery types and fabrics is well understood because sites of many periods have been investigated. This technique has been successfully applied to the dating of many colluvial valley sediments (Bell 1983; Allen 1992) and in a few cases noted below it has been used to date double lynchet trackways. Radiocarbon dating is applicable where organic material such as charcoal or bone is found in a functional relation to a stratigraphic context, i.e. there is clear evidence that it is not residual or reworked. An example would be a defined patch of burning in the fill of a hollow way or below an associated positive lynchet. Also applicable is Optically Stimulated Luminescence dating (OSL; Duller 2008; Rhodes 2011). OSL has been very successfully used to date prehistoric and historic colluvial sediments derived from agricultural fields in Germany (Lang and Wagner 1996; Fuchs and Lang 2009). The writer is not aware of its previous use to date lynchets delimiting or overlying trackways, but an example of this application is provided in Chapter 10 by a case study at Lyminge, Kent. That example also illustrates the application of molluscan evidence derived from apparently known introduction dates (or in other cases extinction dates) and the application of Uranium Series dating of shells (Walker 2005).

On Figure 8.2 stars mark contexts from which dating evidence may be obtained. In Figure 8.2a artefacts or OSL dates from ditch primary fills, or within metalling, may provide a date for trackway use. Artefacts or a radiocarbon date from below metalling or below a bank associated with the ditch will provide a TPQ for the track. In Figure 8.2d artefacts, OSL or biota in the fill of a track hollow will provide a TAQ. Hollow ways (Fig. 8.2c) are a particularly tricky case since they are negative features formed by erosion and lacking datable contexts. However, in the hypothetical case illustrated in Figures 8.2d and e there are earlier phases of hollow way which have filled with eroded sediment and this could be dated using the various methods outlined. Tracks and hollow ways flanked by double lynchet trackways (Fig. 8.2f and g) can be dated by using the techniques appropriate to colluvial sediments, i.e. artefact distributions and OSL; radiocarbon dating and biota may also contribute. In applying these approaches, we need to keep in mind that the dates obtained may in any case be a TPQ for the trackway. For instance, an earlier track may later be flanked by ditches or fields as a landscape becomes more intensively used. It will often be the case that trackways have been used in multiple periods, and field systems may likewise exhibit complex histories, as for instance coaxial fields on Salisbury Plain; some of Bronze Age date were slighted by later linear ditches, then re-established in the later Iron Age and Romano-British periods (Bradley and Fulford 2008). Effective dating will be based, not on one technique, but a combination of field survey, excavation and scientific dating.

Whilst trackways may be well-preserved in places, all too often their survival is over a limited distance. For this reason, identification of longer distance routes in the later Bronze and Iron Ages can be just as challenging as it was in the preceding periods. They may be clear within, or close to, a settlement, or within a group of fields, but then disappear away from the foci of activity. Settlement contexts help to identify trackways where there are linear feature-free areas illustrated in Figure 8.2b. An example is within the extensively excavated Iron Age hillfort at Danebury, UK where internal roads radiating from an elaborately fortified entrance were clearly identifiable from linear strips free of pits and structures (Cunliffe and Poole 1991, fig. 4.1). Outside the hillfort a route continues from the main entrance for some 2.3 km marked in part by a linear ditch which later formed part of a trackway (Palmer 1984). This illustrates a case where, if we start in settlement context where routes are datable, it is sometimes possible to project the evidence onto a wider landscape scale.

Agents of transformation: horses, carts and chariots

Horses, which were present in Pleistocene landscapes, are thought to have become extinct in Britain in the early Holocene. They were domesticated in the Pontic Caspian steppes *c* 4800 cal BC (Anthony 2007) and seem to have been introduced to Britain much later in the middle Bronze Age *c* 1500 cal BC (Bendrey 2012; Bendrey *et al.* 2013). Antler cheek pieces suggest that horses were being ridden and winged sword chapes indicate mounted warriors from the eighth century BC in Britain. Thus it was, from the final Bronze Age that the greater speed and distance facilitated by horse riding transformed communication. That this is likely to have been of some significance in the development of trackways is suggested by the effects of horse introduction in the American North-west *c* AD 1720 which brought about a marked change in communication routes because it became easier to ride along ridge tops and upland slopes than to traverse valley bottoms (Turner 2014, 207). Perhaps this was a significant factor in the development of ridgeway routes in Europe?

In the context of the medieval period, Fleming (2010) has discussed the significance of horse-borne communication as an instrument of elite control, identifying characteristics of routes in Devon which he sees as associated with horse elites (Fleming 2012). In mid-Wales he identifies a long-distance route, the Monks Trod, which he traced for 36.5 km across the Cambrian Mountains linking Cistercian Abbeys. On grounds of historical context he deduces that the route was probably established by *c* AD 1170 (Fleming 2009a). A characteristic of the route adopted was that it represented good going for horses with, in places, longer alternatives for use under less favourable winter conditions. Just as Fleming shows communication by horse was an important element of elite control in the medieval period, so the same argument might be applied to the increased evidence of horse riding in the final Bronze Age and Iron Age. This topic has barely been considered in Britain, although an approach which could be adopted is demonstrated by a GIS based landscape approach in Burgundy, France, which convincingly identified the routes of pre-Roman roads linking hillforts (Crumley and Marquardt 1987; Madry and Crumley 1990).

A second transformative agent in communication is the wheeled vehicle. As we saw in Chapter 6, wheels are more common and earlier in continental Europe than in Britain. The earliest surviving examples in Britain are middle and late Bronze Age. Cart burials are known from the late Bronze Age in Europe and there are cart fittings of this date from Heathery Burn, Yorkshire (Piggott 1983). Some 20 chariot burials are known from the third and second centuries BC, all but two from the Arras culture of Yorkshire (Stead 1979). There are recent finds of an earlier chariot, *c* 535–380 cal BC, well to the north at Newbridge near Edinburgh (Carter *et al.* 2010) and a very late Iron Age example in Pembrokeshire (Dyfed Archaeological Trust, pers. comm. 2019). Wheel ruts also point to the existence of vehicles in the late Bronze Age at Welland Bank, Lincolnshire (Pryor 1988b, pl. 11), and in the Iron Age at Cherbury hillfort and on a limestone causeway at Yarnton, both in Oxfordshire (Lambrick and Robinson 2009). Wheeled vehicles would have increased erosion in hollow ways, where deep wheel ruts are often seen. Mottershead *et al.* (2008) employed a geotechnical and geomorphological approach to demonstrate that deeply incised and extensive cart ruts on soft limestone in Malta were produced by two wheeled carts of gauge 1.4 m. They formed at a time of active soil erosion but are essentially undated. By the close of the Iron Age classical writers provide accounts of great battles between Britains and Romans involving many chariots, a, perhaps debatable, figure of 4000 in the case of the Iron Age leader Cassivellaunus (Piggott 1983). A single farm site at Gussage All Saints, Dorset produced moulds from the casting of an estimated 50 sets of chariot fittings (Foster 1980). If the Cassivellaunus figure is anywhere near numerical accuracy it has major implications for the later Iron Age, both in terms of the availability of horse transport and the existence of viable routeways along which large numbers of vehicles could be marshalled (Spratling 1979, 149).

Hollow ways

Hollow ways (Fig. 1.7) have fascinated many writers who recognise them as ancient routes, their origins lost perhaps in the mist of time. Some have totally gone out of use and represent lost ways of life. One such abandoned way was chosen by Geoffrey Household (1939) as the hiding place for a fictional fugitive in *Rogue Male*. That mysterious place so fascinated the writers Roger Deacon and Robert Macfarlane that they went in search of the track of Household's inspiration and found it in the Chideock valley in Devon (Macfarlane *et al.* 2012). A question to be addressed in this chapter and the next is whether hollow ways are condemned to being places of perpetual mystery, or whether we can find ways of investigating their origins. There are positive signs because increasingly hollow ways are exciting interest from a range of disciplines and internationally. Their study occupies a fertile interface between landscape archaeology and geomorphology. Even so, there are significant challenges, especially in terms of dating, due to their erosive origins.

On bedrocks which were particularly erodible such as chalk, limestone, sandstones and loess, hollow ways can be deeply incised leaving a significant mark on local topography. Frequently, as one route became deeply gullied or muddy, alternative routes were adopted, so that one finds a cluster of parallel hollow ways descending steeper escarpment slopes (Fig. 8.5). They can act as channels for temporary streams and eroded soil from arable fields during high rainfall events, as Boardman (2013; 2014) has demonstrated for some particularly deeply-incised hollow ways in the Rother valley, Sussex, which now act as drainage channels in storms.

Interest in these features is developing in several parts of Europe. In Denmark, where routeways on lowland plateaus descend steeper slopes into valleys, there are fine examples, such as those illustrated at Broskov on Zealand in Figure 6.5, where the multiple ways are both integral with a field system on the plateau and lead to successive excavated causeways across the valley (Kunwald 1962). Similarly, at Kilen in Jutland and elsewhere, Bang (2013) has used LiDAR to document multiple hollow ways

Figure 8.5 Multiple hollow ways descend the escarpment from the Marlborough Downs to the Vale of Pewsey at All Cannings, Wiltshire (photo: M. Bell).

in the form of deep V-shaped incisions in the plateau edge representing routes down to river crossing places. Several examples in Sweden are illustrated by Westerdahl (2006), some converging on river crossings. In the Ljubljana marshes of Slovenia, Cresnar (2016) has identified multiple hollow ways around the wetland basin, one leading to a crossing place of the wetland where there is evidence of Bronze Age ritual deposition. Although the hollow ways themselves are undated, some examples go through the entrances of Iron Age sites. Where areas of European loess support ancient forests, such as the Meeerdaal Forest in Belgium, Holocene erosion gullies are preserved; some are hollow ways and their distribution relates to sites of the Iron Age and Roman period (Vanwelleghem *et al.* 2003; 2006). It is also probable that earlier routeways determined the location of some of the large-scale erosion gullies which have been documented by Bork *et al.* (1998) on the Central European loess. It follows from this that disentangling the role of routeways and large-scale gullying may, more often than imagined, require a combined geomorphological and archaeological approach. On the loess of eastern Austria there are deeply-incised hollow ways (Wiesbauer and Zettel 2014). They are mapped historically and the assumption has tended to be that they date to the Middle Ages. However, LiDΛR survey of these wooded landscapes has provided a fuller picture of their distribution and relationship to sites, demonstrating that some ways lead to Iron Age hillforts (Doneus *et al.* 2011).

Coaxial fields and tracks in moorland

During the later Bronze Age extensive systems of fields, some coaxial, were established in areas of Britain. There is also increasing evidence for coaxial fields on the continental side of the English Channel (Bradley *et al.* 2015). The coaxial plan has one clearly defined long main axis, formed by parallel boundaries, from which fields are divided on a secondary axis at right angles, rather like the face of a brick wall. Figure 8.6a shows a part of the extensively preserved coaxial fields on the moorland edge of Dartmoor. Such systems are often terrain oblivious, *i.e.* the main axis does not follow topography (Fleming 1988, 112). This is opposed to aggregate field layouts characterised by more piecemeal development.

The fringes of Dartmoor exhibit the most extensive survivals of prehistoric fields in Europe; they were discovered and extensively investigated by Fleming (1988), with recent survey by Newman (2011). Extensive surveys of past field systems have also taken place on Bodmin Moor (Johnson and Rose 1994) and west Penwith (Herring *et al.* 2016). Moorlands are today areas of open and rough grazing with impoverished, acidic soils on the higher and wetter parts on which peat has accumulated. Soil acidification followed the loss of woodland, but in the middle Bronze Age the moors were less marginal than today and more intensively used. In the moors, above the present-day limits of farming, low stone banks form parallel coaxial systems with a terminal reave (stone bank) at the moorland edges (Fig. 8.6a). The parallel reaves are divided to varying extents by cross banks which define smaller fields and enclosures, particularly around hut circles which are in neighbourhood groups, as illustrated by the example on Holne Moor (Fig. 8.6a). The reaves were originally constructed *c* 1700 cal BC and appear to have mainly gone out of use by the end of the Bronze Age *c* 800 cal BC, when climate deterioration may have contributed to a reduction of the intensity of land use. It is clear from the palaeoenvironmental evidence that the reaves relate to an open environment which was mainly used for pasture, indeed footprints of cattle and sheep in a reave ditch on Shaugh Moor were noted in Chapter 4. There was some small-scale arable activity attested by pollen analysis and lynchet accumulations against some reaves (Balaam *et al.* 1982).

Despite the extensive traces of prehistoric fields of coaxial form on moorlands, they seldom exhibit the clear evidence of multiple parallel tracks noted below in lowland contexts. The moorland tracks which are visible have not been a major theme in previous surveys which focus more on the fields. Lengths of parallel walls frequently occur, particularly in relation to hut circles representing settlements, but they are often frustratingly short lengths. Occasionally longer lengths of track represented by double walls can be identified. An example highlighted on Figure 8.6a running west–east (A–B) is roughly parallel to the Venford Terminal Reave which was the edge of the open moor in the Bronze Age. Other shorter lengths of parallel walls at right-angles to the Venford Reave probably represent tracks from settlement clusters to the open moor. A rather longer track leading from settlement huts to the open moor is illustrated on Throwleigh Common, where short tracks, also leading to earlier positions of the moorland edge, are highlighted (Fig. 8.6b). Beyond the parallel reave system on open moorland other reaves along watersheds define great tracts of grazing on the open moor, for instance in the Plym valley, where, beyond the coaxial reaves, there are many enclosures and hut circles (Balaam *et al.* 1982). It seems probable that these related to seasonal grazing. Seasonal transhumance characterised Dartmoor in the Medieval period (Fox 2012). The multiple moorland fringe routes associated with this practice were called by Fox 'multiflex routes'. Fleming (2012) has established that some of the lanes used as transhumance

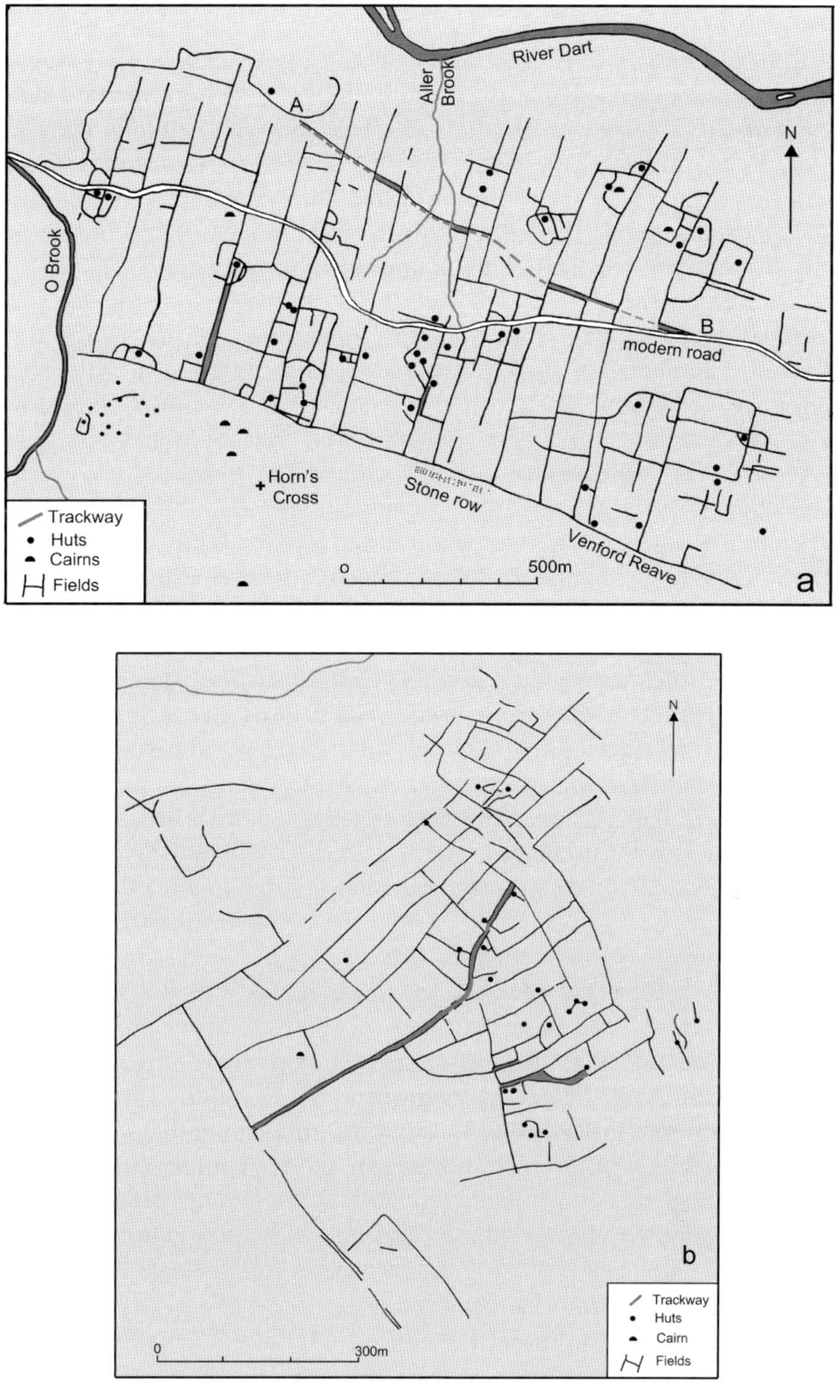

Figure 8.6 Bronze Age landscapes on Dartmoor, Devon: (a) Holne Moor: tracks through fields marked by closely spaced parallel lines, with one discontinuous route marked AB (after Fleming 1988); (b) Throwleigh Common (after Newman 2011) (graphics: J. Foster).

routes go back at least to the immediately post-Roman period, since there are inscribed stones of that date on the routes. This has also been demonstrated in West Penwith, Cornwall where parts of today's landscape of irregular small curved fields can be shown to originate at least as early as the Romano-British period (Herring *et al.* 2017). Probably later in origin are the Trods, or narrow flagstone paths, on the North York Moors (Evans 2008). Some lead to bridges of known medieval origin, some have monastic associations and one leads to a monastic grange; others have a relationship to post-medieval markets. In their present form most seem likely to be medieval or later but the possibility that some formalise earlier routes should be considered.

Yorkshire Wolds

Air photography on the chalk landscapes of the Yorkshire Wolds has revealed very extensive evidence of prehistoric landscapes in which it is possible to trace early routeways over greater distances than in most other areas (Stoertz 1997). Extensive excavations in advance of quarrying have also contributed significantly to the understanding of landscape history of this area. Figure 8.7a shows part of this landscape. A linear feature, which can be traced for more than 12 km, approximately follows the 100 m contour at the boundary between lowland to the south, which is closely divided by many ditches and fields, and, to the north, open upland grazing with some linear ditch divisions. The form of the long linear feature in places is clearly that of a trackway, indicating that at some stages in its life it was a significant routeway. It was followed by later linear boundary ditches *c* 1000 BC (Fenton-Thomas 2005); this is the type of sequence which is illustrated in Figure 8.2i. Other stretches appear more like a linear boundary ditch, demonstrating a rather ambiguous relationship between tracks and linear ditches, which is seen elsewhere, for instance at Danebury and other Wessex sites (Palmer 1984; Bradley *et al.* 1994). Less ambiguity attaches to a trackway which winds from south-east to north-west for 7.5 km. Early origins are suggested by a band of Neolithic and Bronze Age barrows along this line. By the Iron Age it is delimited in part by double ditches with land allotment boundaries laid out at right-angles. The track delimits the extensive middle Iron Age cemeteries at Garton Slack and Wetwang Slack, the latter encroaching in places on its line (Fig. 8.7b; Dent 1982). Notable in both cemeteries is the occurrence of cart burials dated *c* 450–200 BC (Fig. 8.7 d) providing evidence of the vehicles which used these routeways. Perhaps it is not by chance that some of the longest clearly traceable prehistoric routeways in the British Isles are so closely associated with the very localised distribution of Iron Age cart burials of the Arras culture (Stead 1979, fig. 9). It also indicates a relationship between mobility and status in the Iron Age of this specific area (Giles 2012). On the eastern side of Figure 8.7a there are a series of shorter parallel trackways running through enclosed settlement and field areas towards the higher ground, perhaps associated with the seasonal movement of animals from enclosed lowland to open upland grazing. Droving practices are further attested by funnel-shaped entrances leading to trackways, as illustrated by Figure 8.7c.

Coaxial fields and droveways in lowland Britain

Drawing on a wealth of evidence from developer-funded excavations, Yates (2007) has demonstrated that later Bronze Age coaxial fields are widespread in lowland areas along the Thames Valley, in coastal areas flanking the Thames Estuary and along the south coast of England. Yates has shown convincingly that these lowland fields are associated with intensive animal husbandry. Where excavation has been on some scale, droveways often lie between fields and are flanked by ditches. Often

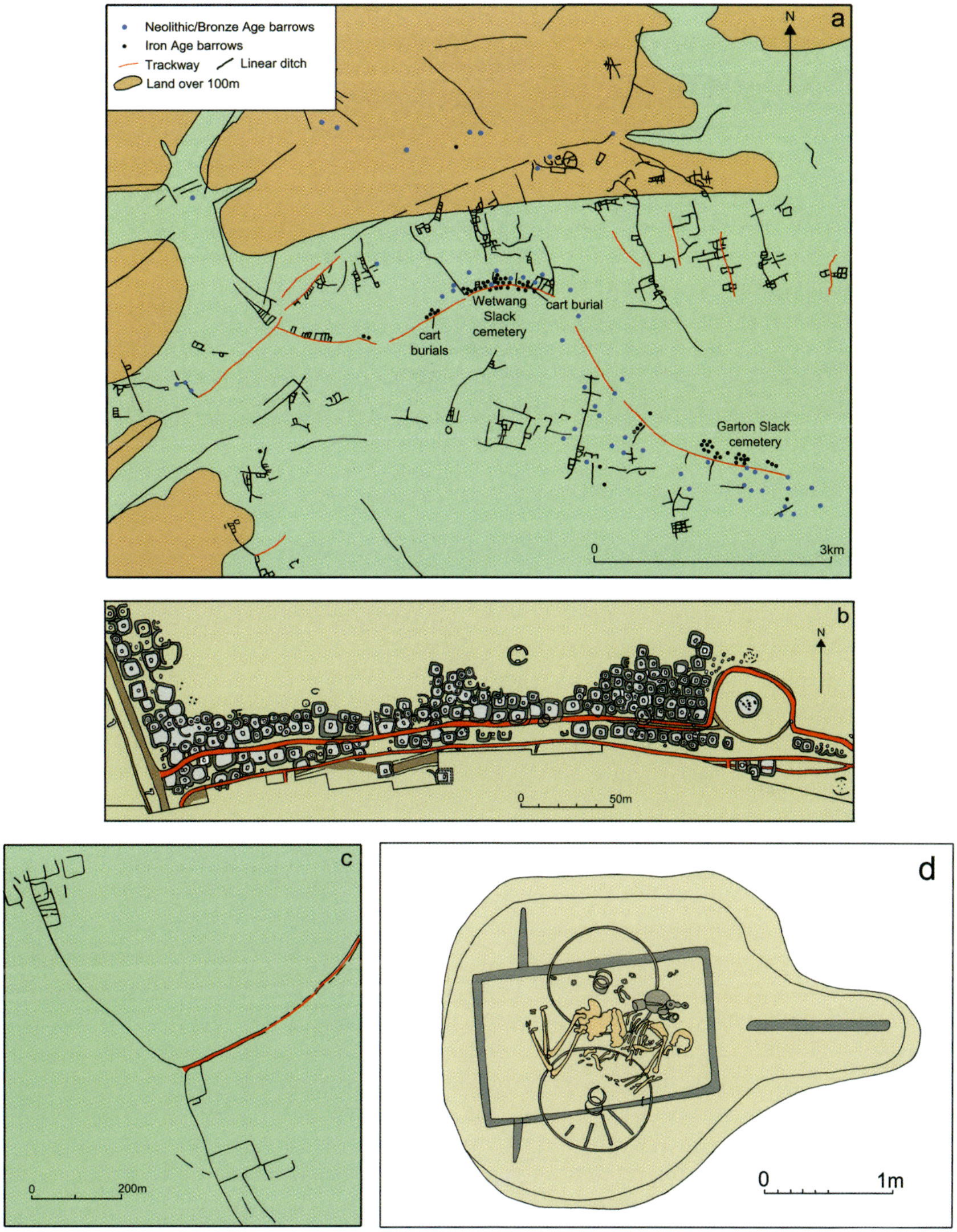

Figure 8.7 Yorkshire Wolds: (a) Trackways, linear ditches and barrows in the Wetwang Slack and Garton Slack areas (after Stoertz 1997, fig. 36); (b) Wetwang Slack Iron Age cemetery in relation to trackways and linear ditches (after Dent 1982, fig. 3); (c) Wetwang trackway with funnel entrance, location bottom left on (a) (after Stoertz 1997, fig. 23); (d) Wetwang Slack, cart burial (after Dent 1985, fig. 3) (graphics: J. Foster).

the droveways are aligned towards rivers such as the Thames, suggesting that animals were being driven down to seasonal pasture on the floodplain and saltmarshes. Gravel causeways and wood trackways carried routes of this type onto the wetland itself (Meddens 1996). An important example of coaxial fields at Fengate dated to the earlier Bronze Age has been discussed above, in the context of dryland–wetland communication routes (Fig. 6.10). An example of aligned drove ways was found at South Hornchurch, Essex where a droveway flanked by fields is beside a late Bronze Age ringwork of the ninth–eighth century BC (Guttman and Last 2000). The scale of these Bronze Age field systems is apparent from the very extensive excavation of an area 1.3 × 1 km in advance of construction of Terminal 5 at Heathrow Airport, London (Lewis *et al.* 2010). Here 12 trackways were identified, the most prominent seven were oriented north–south and spaced at intervals of between 270 m and 100 m. The trackways formed the main axis of an extensive area of coaxial fields (Fig. 8.8). Twelve farmsteads lay within the fields and there were 48 watering holes, which are very much a feature of the Thames valley fields, and, together with environmental evidence, such as dung beetles and pollen, these strengthen the view that the landscape was mainly associated with animal husbandry. The river Thames is 2 km south of this site and it is probable that animals were being driven down to floodplain pastures, just as has been suggested at Fengate and sites in the Severn Estuary (Chapter 6).

Survival of coaxial field systems

There is also evidence for extensive coaxial field systems surviving in some areas in the pattern of fields today. Whether some of these originate in the previously described coaxials of the later Bronze and Iron Ages remains to be established. One of the clearest examples of surviving coaxials is in the Scole–Dickleburgh area of East Anglia where a landscape of fields and tracks documented from early maps appears to be cut across by the Roman road from Colchester to Norwich (Williamson 1987). Environmental evidence from the adjoining Diss Mere indicates extensive clearance and agriculture from the Iron Age (Peglar 1993). In the Lea Valley, Hertfordshire coaxial fields are bounded by long east–west tracks running down to the River Lea and pre-dating Roman Ermine Street. Excavation of a ditch suggested the system originated in the pre-Roman period (Fig. 8.4c; Bryant *et al.* 2005). Identification of a coaxial pattern of roads and tracks on the Chiltern Hills (Bull 1993) may have early origins as suggested by the discovery of an Iron Age track at Aston Clinton below a later parish boundary (Lambrick and Robinson 2009). At Shapwick, Somerset the boundary of a surviving coaxial field system was shown to directly overlie medieval, Roman and Bronze Age ditches (Aston and Gerrard 2013). Whilst not directly related to documented coaxial systems, parallel droves have been mapped connecting the chalk downs and lowland in south east Hertfordshire and Cambridgeshire (Harrison 2003) and extensively in the Wealden landscape investigated as a case study in Chapter 10. The first detailed synthesis of the extent to which Romano-British field orientations survived into recent landscapes came to the very significant conclusion that 64% of the excavated Romano-British field systems studied shared a common orientation with medieval landscapes, though the proportion varied regionally (Rippon *et al.* 2015). Furthermore, pollen analysis showed that the oft postulated abandonment of agricultural land and woodland regeneration at the end of the Roman period was far less than generally assumed, in the south east of England, attested by only a 5–6% increase in woodland.

To date, discourse about coaxial landscapes has been largely framed around the fields themselves, but there are signs of an emerging view that it is not so much the individual fields that

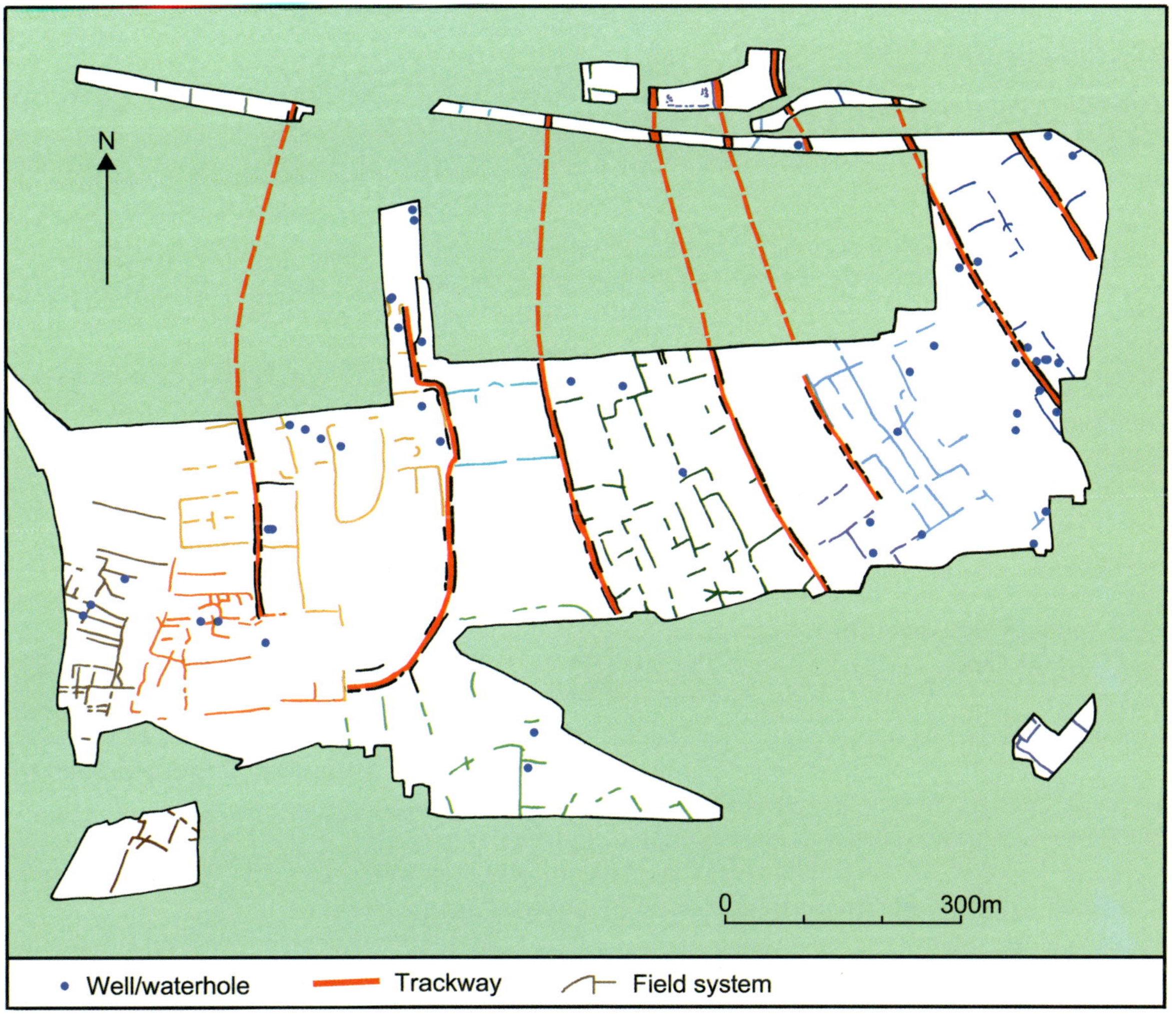

Figure 8.8 Bronze Age landscape of fields and trackways revealed by excavations at Heathrow Terminal 5 (graphic: J. Foster after Lewis et al. *2010).*

demonstrate continuity, as the systems of parallel droveways connecting environmental zones which are the key to landscape structuration (Williamson 2008). Within these patterns some individual field boundaries show continuity and there are others which clearly do not (Rippon *et al.* 2015). What is emerging as controversial is the extent to which continuity of tracks and fields may be interpreted in terms of social continuity of land-use practices, such as common property rights, commons and field system organisation, for which Oosthuizen (2011; 2013) has argued. Rippon *et al.* (2015, 320) do not accept the arguments for long-term continuity of social practices. It could be argued that the existence of bands of landscape offering complementary resources, connected by drove ways, in places deeply incised as hollow ways, created structures within which similar practices of resource connectivity and lesser transhumance might have been practised, whether or not there was any actual continuity in the social arrangements which articulated these practices.

Ridgeways

This brings us to perhaps the most contentious topic in the study of prehistoric routeways, the so-called ridgeways. These are supposed routes along the crests of escarpments and ridges, particularly those of chalk and limestone, in lowland England. They were regarded by early prehistorians as routeways. Some were marked on the Ordnance Survey maps of Iron Age and Roman Britain (Ordnance Survey 1956; 1962) on the basis that they were 'natural routes' marked by concentrations of sites (Fig. 8.1). Those marked were: the Wiltshire/Oxfordshire Ridgeway and its continuation into East Anglia to the Wash as the Icknield Way; the North Downs Way to Dover; the Jurassic Way, shown as running from the Cotswold Hills to the Humber; and a route called the High Street along the Lincolnshire Wolds. Many of these are now long-distance walking paths of great recreational importance. In Chapter 1 it was emphasised that routeways play a seminal role in how people encounter and perceive particular landscapes, so it follows that the long-distance ridgeway paths are an important aspect of a twenty-first century walker's encounters with landscape, whatever the history of their origins. Many have acquired a sort of iconic status providing a veneer of great antiquity and were beloved by many of the landscape writers celebrated in Chapter 1. The poet Edward Thomas (1916) wrote on the Icknield Way. Stroud (2017) has described the magnetic attraction of the Oxfordshire Ridgeway and its Uffington White Horse. Hilaire Belloc (1904) wrote lyrically of the North Downs Way; illustrations in his book include a map of six ridgeways which in a very general way he saw as converging on Stonehenge. Their assumed prehistoric origins are, however, open to debate. Sherratt (1996) regarded them as a myth, proposing instead the significance of riverine transport, a theme returned to in Chapter 9.

A key factor which originally led to their identification as routes was that they seemed to be associated with concentrations of prehistoric monuments such as barrows, hillforts and other settlements along the ridges. At the time it was assumed that the lowlands and clay vales remained wooded and impenetrable. The situation changed when air photography, and increased archaeological survey and excavation, demonstrated there were equally dense concentrations of barrows, prehistoric ritual sites and settlements in the surrounding lowlands, such as the Thames valley north of the Oxfordshire Ridgeway (Hey *et al.* 2011). Grimes (1951) discussed the Jurassic Way across England from Mendip to the Humber but frankly does not produce any convincing evidence for its former existence. Taylor (1979) showed that the part of that route which crosses Northamptonshire was neither a 'natural' route, since it is crossed by multiple wet valleys, nor one that was associated with a particular concentration of prehistoric sites, once the many newly discovered sites in the lowlands and valleys were taken into account; consequently, the existence of the Jurassic Way in prehistory seems especially questionable.

Where we have environmental evidence from barrows, both long (Neolithic) and round (mainly Bronze Age), they were constructed in open short grassland environments with little evidence for scrub and woodland (Evans 1972). For such an environment to have been maintained, active grazing must have occurred. Thus, environmental evidence strongly supports Fleming's (1971) view of the ridge crests as upland pasture and it seems reasonable to infer that concentrations of barrows are found in ancestral grazing lands.

Case Study: the Wiltshire and Oxfordshire Ridgeway

Two recent landscape archaeology projects have contributed to an understanding of the history of the Ridgeway: one in Oxfordshire, the other in Wiltshire. The Oxfordshire project was focused on the Iron Age hillfort of Uffington which is right next to the present Ridgeway as Figure 8.9a shows (Miles *et al.* 2003). Two kilometres west of Uffington, beside the Ridgeway, is the Neolithic tomb at Wayland's Smithy (Fig. 8.9c), which has often been cited as evidence for the route's early origins. Dominating the Vale below the escarpment and just 160 m from the hillfort is the giant hill figure of the White Horse of Uffington which OSL dating showed originated in the late Bronze, or early Iron Age (Miles *et al.* 2003). The hillfort originally had two opposed entrances, so any route along this line may have passed through that axis; however, one entrance was blocked *c* 400 BC. The landscape south of the escarpment is dissected by a series of north-south dry valleys with the result that the Ridgeway can be seen as a 'natural route'. Acceptance of the Ridgeway as a route gains support from computer modelling of least-effort paths of movement, which did nonetheless involve certain assumptions about sticking to higher ground (Bell and Lock 2000). The computer model shows that the least cost path passes through three of the four hillforts on the ridge and within 80 m of the fourth, that at Uffington.

However, early origins for the Ridgeway are called into question by a linear ditch of probable later Bronze Age date which runs up to the hillfort from the south at right-angles to the Ridgeway (Fig. 8.9a). Excavation showed that the ditch was infilled by the Romano-British period and that the present Ridgeway, which crosses it here, is not earlier than this date (Miles *et al.* 2003). The project concluded that there was no certain evidence for the prehistoric origins of the Ridgeway; however, a route could have taken somewhat different line in some periods, e.g. through the hillfort, or to its north. Other prominent features of this landscape are the hollow ways, which descend the steep escarpment and form north–south routes to the clay vale below; these are undated. Multi-isotope studies of herbivore teeth from sites on the Ridgeway and the clay vale to the north demonstrate that sheep were mainly raised on the chalk, whilst cattle spent their first two or three years in the isotopically-distinct vale but subsequently spent time on the chalk (Schulting *et al.* 2019). That could be consistent with a model of seasonal movement of older cattle up the hollow ways to the downs in summer but if so they were not accompanied by sheep.

One hollow way curves round the slope behind the back and tail of the White Horse (Fig. 8.9b) raising the intriguing possibility that what inspired the horse image were the curving lines of an early chalky hollow way showing up against the green of the escarpment. Either way, the horse symbolism is highly significant given the transformative effect of horse riding in facilitating far greater mobility from the late Bronze Age. It has furthermore been proposed by Pollard (2017) that the white horse, which is clearly running from east to west, reflects the mobility-related sun horse imagery, depicted on razors and rock art, which is such a feature of Bronze Age Scandinavian cosmology (Chapter 9).

The second project which has proved important for an understanding of the Ridgeway is on Overton Down, Wiltshire. Here the Ridgeway runs along the top of the escarpment 1.7 km east of

Figure 8.9 Uffington hillfort, the Ridgeway, hollow ways and the White Horse: (a) plan (graphic: J. Foster after Miles et al. *2003); (b) the White Horse, Uffington hillfort and hollow ways (left) descending the escarpment (photo: C. Speed); (c) the Ridgeway near Wayland's Smithy with Uffington hillfort in the distance (photo: M. Bell).*

the great late Neolithic henge at Avebury (Fig. 5.5). A detailed survey based on air photographs of the field systems and settlements of Overton and Fyfield Down, and selected excavation of settlements and fields, though not the Ridgeway itself, enables the Ridgeway to be seen in the context of its wider landscape (Fowler and Blackwell 1998; Fowler 2000). The Sanctuary on Overton Hill lies beside the Ridgeway and along it running north for 800 m is a line of barrows. However, detailed mapping of the early landscape shows that the present Ridgeway line overlies 'Celtic' fields and must have been established after these went out of use in about the fifth century AD. The present line appears to have been established by about the tenth century when the Ridgeway formed the boundary of several parishes. Four kilometres to the south of Overton the Ridgeway was in existence when the long distance defensive dyke, Wansdyke, was laid out in the post-Roman period (fifth–tenth centuries), because there is a defended gateway in the Dyke where the Ridgeway crosses at right-angles (Fowler 2000; Reynolds and Langlands 2011). The Ridgeway as currently defined does not accord with the track systems of the prehistoric and Romano-British landscape which Fowler (1998; 2000) identified, at Overton and Fyfield, but he argues that rather than one ridgeway route, landscape analysis reveals multiple, roughly parallel routes forming an axis of movement. He suggests that this is of at least Romano-British date and probably had its origins in later prehistory with the seasonal movement of animals from the lowland to the south up to the high downland in summer (Fowler 1998).

The relationship between the Ridgeway and the great Neolithic Henges remains enigmatic. The West Kennet Avenue (Fig. 5.5) leads from Avebury to the Sanctuary on the present ridgeway, but there is no obvious connection (e.g. entrance) linking the two. The east entrance to Avebury leads up the Herepath crossing the Ridgeway and continuing along the north flank of the Kennet Valley. The great sarsens used to construct Avebury are likely to have travelled down this route from Fyfield Down where the greatest concentration of sarsens is found today. It has also been assumed that the sarsen stones used at Stonehenge may have come from the Overton and Fyfield areas. If so, they are likely to have been transported 30 km south along a continuation of the Ridgeway across the Vale of Pewsey to Salisbury Plain. Figure 8.5 shows one of two places where multiple incised hollow ways descend into the Vale of Pewsey in the general direction of Stonehenge. The other hollow way route descends from a prominent col between the Neolithic tomb of Adam's Grave and the causewayed enclosure on Knap Hill.

The Icknield Way

The Wiltshire-Oxfordshire Ridgeway continues east through Buckinghamshire, Cambridgeshire and into East Anglia as the Icknield Way, a name sometimes applied to the entire route. Together the Ridgeway and the Icknield Way are probably Britain's most widely acknowledged 'prehistoric' route. Fox (1923) described it as a superb 'natural' routeway with many barrows and other archaeological sites along its line. Today, however, it is appreciated that archaeological sites are also very abundant in the surrounding lowlands, for example along the Fen edge to the north (Taylor 1979, fig. 17). In Norfolk the Icknield Way runs to the east of the Fenland; it is paralleled by another route, the Pedders Way, which was a road in the Roman period (Fig. 8.10a). Neither

a

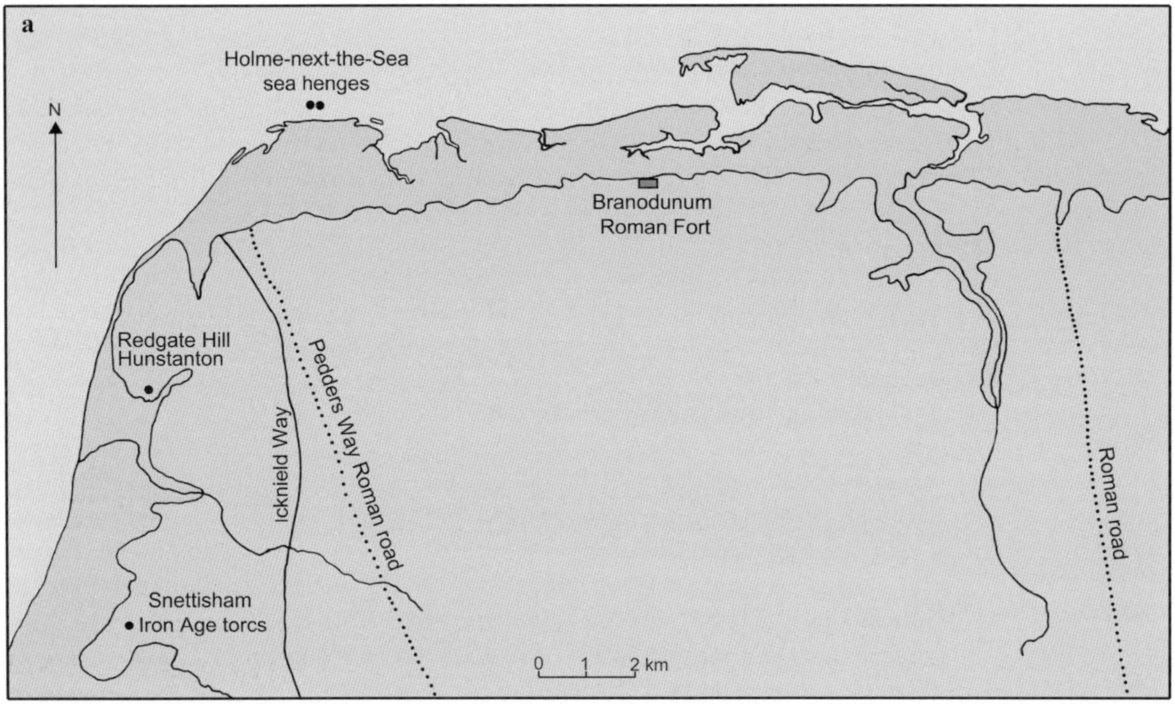

b

Figure 8.10 Holme seahenges in relation to the line of the Icknield Way: (a) Map of the seahenges in relation to early routes (graphic: J. Foster); (b) Seahenge on the foreshore (photo: Francis Pryor).

has been the subject of very detailed field investigation. The Icknield Way and Pedders Way converge on the North Norfolk coast close to Holne-next-the-Sea which is a natural crossing point by sea over the Wash (Fig. 8.1). At Holme two circular wood structures known as 'sea henges', together with trackways and other wood structures, have been revealed by coastal erosion on the foreshore (Fig. 8.10b). Both are dated dendrochronologically to 2049 BC (Brennand and Taylor 2003; Robertson 2016). The henges were erected in a saltmarsh environment. It seems possible that the location of the sea henges reflects the landward end of a significant routeway, but whether this represents the long-distance route envisaged by previous generations, or a more localised route is open to question.

Recent writers have grown more sceptical of the Icknield Way's prehistoric reality. In Buckinghamshire cross dykes of the first millennium BC cut across the ridge (Bryant and Burleigh 1995). In Cambridgeshire the bundle of Ridgeway routes seem to postdate apparently defensive dykes of the post-Roman period and Harrison (2003) calls into question the very existence of a long distance route in prehistory. Post-medieval maps suggest an escarpment foot route of the Icknield Way was at that time more significant and at Aston Clinton that route has been shown by excavation to post-date features of Iron Age to Anglo-Saxon date.

The origins of Roman roads in Britain

The Roman roads of Britain were comprehensively reviewed in a pioneering study by Margary (1955). They were mapped on the *Ordnance Survey Map of Roman Britain* (1956) and the network is constantly being extended by new discoveries. They are one aspect of past routeways which has long been the subject of quite lively archaeological research, in contrast to prehistoric routes. Roman roads are readily recognisable since they run more or less straight between towns and other sites, are generally flanked by ditches and have well-constructed metalled surfaces. The more important routes are described in written sources: the *Antonine Itinerary*, the *Tabula Peutingeriana* and the *Ravenna Cosmology* (Jones and Mattingly 1990). The general assumption is that they represent a radical, efficient and entirely new imposition on the British landscape, an imposed system of elite control.

Roman roads themselves are outside the prehistoric scope of this book but a pertinent aspect, which has received little attention, concerns the extent to which some of those routes may have appropriated pre-Roman antecedents. Bishop (2014) in reviewing the Roman roads of Britain argues that although we know virtually nothing of the pre-Roman road network, it is inherently likely that the invading Romans made at least some use of existing routes on arrival. The last two decades of research have been notable in demonstrating the extent to which some of the trappings of a Roman way of life were introduced as a result of contact and trade in the century before the conquest in AD 43 (e.g. Creighton 2005). *Oppida* settlements developed and were defended by systems of ditches. Excavations have revealed that within the *oppidum* at Silchester a system of laid roads and buildings was established from the last two decades BC (Fulford 2012). A series of *oppida* existed in lowland Britain and, given the importance of trade and continental contact in the genesis of these sites, it seems probable that they would have been connected by roads which are likely to have been on a similar line to the Roman roads which connected towns that developed from *oppida*. The dykes round the Silchester *oppidum* seem to have had entrances in

Figure 8.11 Sharpstone, Shropshire: Roman road overlying roads of Iron Age date (graphic: J. Foster after Malim and Hayes 2011; photo: T. Malim).

the same position as the later Roman gates, suggesting roads radiating in the same directions. A late Iron Age mirror burial dated 75–25 BC 2 km south of Silchester (Fulford and Creighton 1998) lies close to the point where roads to Winchester and Chichester, both pre-Roman foundations, diverged. At the Winchester end of one of these roads, 48 km to the south-west, the Roman road enters the town through an entrance which was also that of the Iron Age defended site of Orams Arbour which was on a different plan to the Roman town (Cunliffe 1965). Within Orams Arbour a hollow way has been excavated linking this entrance to another. It is argued that the site originated in the middle Iron Age to control a network of routes at a point where there was a river crossing (Ford and Teague 2011; Ottaway 2017).

The Roman road between Colchester and *Verulamium*, which both originated as Iron Age *oppida* and became Roman towns, has been argued to be a development of a late Iron Age route (Witcher 1998). Copeland (2009) reviewed this question in the case of the Roman Akeman Street between the late Iron Age *oppida* at *Verulamium* and Bagendon but without identifying any conclusive evidence that the road had Iron Age origins. Roman roads also meet at Iron Age hillforts, at Badbury Rings and Old Sarum, suggesting the possibility of earlier origins (Bishop 2014). Rackham (1986) advocates a pre-Roman origin for many routes in areas of ancient British countryside, in support of which he draws particular attention to Wool Street which, close to Cambridge, is a typical Roman road, whereas 14 km out it no longer has a typically Roman form and its route is less direct; this, he maintains, is suggestive of an earlier route, only part of which was adopted and improved by the Romans. In cases where the end points of a late Iron Age road appear to be known it should not be impossible to identify by survey or excavation traces of an early route between them.

Evidence for the pre-Roman origins of a Roman road is most clearly demonstrated by excavation in advance of quarrying at Sharpstone, west of Wroxeter Roman town (Malim and Hayes 2010). Several phases of earlier tracks were dated using a combination of radiocarbon, Optically Stimulated Luminescence and Bayesian statistical analysis of the stratified dated sequence (Fig. 8.11). The phases of road construction in the last two centuries of the Iron Age from about 200 cal BC involved initially brushwood and later sand, pebbles and gravel. The Iron Age road was straight, had been carefully engineered and was well maintained. Phases of road construction were interleaved with colluvial sediments. On the surface of one phase were wheel ruts, and another phase had spherulite particles indicative of herbivore dung and dung beetles, indicating that animals were moved along the route. Cattle footprints below the earliest brushwood phase indicated it originated with animal droving. An underlying post was dated to the Bronze Age suggests the possibility of still earlier origins.

Whilst Sharpstone was perhaps unexpected in terms of its remoteness from the main late Iron Age elite centres, there is evidence that it is not an isolated example. In this and earlier chapters we have identified several cases where earlier routes appear to be succeeded by Roman roads. In some cases, these may justifiably be described as 'natural' routes, because they are by far the most suitable way to access an area. One example is the Roman road linking forts up the Usk Valley, Wales which we have suggested originated as a Neolithic route (Fig. 5.2). The presence of Roman forts also carries with it the implication that it was also a significant route in the Iron Age. Another example is the main east-west Roman road across upland in north Wales at Tal-y-fan which follows a route marked by Bronze Age monuments (Fig. 7.4). There is also the Roman Fen Causeway which crosses the Flag Fen embayment at the same natural crossing point as the Flag Fen late Bronze and Iron Age post alignment (Fig. 6.10). Other examples are less obviously 'natural' routes. A Roman road descending the North Downs escarpment at Whitehorse Stone has been argued to be on the same alignment as a probable Neolithic route (Fig. 10.11).

Conclusions

With the appearance of extensive agricultural landscapes of fields and settlement in the middle Bronze Age evidence for trackways becomes far more extensive. So much so that in this chapter we have had to focus down and concentrate on the evidence from Britain with only occasional references to elsewhere in Europe. The various ways in which trackways may be preserved, recognised and dated in relation to fields have been identified. The potential transformative effects on routeways of the horse and wheel have been suggested but this is a topic awaiting investigation in prehistory. Two types of routeway have been given special emphasis in this chapter. One is the hollow way, important because once deeply incised they are virtually permanent features of the landscape and are likely to have influenced subsequent long-term human mobility. The second main topic has been coaxial field systems which in lowland and river valley areas provide evidence for multiple parallel routes linking areas of contrasting landscape. These survive in the present-day landscape and some may have had prehistoric origins.

As for the Ridgeways, on which archaeologists pinned their faith for generations, we have seen there is surprisingly little evidence of prehistoric use. Burials and historical sources show that some, such as the Wiltshire-Oxfordshire Ridgeway and the Pilgrims Way in Kent were established routes by mid-Saxon times (Reynolds and Langlands 2011). Notwithstanding these uncertainties there remain grounds for anticipating longer distance routes from at least the later Bronze Age, as the evidence for wetland causeways/bridges discussed in Chapter 6 suggests. Evidence for wheeled vehicles from the late Bronze Age and Iron Age strengthens that inference. However, to date there is surprisingly little evidence for prehistoric routes covering any distance in Britain, compared with the evidence reviewed in Chapter 7 from the Netherlands, north Germany and Denmark. That this is a puzzling conundrum is highlighted by growing evidence for the movement of artefacts, people and animals, the last two especially attested by recent isotopic evidence. The next chapter argues that riverine and maritime communication provides a partial solution to this conundrum.

9. Maritime and riverine connectivity and the allure of the exotic

Introduction

Preceding chapters have found limited evidence for long-distance prehistoric communication routes on dryland, particularly in the British Isles. Long accepted prehistoric origins for the ridgeways have been shown to be limited and questionable. Thus far the focus has been on movement and routes in terrestrial landscapes, but that is only part of the story. Terrestrial routes will have been linked by boat on rivers and the sea and may often only make sense when considered in that wider context. In some areas, and during some periods, it is likely that most journeys were by water. Archaeologists have tended to treat movement on land and by water as separate specialist categories whereas the two must often have been complementary, as Westerdahl (2006) has argued in a Scandinavian context. Most early journeys would have been along the coast and people would have navigated using many of the markers which we considered in a terrestrial context: cairns, standing stones, local clearings, modified trees; all, along with natural features, are likely to have played a part and connected maritime with terrestrial routes.

The purpose of this chapter is to complete the story by highlighting the complementary role of transport by water. The comparatively brief treatment here is justified in two ways. Firstly, in relation to riverine transport, although many log boats are known, there has been comparatively little synthetic treatment of the role which rivers and lakes played in movement. This is a challenging topic beyond the scope of this book. However, one or two case studies provide pointers to possible ways forward. Secondly, in relation to maritime transport, the situation is almost the opposite; there has been much recent research and synthesis which has advanced understanding significantly, so what is required here is to signpost this work and consider how it may have linked to patterns of riverine and terrestrial movement.

The advantages of water in facilitating rapid movement over distance are evident; however, such practical issues are not the only reasons prehistoric communities were interested in watery places. They were also sometimes places of ritual and symbolic significance, liminal places where ritual deposition took place in many periods from the Mesolithic to the Iron Age and later. Certain rivers are particularly associated with the deposition of selected artefact types such as weapons and also human skulls, as for instance in the Thames during the Bronze Age and Iron Age (Bradley 1990; 2016). The deposition of metalwork in association with bridge and trackway crossing places has already been noted in the context of Flag Fen and several similar sites in the UK and La Tène in Switzerland (Chapter 6; Schulting and Bradley 2013). On the Atlantic fringe and the Baltic travel by sea was the connecting factor. In much of continental Europe the major river arteries such as the Danube and the Rhine would have served as the main conduits. Thus there is a need to extend recent advances in the understanding of maritime connectivity to riverine environments.

Riverine transport

Riverine movement is not particularly easy to identify. We have to consider the character of the river and its suitability for transport at relevant periods in the past. Many channels may have been very different prior to the human modifications in recent centuries. Increasingly their past character can be reconstructed using geoarchaeological techniques (Brown 1997). We can be reasonably confident of the major rivers being navigable at most times, whilst minor rivers may only have been seasonally navigable. In some cases it may be the river valley itself which acted as corridors of movement rather than riverine transport. Where drainage was disrupted by the effects of glaciation, or the activities of beavers (Coles 2006), extensive early Holocene water bodies would have facilitated local movement, although, in the more minor rivers and streams, beaver dams may have blocked routes, so that easily carried lightweight canoes were more suitable than dugouts fashioned from trees. There are finds of dug-out canoes demonstrating water transport but only rare occurrences of artefacts in them providing evidence for their purpose, or where they came from.

This highlights the importance of making the connection between dry ground and wetland landscapes. Such links have been most effectively demonstrated in the Stonehenge landscape, as discussed in Chapter 5. The Avenue at Stonehenge leads down to the River Avon leaving no doubt about the relationship here between a route and riverine movement. This case has been significantly strengthened by the recent discovery of a short avenue leading from the entrance to Durrington Walls henge down to the same river. It has generally been considered that the bluestones at Stonehenge completed their journey from west Wales on the Wiltshire River Avon; the much larger sarsen trilithon stones could also have completed their journey from the Marlborough Downs on the same river. Both journeys would also have involved overland tracts and, in the case of the bluestones, transport by sea, probably up the Bristol Channel, then the Bristol River Avon as the least risky route (Parker Pearson 2012).

The recent evidence from the Stonehenge landscape demonstrates the perspicacity of an earlier paper by Sherratt (1996) which encouraged prehistorians to consider the role of rivers as arteries of communication in prehistory. Sherratt argued that the rivers were much more important than the ridgeways which archaeologists had previously emphasised, but which he regarded as a myth. Sherratt particularly emphasised the role of the Wiltshire and Bristol Avon rivers in providing a key communication route across southern Britain. He also suggested that both rivers may share the name Avon, the Celtic word for river, because the corridor that they together provide was seen as part of one conceptual entity. Coles (1994) has extended this argument by proposing that other river names such as Trent and Tarrent, also of Celtic derivation, may point to their role as routes in later prehistory; several such names occur on the Avon/Severn route. The distribution of amber artefacts of early Bronze Age date particularly highlights the key role of maritime connections into the Poole Harbour area and riverine transport up the Wiltshire Avon to Stonehenge (see Fig. 9.5).

Log boats

Many types of boat will have proved suitable for river and lake transport under certain conditions. Log, hide and bark boats can be inferred to have been used in the Mesolithic and Neolithic periods, though only log boats are represented in the prehistoric archaeological record and these are likely to have been used principally for riverine and lacustrine travel, as indeed the findspots of most examples indicate (McGrail 1978; Mowat 1996; Gregory 1997). More sophisticated forms extended, or improved, the shape with separate fitted transoms, bows and plank extensions at the side to create larger, more complex craft. A database of European log boats compiled by Arnold (1995; 1996) comprises some 2400 examples. Many are old discoveries of the nineteenth and twentieth centuries and do not survive. Scientific dating programmes provide radiocarbon dates for 571 examples and dendrochronological dates for 58 (Lanting 1997–8). The earliest dugout in Europe seems to be from Pesse, Netherlands, dated 8010–7510 cal BC (Menotti 2012). Excavations of a major waterlogged Mesolithic site on the edge of a river dune at Hardinxveld De Bruin, Netherlands revealed a dugout at the dune edge and dated *c* 5400 cal BC (Fig. 9.1; Louwe Kooijmans 2001b). The greatest concentration of Mesolithic dugouts is in Denmark where they are known from 15 sites,

Figure 9.1 Dugout boat under excavation at the Mesolithic site of Hardinxveld De Bruin, Netherlands (photo: Betuweroute project).

all of them later Mesolithic in date; wooden paddles are also recorded from 14 sites (Fig. 9.2b–c). This distribution suggests that the dugouts were important not just in riverine communication but along the coast and between the mainland and islands of the Baltic. A dug-out from the submerged settlement at Tybrind Vig (Figure 9.2a; Andersen 2013) had a clay hearth, which recent folk practice suggests was used in night fishing to attract eels. Tybrind Vig also produced richly decorated wooden paddles (Fig. 9.2c) highlighting the role which boats are likely to have played in social communication both within and between communities. Excavations in advance of the fixed link bridge and tunnel across the Storebaelt, Denmark revealed fragments of 19 dugouts in a mass of drifted wood, much of it derived from fishtraps, at Magrethes Naes, Zealand; these dated between *c* 5440 and 4950 BC (Pedersen *et al.* 1997).

From the British Isles the evidence for early logboats is much more limited. There is an early Mesolithic paddle from Star Carr but that could have been used on a boat of hide or bark. Two Mesolithic log boats are known from Ireland, at Brookland dated 5407 BC and at Lough Neagh dated 5490–5246 BC, and there are three Neolithic examples from Ireland (Gregory 1997; McErlean *et al.* 2002). In the rest of the British Isles there are no certain Mesolithic dugouts, although two old finds from Scotland at Friarton on the river Tay and the river Carron were described respectively as coming from under, and within, a Carse clay deposit which was laid down in the Mesolithic.

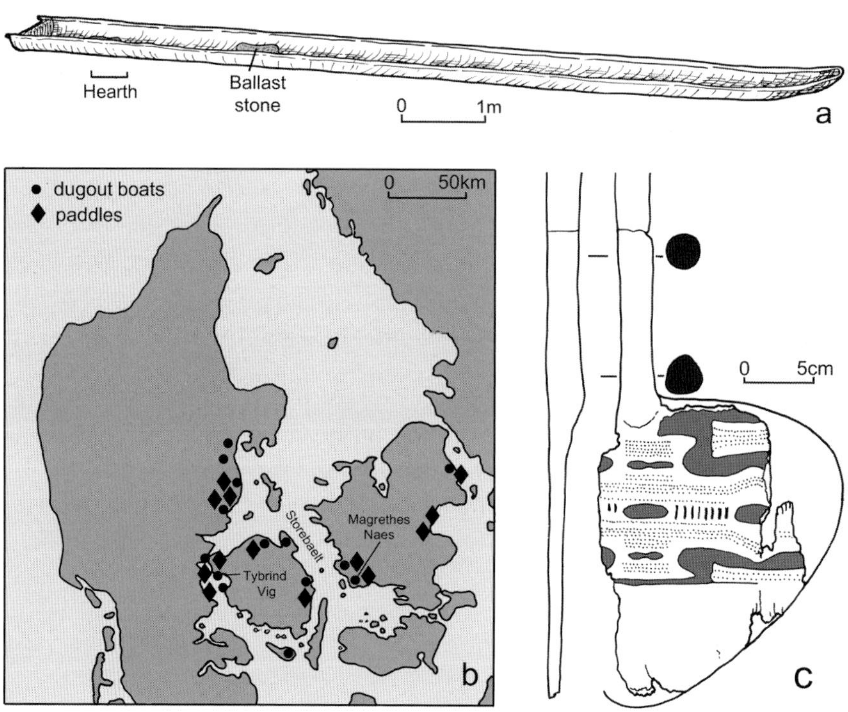

Figure 9.2 (a) Reconstruction of a Danish Mesolithic dugout canoe of the type found at Tybrind Vig, Denmark; (b) the locations of dug out canoe finds and paddles in Denmark; (c) decorated paddle from Tybrind Vig (graphics: J. Foster, all after Andersen 2013).

However, without direct dating this remains uncertain. Three Scottish log boats are dated to the Bronze and Iron Ages. No Neolithic log boats are known from England where the earliest is early Bronze Age. Several British early Bronze Age log coffin burials are boat-shaped and interpreted as either metaphorical, or reused, dug-out boats, for example the burial of Gristhorpe Man in Scarborough, Yorkshire (Melton *et al.* 2013). Boat-shaped stone settings and graves also occur, mostly close to the coast in Scotland with some round the Bristol Channel; these are argued to highlight the significance of connectivity by water in the early Bronze Age (Bradley 2018). Six log boats dated between 1300 and 700 cal BC have recently been found in a palaeochannel at Must Farm in the English Fenland, near a timber platform settlement and with evidence for some very well-preserved basketry fishtraps (Gibson *et al.* 2012; Knight and Murrell 2012).

Although experiments have shown that log boats can be used at sea under certain conditions (Dunkley 2014), in practice their use is likely to have been restricted to the more sheltered locations. We may suppose log boats were the general workhorse of communication in many inland waters. At Shardlow a log boat dated 1430–1190 cal BC was used to transport large blocks of sandstone and was found beside a causeway of wood and sandstone in a palaeochannel of the river Trent (Garton *et al.* 2001). An especially magnificent dugout, which provides some pointers to its use, is the Hasholme boat from Yorkshire (Millet and McGrail 1987). It was of sophisticated construction and was fashioned from a 800 year old tree, 14 m long and 1.4 m in diameter. It was capable of carrying a crew of 20, or, with a crew of five, could carry 5.5 tonnes of cargo. Its final cargo included sides of beef and timber. The boat had a carved oculus or eye, a feature seen on boats of many periods especially in the Mediterranean.

Hide boats

Sea-going boats are described in written Classical sources relating to Iberia, Brittany and Britain from the later first millennium BC (Cunliffe 2001a; McGrail 2004, 181); they were of wooden frames covered with animal hides. Representations in the form of artefacts dated to the Bronze and Iron Ages mean we can be confident that such craft existed. The Caergwrle bowl is a wetland find from north Wales; it is an elaborate oval bowl fashioned from Kimmeridge shale, decorated with sheet tin and gold apparently depicting shields or sun motifs along the side, with oars and a wavy sea; it is dated *c* 1200 BC (Green 1985; Rednap 2011). Though the boat attribution may seem speculative, a clinching piece of evidence is the presence of pairs of oculi at either end. These eye motifs, seen also at Hasholme, are found on boats of many periods. A second artefact, the Broighter sheet gold boat from Ireland, is dated by associated objects to the first century BC. This certainly represents a boat which, from its oval shape and curved profiles, was hide and probably seagoing. It was propelled by nine oars on either side and a steering oar is also represented; of special significance is the presence of a mast demonstrating the use of a sail (McGrail 2004). Its proportions suggest a boat maybe 12–15 m long. Hide boats known as currachs have been used in areas such as the Aran Islands in the west of Ireland into recent times, demonstrating their suitability for challenging sea conditions (Synge 1979). McGrail (2004) is of the view that hide boats were more seaworthy than the types of prehistoric plank boats discussed below. From Nors in north-west Jutland, Denmark there are 100 miniature gold leaf canoe-shaped boats, some decorated with solar symbols. Their form is suggestive of skin rather than log boats (National Museum, Copenhagen, Denmark). A much more modest type of hide boat, which has survived to the present day in Wales, is the coracle, traditionally a bowl-shape of woven wickerwork with a hide covering and seat for one person, which was used on rivers for fishing (Jenkins 1988). One such fisherman, Mr Edward Thomas, crossed the English Channel in a coracle in 1974 in 13.5 hours (Dunkley 2014) but their historical use was in riverine and very sheltered estuarine locations. A key point is they were lightweight, easily carried on a person's back, and are thus illustrative of the intimate connection between dryland and water mobility. An oval, bowl-shaped burial with traces of organic lining from Fife, Scotland has been interpreted as a hide coracle reused for burial in the early Bronze Age (Watkins 1980).

Sewn plank boats in the British Isles

In the Bronze Age and Iron Age, the one boat type which is very well represented in the archaeological record is the sewn plank boat. In all probability, these were the vessels which provided the connections between cultural groups both in North-west Europe and the Baltic, the means by which metals and other commodities of great social significance were obtained. The archaeological evidence for these boats contrasts geographically: in England and Wales there are finds of the boats themselves mainly of Bronze Age date, while in Scandinavia there is very abundant evidence for Bronze Age boats in rock art and images on metalwork but the only boat finds are of Iron Age date. We will consider first the boats themselves, then the Scandinavian iconographic evidence.

The sewn plank boats are typified by the most complete example found at Dover in 1992 and now magnificently displayed in its own gallery in Dover Museum (Clark 2004a). There is another group of finds from the river Humber, especially the discoveries from Ferriby made by the pioneering boat archaeologist Ted Wright from 1937 onwards (Wright 1990). These boats are sophisticated and complex pieces of carpentry (Fig. 9.3a and b) which demonstrates the importance of maritime communication during the Bronze Age. The boats were made from hewn planks sewn together with withy ties (twisted thin roundwood) through individual holes on the planks. The ties were generally rebated into the wood to reduce abrasion on beaching. The planks have internal raised ridges and semi-circular cleat bosses which were perforated for transverse timbers providing lateral strength. The joints between planks were caulked with moss and covered with laths to create a more effective seal between planks. None of the sewn-plank boats has produced any evidence for a mast, sail or fitments for oars and they are presumed to have been paddled.

Eleven or 12 sewn plank boats are known in the British Isles (Fig. 9.3c): 4 or 5 from Ferriby, 1 from Brigg and a fragment from Kilnsea, all in the Humber wetlands; 3 fragments of 2 or 3 phases from Caldicot, and 2 planks of the same date from Goldcliff, both in the Severn Estuary; the two-thirds complete boat from Dover; and a fragment from Testwood Lakes inland of the Solent. The dates of these finds are shown on Figure 9.3d which demonstrates they are mainly Bronze Age, with fragments from Ferriby of Iron Age date. The earliest are very close to the beginning of the Bronze Age. Van de Noort (2011) has argued that such complex carpentry became possible with Bronze Age technology and that the sewn plank boat was of Beaker origin, associated with accelerated maritime contacts which occur at this time, although the earliest radiocarbon dates are a few centuries later (Fig. 9.3d). As reconstructed, Ferriby 1 is the largest of the sewn plank boats, 15.4 × 2.6 m and 0.7 m high, a boat capable of carrying about 3 tonnes including a crew of up to 12. When complete the Dover boat would have been *c* 11 m long and was 2.2 m wide and capable

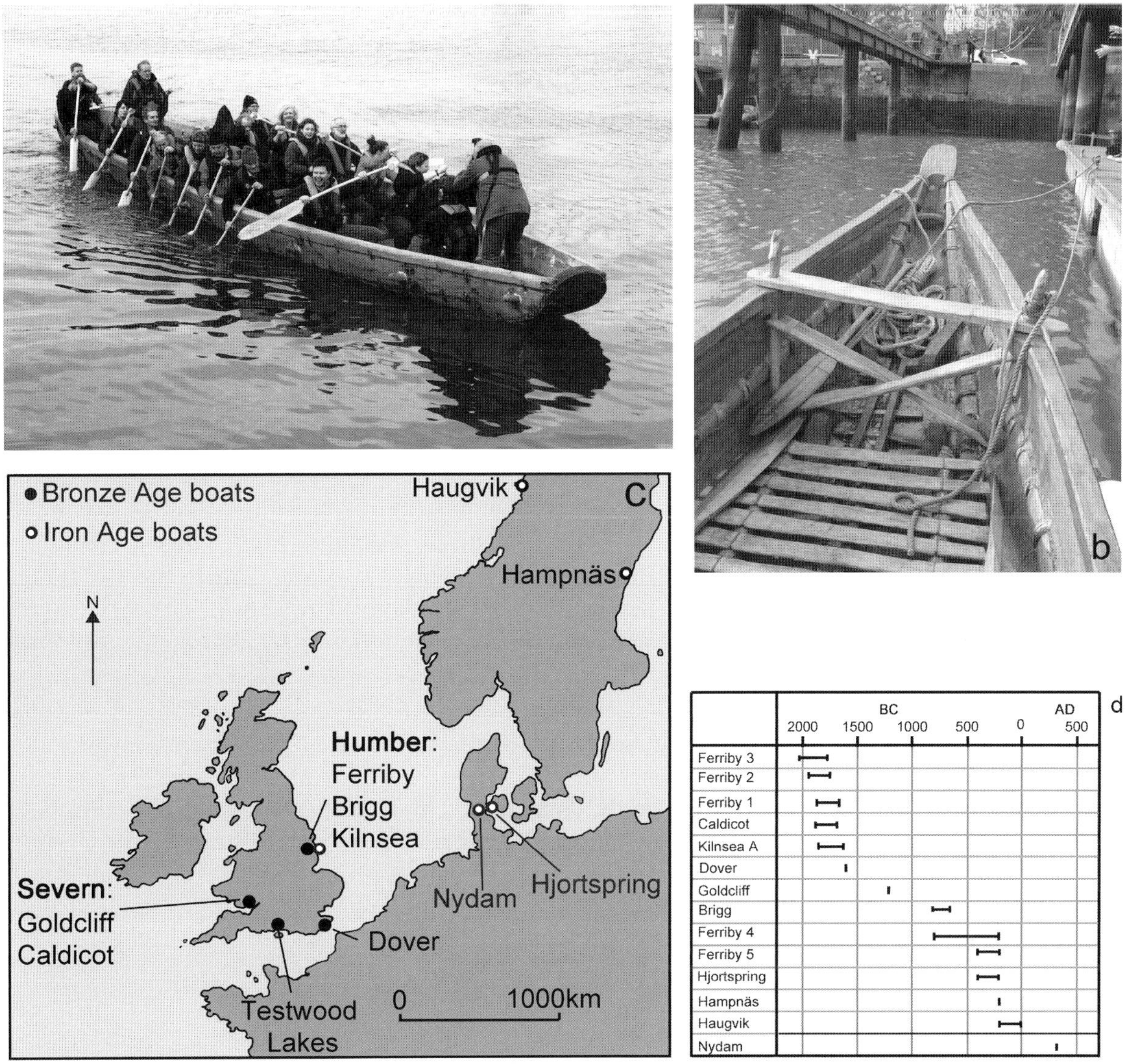

Figure 9.3 Sewn plank boats: (a) a fullsize reconstruction of the Ferriby boat built at the National Maritime Museum, Falmouth (photo: P. Marsden courtesy of Prof R. van de Noort); (b) Half size reconstruction of the Dover Boat (photo: M. Bell); (c) the distribution of prehistoric sewn plank boats (and Nydam); (d) the dates of North-west European sewn plank boats (and Nydam) (graphics c–d: J. Foster).

of carrying 2 tonnes. It was calculated that with 16 paddlers it could travel at 5 knots and cover up to 56 km a day.

There have been conflicting views as to the environment in which the sewn plank boats were used. McGrail has argued that they were boats for river and estuary use but not seagoing, a purpose for which he favoured the use of hide boats. The locations in which they were found are mostly estuarine, with Dover, Brigg and Testwood being riverine (Van de Noort 2011, fig. 40) but the finding of a boat in a river need not indicate where it was habitually used. Environmental evidence showed that the Dover boat lay in freshwater conditions yet the presence of glauconitic sand and beach-rounded marine shells within the boat suggest it travelled along the coast (Green 2004), whilst a fragment of Kimmeridge shale from the boat may indicate longer sea journeys some 220 km west as far as Dorset. The River Dour on which it was found is such a minor drainage of *c* 6 km that it scarcely seems to have justified such a boat and the same can be said of the Nedern Brook where the Caldicot fragments were found, at the tidal limit in this inlet of the Severn Estuary (Nayling and Caseldine 1997).

Understanding of the capability of these boats has recently been significantly advanced by the making of two experimental reconstructions. A half scale replica of the Dover boat was launched in 2012 (Fig. 9.3b) and has been trialled in waves in the outer harbour at Dover (Lehoërff *et al.* 2012). A full-size replica of a Ferriby boat (Fig. 9.3a) was made in 2012 at the National Maritime Museum, Falmouth and has undergone trials in Falmouth Harbour (Van de Noort *et al.* 2014). Though neither boat has yet been trialled in the open sea, mainly due to modern health and safety and training issues, those responsible for the trials are confident of the boats' sea-going capabilities. It is reasonable to suppose that Bronze Age people would have chosen the time to travel very carefully and that journeys would generally have been made under the calmest conditions of summer. Even so, journeys in the open sea would have carried a high risk to set against the substantial rewards in terms of material goods and status for those that succeeded. Boats, and the distant contacts which they facilitated, can be seen as having an pivotal social role by providing access to exotic raw materials, especially the metals which were only available in very restricted areas. Production of the most prestigious artefacts involved the combination of alloys and raw materials available in very different locations. In that sense long distance communication was central to Bronze Age society. Voyages would also have provided access to exotic and esoteric knowledge and are likely to have been rites of passage for young elite crews. Van de Noort (2011) has argued that the sewn plank boats were seen as such powerful symbols on account of their exotic contacts that boat pieces were ritually incorporated in structures such as bridges and tracks at Caldicot 2, Testwood Lakes and Goldcliff.

Possible wrecks round Britain

Support for the transport of exotic goods by boat may be provided by two collections of Bronze Age metalwork from the sea bed which are both interpreted as the sites of wrecks (Needham *et al.* 2013). Both are from high energy exposed sea beds, so no trace of a boat itself can be expected to survive. Both collections of metalwork date to the thirteenth century BC on metalwork typology. The most productive of these sites is at Langdon Bay just outside the east harbour wall at Dover, 0.5 km from the present shore and just 2.25 km east of the find spot of the Dover Boat. The find comprises some 360 bronzes, many of them types which are normally found in continental Europe, though some are of types thought to have been made in Britain (Lehoërff *et al.* 2012; Needham *et al.* 2013). The second collection is from Salcombe, Devon and was made 450 m seaward of a rocky cliffed shore. It comprises 29 objects, mostly bronzes but also gold, tin, iron and bone.

Landing places in Britain

Making connections between waterborne and terrestrial routes is made more difficult by the fact that the dugouts, skin boats and other lightly constructed boats of prehistory could have been beached at any gently shelving, sheltered location with no need for a landing stage. No recognisable dock arrangements are represented in North-west Europe until the Roman period. Nonetheless there are indications of landing places. In a riverine context, it was noted in Chapter 5 that the intersections of the Stonehenge avenue with the river Avon and those of routes at Durrington Walls indicate the locations of landing places. In marine contexts a series of sites have been identified, mostly in Scotland. They are protected havens at river mouths, often associated with sand dunes and characterised by artefacts dated between 4000 and 2000 cal BC, including the use of exotic lithic raw materials, evidence of artefact production and the implication that they were places where traders arrived, and goods were produced and exchanged (Bradley *et al.* 2016). The presence at Ferriby, Humberside of fragments of plank boats spanning some 1500 years right through the Bronze Age and into the Iron Age can be interpreted as both a frequented landing place and somewhere where boats were also dismantled, repaired and rebuilt. The case is strengthened by the presence of trackways and other wood structures in the Ferriby intertidal area (Van de Noort and Ellis 1999) and by the presence of a long established routeway which later became a Roman road, known today as High Street, along the Lincolnshire Wolds to south Ferriby on the opposite bank of the Humber (Fig. 8.1; Van de Noort and Ellis 1999). Also in the Humber wetlands, at Brigg, a natural crossing place where the Ancholme valley narrows, is the findspot of a middle to late Bronze Age dug-out canoe, an early Iron Age trackway and the Brigg sewn plank boat of the late Bronze Age. The presence at Caldicot, Wales of three sewn plank fragments, one early and two late Bronze Age, also implies a frequently used landing place, an idea supported by other wood structures including a bridge/jetty of the late Bronze Age (Nayling and Caseldine 1997). In the intertidal zone of the Essex coast a number of wood structures were interpreted as possible jetties and landing places (Wilkinson and Murphy 1995). The possibility also needs to be considered that some of the double post alignments on the river Thames at London, for instance that of middle Bronze Age date at Vauxhall (Cohen and Wragg 2017), may represent landing places rather than bridges, or perhaps a combination of the two, since the place where riverine and terrestrial routes cross are natural landing places. In the Thames Estuary trackways, of both wood and stone, are reported leading in the direction of the estuary, some, as noted in Chapter 6, continuing dryland drove ways. The destinations of these are potential landing places and potential locations for boat finds. The clearest evidence for prehistoric harbour installations comes from Poole Harbour, Dorset where two substantial timber and stone jetties were constructed in the later Iron Age (Wilkes 2007). Poole is one of a series of natural harbours on the English Channel Coast associated with cross channel trade at this time.

Portages are places where boats, or goods, can be transported across an isthmus or promontory to avoid long, or dangerous, sea crossings, or river rapids; they are obvious focal points for future investigation. There are many examples in Scandinavia and Scotland which are documented using historical and place-name sources. Some Scandinavian examples have hollow ways leading to them and it has been suggested that such places might be associated with ballast dumps, though no such examples are yet documented (Westerdahl 2006; Bradley *et al.* 2016).

Artefact distributions in Scandinavia

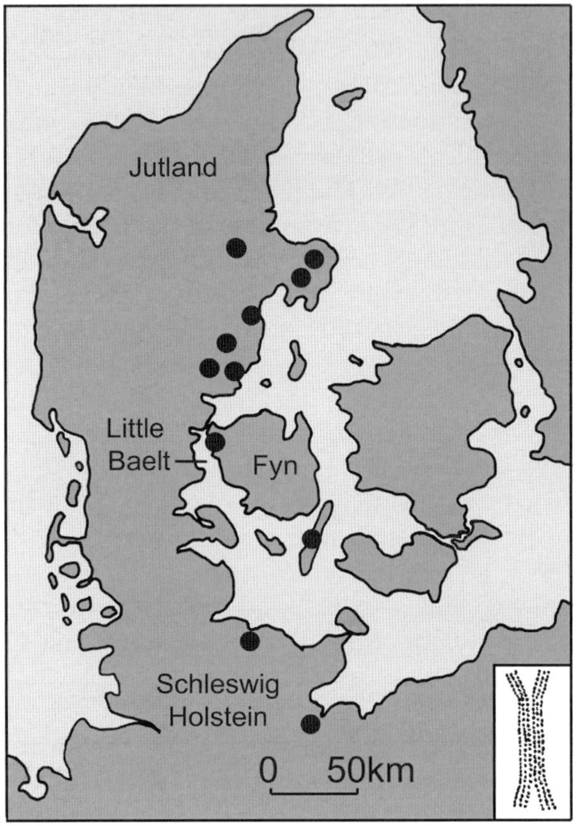

Figure 9.4. Denmark showing the distribution of 'wheatsheaf' motif on Mesolithic bone objects (graphic: J. Foster after Terberger 2006, fig. 39).

In the identification of patterns of maritime communication we can also deploy evidence provided by artefact distributions. As early as the late glacial interstadial (Allerød, *c* 13,000 BP) it has been inferred that upper Palaeolithic groups colonised the west coast of Norway by boat since there seems to have been no ice-free corridor on land at that time (Coles 2000). In the Mesolithic of Denmark, the distribution of distinctive axe styles has been used to suggest Mesolithic territories spanning 50 km at most (Fischer 2003). During the Danish Kongemose *c* 6500–5400 cal BC, the distribution of a distinctive so-called 'wheatsheaf' decorative motif on bone artefacts points to maritime connections around the Little Baelt, between the west of Fyn, the east coast of Jutland and south to Schleswig-Holstein (Fig. 9.4; Terberger 2006). This argument is strengthened by similarities of material culture between what is argued to be a year-round site at Tybrid Vig and other sites in this area which may have been visited as a result of logistical (i.e. task specific) mobility (Andersen 2013). Close similarities in material culture between Scania in southern Sweden and the island of Bornholm, 37 km to its south, point to the maintenance of connections across this open sea (Fischer 2002). In the final Mesolithic Ertebolle period, much more distant connections are manifest with the appearance of stone 'shoe last' axes obtained from Danubian farmers and originating from outcrops 1000–1500 km to the south in Bulgaria and Slovakia (Fischer 2002). These probably do not reflect the actual movement of people but of objects. However, the significant point is that they are found on Baltic Islands well to the north of contemporary farming communities and must have arrived there by boat.

Transported things in Britain and Europe

Britain was recolonised *c* 12,700 cal BC but that was overland at a time when low sea level connected it to continental Europe by Doggerland in the North Sea. Britain finally became an island as a result of sea level rise *c* 6000 cal BC and subsequent contact required a boat crossing of at least 38 km. Ireland was colonised by Mesolithic communities about 8000 cal BC and that must have been by boat since it became an island *c* 12,000 cal BC. Recent evidence shows that Orkney, Shetland, the Outer Hebrides and Isles of Scilly were visited in the Mesolithic. There is increasing evidence from the western seaways of the British Isles of the facility with which people moved significant distances, both in the Mesolithic and Neolithic (Garrow and Sturt 2017). Domesticated animals and plants reached the British Isles by boat. Domestic cereals and sheep/goat are non-native and bone measurements also show that domestic cattle and pigs were from introduced stock rather than local domestication (Davis 1987). The scale of human migration as opposed to acculturation of Mesolithic populations remains the subject of active debate, nonetheless it is clear that there was active maritime communication through the Neolithic. Curiously, the earliest dated domesticate, a bone of *Bos* dated 4495–4165 cal BC, is from the west of Ireland on a Mesolithic site at Ferriter's Cove (Woodman *et al.* 1999). *Bos* was non-native in Ireland and this suggests the possibility of precocious Neolithic contact up the Atlantic fringe, for which Sheridan (2003; 2004) has argued on the basis of tomb and pottery typology. In England the earliest Neolithic dates are from Kent and the Thames valley, *c* 4100 cal BC. From there Neolithic attributes spread widely and rapidly west and north as far as Wales and Scotland by *c* 3800 cal BC (Whittle *et al.* 2011). West to east Irish Sea connections are demonstrated by links of megalithic tomb type and art between the great tombs of the Irish Boyne Valley and north Wales, and from south to north between the Boyne valley and the tombs of Orkney. Cunliffe (2001a, 199) saw this as reflecting a common Atlantic cultural province. Noble (2006) has likewise argued that travel by sea was a significant factor in the positioning of Orkney Neolithic tombs. Even Britain's most remote island, St Kilda, has produced evidence of Neolithic material culture (Fleming 2005a).

With the arrival of metals, maritime connections are even more clearly demonstrated. Maritime Beakers, among the earliest of the type, have, as the name suggests, a strongly Atlantic distribution from Iberia to Britain (Cunliffe 2001a, fig. 6.11). Sheridan (2008) has pointed to close similarities between early Beakers and associated burial practice in Holland and Scotland. Similarly, gold lunulae, which are mostly found in Ireland, also occur on the Cornish coast where one example is almost identical to a find at Kerivoa on the French Cotes d'Amour (Fig. 9.5). Figures on stone menhirs in Brittany and Guernsey are shown wearing lunulae (Cunliffe 2001a, fig. 6.13). Recently a combination of lead isotope and elemental analysis has concluded that much of the Chalcolithic and early Bronze Age gold, originally thought to come from Ireland, may derive from south-west British alluvial, or other, as yet unidentified, sources (Standish *et al.* 2015). Many metal artefacts demonstrate connections between Ireland, the Wessex region and continental Europe, as shown for instance by gold hair rings, neck rings and bar torcs (Eogan 1994), and Cunliffe (2001a) includes many distributions of bronze artefacts illustrating these Atlantic connections. Some of the most compelling evidence for maritime connections has been assembled from the English Channel area, leading Needham to refer to this as a 'maritory', meaning a maritime cultural area (Needham and Parfitt 2006). Connections are demonstrated by a series of precious cups of gold, silver, amber and shale, which span the period 1850–1550 cal BC and occur in the south coast of England, the Netherlands and Brittany (Fig. 9.5c). Needham proposes that these cups may have had a role in propitiatory rites associated with maritime travel and related exchange networks. The distribution of amber in southern Britain, much of it probably of Baltic origin, indicates transport along the Channel coast and then, as previously noted, riverine transport to the Stonehenge area (Fig. 9.5d). Amber spacer beads and discs mounted in gold, an artefact type distinctive of Wessex, then found their way to the shaft graves of Mycenae and to Crete respectively, perhaps via an Atlantic, French and Mediterranean route (Bouzek 1993). Amber in central Europe seems to have moved south via riverine exchange networks. This is part of a network of amber trade between the Baltic coast of Jutland, where vast numbers of amber beads are found in Bronze Age contexts, and Europe as far as

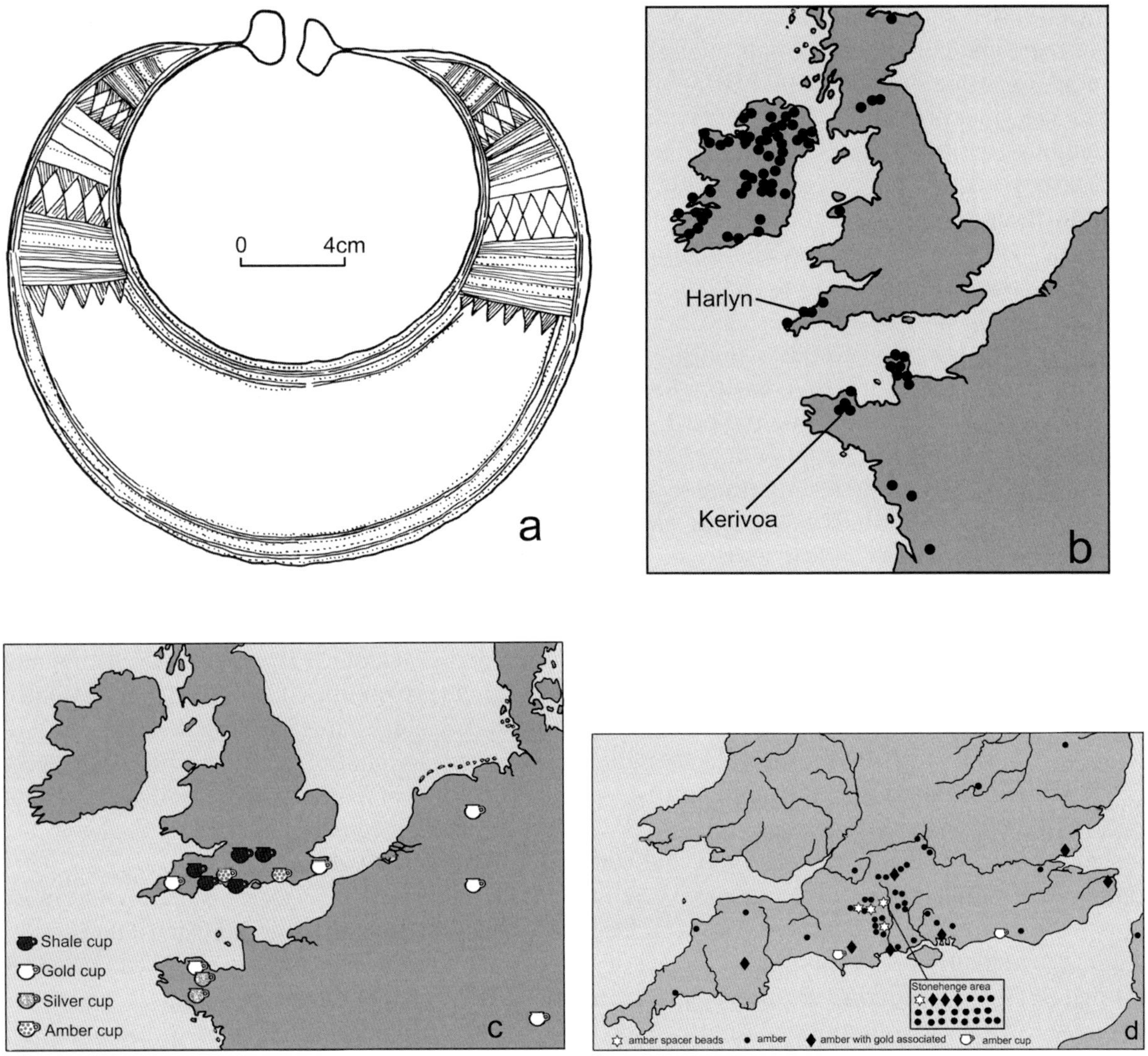

Figure 9.5 Maritime connections indicated by metalwork: (a) gold lunula from Harlyn, Cornwall (after Hencken 1932); (b) Atlantic distribution of gold lunulae (after Cunliffe 2001a); (c) distribution of gold, silver, amber and shale cups and embossed gold ornaments; (d) Distribution of amber in southern Britain (c–d after Needham and Parfitt 2006) (graphics: J. Foster).

the Mediterranean (Beck and Shennan 1991). In reviewing the rock art of Atlantic Europe, Bradley (1997) has observed that from Scotland to Iberia similar motifs are arranged in similar ways at a similar chronological horizon between *c* 3000 and the early second millennium BC suggesting that these communities were linked by maritime communication.

In the middle Bronze Age the volume of metalwork as represented by hoards and other finds increases markedly and there is abundant evidence for occurrence in British hoards of types characteristic of production on the continent, and vice versa (Lehoërff *et al.* 2012; Needham *et al.* 2013). Combining materials of high status to create artefacts of probable cosmological significance contributes to the evidence for exotic maritime contacts. Pre-eminent among them is the Nebra sky disc of *c* 1600 BC, found in central Germany, and apparently depicting the phases of the moon, sun and stars, perhaps used in navigation (Meller 2004). Geochemistry has shown that the gold used was of Cornish origin (Ehser *et al.* 2011). Another example is the Caergwrle bowl combining materials of diverse geographic origin: shale, gold and tin (Green 1985; Rednap 2011).

In recent years, evidence of distant metalwork connections has been supplemented by other forms of artefactual and settlement pattern evidence which suggest the connections extended well beyond simple concepts of trade, as indeed Needham's 'maritory' label implies. Among the most telling evidence is the discovery of a very distinctive large pot in the Cornish Trevisker style which was made of Gabbroic clay from Cornwall but was found at the other end of the English Channel on the Isle of Thanet, Kent (Gibson *et al.* 1997). Pottery in Trevisker style also turns up on the French side of the English Channel. That pottery also has a coastal distribution up the Bristol Channel and Severn Estuary into south Wales (Bell 2013, fig. 17.4) which may suggest that the Trevisker-producing communities of middle Bronze Age Cornwall played a significant role in maritime communication in western Britain and the English Channel. Pots made in Finistère, France are also found in the Solent region of England, for instance on the Isle of Wight (Murphy 2009). Coaxial field systems, discussed in Chapter 8, which are widespread in coastal areas of south-east England have been found in the Cherbourg peninsula, Normandy and elsewhere on the continental side of the Channel (Bradley *et al.* 2015). A range of evidence has been assembled in terms of pottery, fields and house forms for strong cross-Channel connections in the middle and late Bronze Age (Marcigny *et al.* 2007; Lehoërff *et al.* 2012).

During the Iron Age metalwork continued to reach Britain from continental Europe via both Atlantic connections and cross-Channel links to central Europe (Cunliffe 1991). In the later Iron Age cross-Channel trade appears to have particularly focused on favoured harbours at sites such as Hengistbury Head, Mount Batten and Poole Harbour. Particularly illuminating are the writings of Pytheas of Massalia some time before 320 BC whose book *On the Ocean* is only known from references in later works. Pytheas appears to have circumnavigated Britain, voyaging as far as Iceland since volcanos and frozen seas are described (Cunliffe 2001b). In the final centuries of the Iron Age other Greek and Roman writers such as the Massaliote Periplus, Timaeus and others provide tantalising glimpses of voyages up the Atlantic coast to Britain and beyond, often for trade in tin (Cunliffe 2001a; 2001b). From the north coast of the Isle of Wight, pottery and foreign stone from Brittany and south-west England, inferred to be ships' ballast, indicate increasing cross-Channel and along Channel shipping in the later Iron Age (Tomalin *et al.* 2012). By the late Iron Age pottery was reaching the Solent area from Brittany (Cunliffe 1991, 437) and wine amphorae of Mediterranean origin were reaching Britain, at first via Hengistbury Head and later, when Caesar's conquest of Gaul disrupted Atlantic routes, the focus of trade shifted to a cross-Channel route into eastern England for the century preceding the Roman conquest in AD 43.

As regards the ships which carried this Iron Age trade, we have only indirect evidence. Ships illustrated on Celtic coins of Cunobelin (*c* AD 10–43) show a mast, square sail and side rudder (McGrail 2004). Anchors of Iron Age date are known from Bulbury, Dorset and Porth Felen, Wales, the latter apparently a Graeco-Roman type of the second or first century BC (Cunliffe 2001a). Caesar, when writing of the Veneti of western Brittany, describes ships with leather sails, the hull fashioned from flush-laid oak planks fastened to thick framing with substantial iron nails, vessels more suited to Atlantic conditions than the ships at the disposal of the Romans. Thus, at least by the later Iron Age, there were larger and more substantial ships than sewn-plank boats. Further evidence of their character is provided by a series of so-called Romano-Celtic boats which share some of the characteristics Caesar described but which were made after the Roman conquest and have been found in the Severn Estuary, the Thames at London, the Channel Islands, the Rhine basin and in Switzerland (Nayling and McGrail 2004). The construction of these ships differs markedly from Roman vessels in the Mediterranean hence the view that the type is of indigenous Iron Age derivation.

Scandinavia: ships and rock art

Sewn-plank boats are represented in continental Europe only by a more-or-less complete example from Hjortspring, Denmark, displayed in the National Museum, Copenhagen. It is a light slender war canoe with bevelled lap planks with raised cleats lashed to frames, some 19 m long and 2 m wide with distinctive upturned projections (horns, i.e. keel and prow extensions) at either end. It would have carried 18 paddlers. It was found in a tiny peat bog 3.4 km from the sea, in which it had been ritually deposited together with the weaponry of 70–100 warriors, perhaps the booty of a defeated raiding party (Crumlin-Pedersen and Trakadas 2003). An experimental full size replica performed effectively as rapid marine transport capable of covering *c* 100 km a day. Other probable fragments of sewn plank boats are reported from Haugvik, northern Norway dated to the first or second century BC, and Hampnas, Vasternorrland, Sweden, dated *c* 220 BC (Sylvester 2009; Janssen 1994).

For the Bronze and Iron Ages the rock art of Norway and Sweden, and depictions of boats on bronzes from Denmark, provide a complementary perspective on later prehistoric maritime communication which is especially important in illuminating the social and cosmological aspects of exotic contacts. The rock art of southern Scandinavia is dominated by pictures of boats of which 19,322 images are documented (Nimura 2016). Many examples are remarkably informative as to boat type; they apparently show sewn-plank boats with the distinctive horned prow as at Hjortspring. The prow often has an animal head, usually horse.

Figure 9.6 is an example from Soletorp, Sweden depicting three boats, the upper with 13 crew, larger figures fore and aft with horned helmets and axes, horse head terminals at either end of the boat and an acrobat above the boat. A few examples with deeper hulls are thought to depict hide boats (Crumlin-Pedersen 2010). Many of the plank boats depict rows of paddlers and occasionally a steering oar is shown. They do not generally have obvious sails although Bengtsson (2017) has recently argued that aspects of the iconography should be interpreted as sails. Among the crew there is often one or more much larger figure interpreted as leader(s). These prominent figures are often phallic and the overall imagery with the emphasis on weapons is strongly male with a near absence of female figures and obvious children (Hygen and Bengtsson 2000). Prominent figures are sometimes depicted with distinctive types of artefacts which can be dated typologically (Coles 1990), for example, sword scabbards with forked terminals and swords with curved blades (early Bronze Age), horned helmets (late Bronze Age), shield types (round Bronze Age, rectangular Iron Age), *lurs* (late Bronze Age blowing horns), and crescentic axes (middle Bronze Age). Sometimes, as in Figure 9.6, mysterious acrobatic figures are shown leaping over the boat.

Figure 9.6 Boats in rock art: Soletorp, Tanum, Sweden (photo: M. Bell).

The Scandinavian rock art is pecked on prominent glacially smoothed rock faces which often sit at the edges of level plains. Ling (2004; 2008) has shown that many examples sit around the margins of former shorelines, 15–20 m above today's sea level, as a result of still continuing isostatic readjustment which followed removal of the huge weight represented by the Pleistocene ice sheets. Many boat images are now kilometres inland but at the time they were made they overlooked the beach and would have been visible from the sea (Hygen and Bengtsson 2000). Isostatic readjustment also provides some guide to dating since earlier images and types of boat relate to higher former shorelines. Sites where ship representations are abundant may represent places where boats from across the Baltic habitually landed for seasonal gatherings involving social and economic exchange. They are concentrated in areas where the effects of shoreline displacement were especially marked, and it has been argued by Nimura (2016) that the art may have had active agency in enabling communities to cope with dramatic environmental changes.

The case for cross-Baltic connections is strengthened by links between the belief systems, symbolism and cosmology of the art in southern Norway and Sweden and those represented iconographically on bronzes in Denmark. Indeed the significance and meaning of the ships, their horse head prows, the acrobats and other features, is greatly augmented by depictions of ships on bronze objects, mainly razors, of which there are 800 from Denmark (Kaul 1998). Kaul has shown that the ships on bronzes are a central part of a sophisticated cosmological scheme representing the solar cycle and the passage of day and night and probably also the seasons, life and death. A significant clue in deciphering this scheme was the Chariot of the Sun from Trundholm, Denmark (*c* 1400 BC), a bronze wheeled model of a horse, drawing a solar disc, with gold on one side, representing day, and silver grey on the other representing night. In Denmark the mysterious acrobatic figures, horse head terminals, figures with horned helmets and axes are also represented as cast bronzes, probably originally fitted to models of ships (Kjaerum and Olsen 1990). All this seems to point to a common cross-Baltic cosmological belief shared by southern Norway, Sweden and Denmark, in which boats played a central part and to which other means of transport, the horse and wheeled vehicles, also contributed. The significance of the means of transport, and particularly the boat, is partly explained by the fact that the metals and ostentatious metal artefacts used in Norway and Sweden were imported from central Europe, as lead isotope and elemental analysis shows (Ling *et al.* 2014).

Conclusions: maritime connections and cultures

A theme of this chapter has been the role of movement by water, rivers and the sea in connecting rather than dividing communities (Van de Noort 2011). The significance of this from an early date is demonstrated by the Palaeolithic and early Holocene colonisation of islands in the Mediterranean and South-east Asia (Gamble 1993; Broodbank 2013). It is also increasingly apparent from material cultural evidence that many islands had more connection with outside communities than previously imagined and that their distinctiveness sometimes relates not so much to isolation as to the establishment of a cultural identity in the context of wider contacts. This case has been effectively made by Fleming (2005a) in the case of Britain's most remote island, St Kilda. A similar case can be made for Orkney and the Isles of Scilly, which are now known to have been more connected to other areas in the Mesolithic to Bronze Age than previously imagined (Garrow and Sturt 2017). The implication of this is that archaeologists have tended to underplay the social significance and probably the frequency of maritime communication. Maritime archaeology has suffered from pigeonholing as a separate category of evidence disconnected from the mainstream. Fortunately, that tendency is now being very actively counteracted by Cunliffe's (2001a) approach to the archaeology of the Atlantic and that of Broodbank (2013) in the Mediterranean. Such studies take account of currents and winds and the natural patterns of movement which are observed today and would have been utilised in the past. These help to explain the patterns of artefactually attested connectivity between Aegean islands and the routes used to cross the English Channel (McGrail 1983).

The challenge is to develop an approach to maritime cultural landscapes or seascapes which integrates the archaeology of the sea and land (Westerdahl 1992; 2006; Hunter 1994). A move towards this has been the development of a seamless approach to coastal archaeology which incorporates submarine, intertidal and terrestrial evidence, as pioneered on the Langstone Harbour Survey, UK (Allen and Gardiner 2000) and the Strangford Lough Survey in Northern Ireland (McErlean *et al.* 2002). The subject has certainly benefited from greater attention in recent decades, because heritage agencies in many countries have placed increasing emphasis on coastal archaeology. In the UK this has dramatically increased our knowledge of the coastal and maritime heritage resource (Murphy 2014). In seeking to understand distinctive maritime ways of life we can draw not just on the direct archaeological evidence, but on wider studies of human engagement with the sea from a diversity of periods and perspectives including anthropology, literature and history (Mack 2011; Gillis 2012). Examples where ethnohistorical perspectives have proved especially illuminating include an understanding of traditional fishing practice (O'Sullivan 2003c) and the performance and role of skin currach boats in the Aran Islands on the west of Ireland (Synge 1979). As Van de Noort (2011) has emphasised, people's relationship with seascapes defined them in terms of their livelihood, identity and place in history.

So why do coasts often appear to be seen as special places and what motivated distant maritime communication? It has been widely recognised that coasts have been seen as liminal places representing transitions, for instance from life to death. Mesolithic Ertebolle burials in dugout canoes have been interpreted in these terms (Bradley 2000), as have ship images on Bronze Age razors from Denmark and the related rock art of Scandinavia. The liminal properties of the coast may also account for the concentration of megalithic tombs on the coastal fringes of Atlantic Europe, for instance in Brittany (Scarre 2002) and Pembrokeshire (Tilley 1994). An extension perhaps of this liminality or transitional concept concerns the particular properties of the ship itself. Van de Noort (2011) has highlighted the notion of the ship as hererotopia, a space where the conventional is contested. He further argues that, through the significant social role that ships played, the ships themselves acquired iconic status and were ritually deposited. That is a somewhat different emphasis from the ritual deposition of the boats of defeated raiders which has been proposed at Hjortspring in Denmark. Regarding the motivation of distant contact, valuable insights are provided by Homer's (ND) *Odyssey* written in the ninth century BC but set in the time of the Trojan War in the thirteenth century BC. It describes a distant, immensely challenging voyage by which the status and character of those involved were built. Helms (1988) has developed this idea from the perspective of anthropology, demonstrating that by distant travel people acquired esoteric knowledge and privileged access to exotic objects; this exoteric knowledge became part of the validation, or legitimation, of elite status (Kristiansen and Larsson 2005). Amber can be seen in these terms (Beck and Shennan 1991) and the same can be said of many of the prestige goods of later prehistoric North-west Europe. It is striking how many of them combine metals and geological material derived from diverse geographical sources and are themselves icons of mobility, for instance the Caergwrle bowl in the form of a boat, the Broighter boat, the Nebra Sky Disc and the Trundholm 'Chariot of the Sun'. Clearly acquisition and display of exotic objects was of central importance to the maintenance and reproduction of elites who, through distant travel, had privileged access to other worlds and the esoteric knowledge which they provided.

This chapter has highlighted the importance of understanding relationships between terrestrial, riverine and maritime transport and the significant social role which these relationships played in prehistory. These relationships will be further touched on at the end of the following case study chapter and in the conclusions.

10. A case study of the Wealden district in south-east England

Introduction

In previous chapters we have taken a broad-brush approach drawing on evidence where it was available in many different regions. This selective, prehistoric focused approach has served its purpose in identifying the wealth of evidence which, contrary to previous expectations, exists for work on past routeways and mobility. Such an approach is inevitably less effective in the identification of long-term continuities, especially into the historic periods, and the contribution of routeways to the wider structuration of landscapes at district and regional scale. Accordingly, this chapter takes a regional approach with a case study of the Weald in south-east England. Here the evidence for connectivity seems particularly strong but previous studies have seldom explored this from a prehistoric perspective. As observed in Chapter 1 the detailed investigation of landscapes can, and should, seldom focus on prehistory alone, as the routeways of the Roman and medieval periods and present landscapes may often provide clues to earlier patterns.

The essential background to this is the Wealden District's geological structure (Wooldridge and Goldring 1953; Gallois and Edmunds 1965), shown in Figure 10.1, which also illustrates the location of key trackway sites. The Weald was originally an anticline, a dome of rock flanked to north and south by synclines, i.e. basins. The central part of the dome has been eroded to expose an area of Lower Cretaceous clays and sandstones. These are flanked by the steep broadly east to west escarpment ridges of the Upper Cretaceous North and South Chalk Downs. To north and south of the two escarpments are gentler dip slopes dissected by dry valleys separating spurs running north/south. To the north of the Weald is the London Basin with Tertiary beds and the Thames Estuary. To the south is the Coastal Plain with Tertiary beds overlain by Quaternary raised beaches.

Historic settlements in this area often have long strip parishes which cut across the grain of the geology, facilitating the integration of resources from upland and lowland. From at least medieval times there is evidence for multiple parallel drove roads at right angles to the topography, along which animals were moved seasonally, from parent settlements on the Chalk and the Tertiary basins to south and north, out to daughter settlements in the central Weald (Brandon and Short 1990). Thus, Tenterden in the Weald of Kent is the den or swine pasture of the people of Thanet, 45 km to the north-east (Fig. 10.1). Some drove roads have been mapped in Sussex by Brandon (1974), and more extensively in Kent by Everitt (1986). The economic and tenure system of the parent and daughter

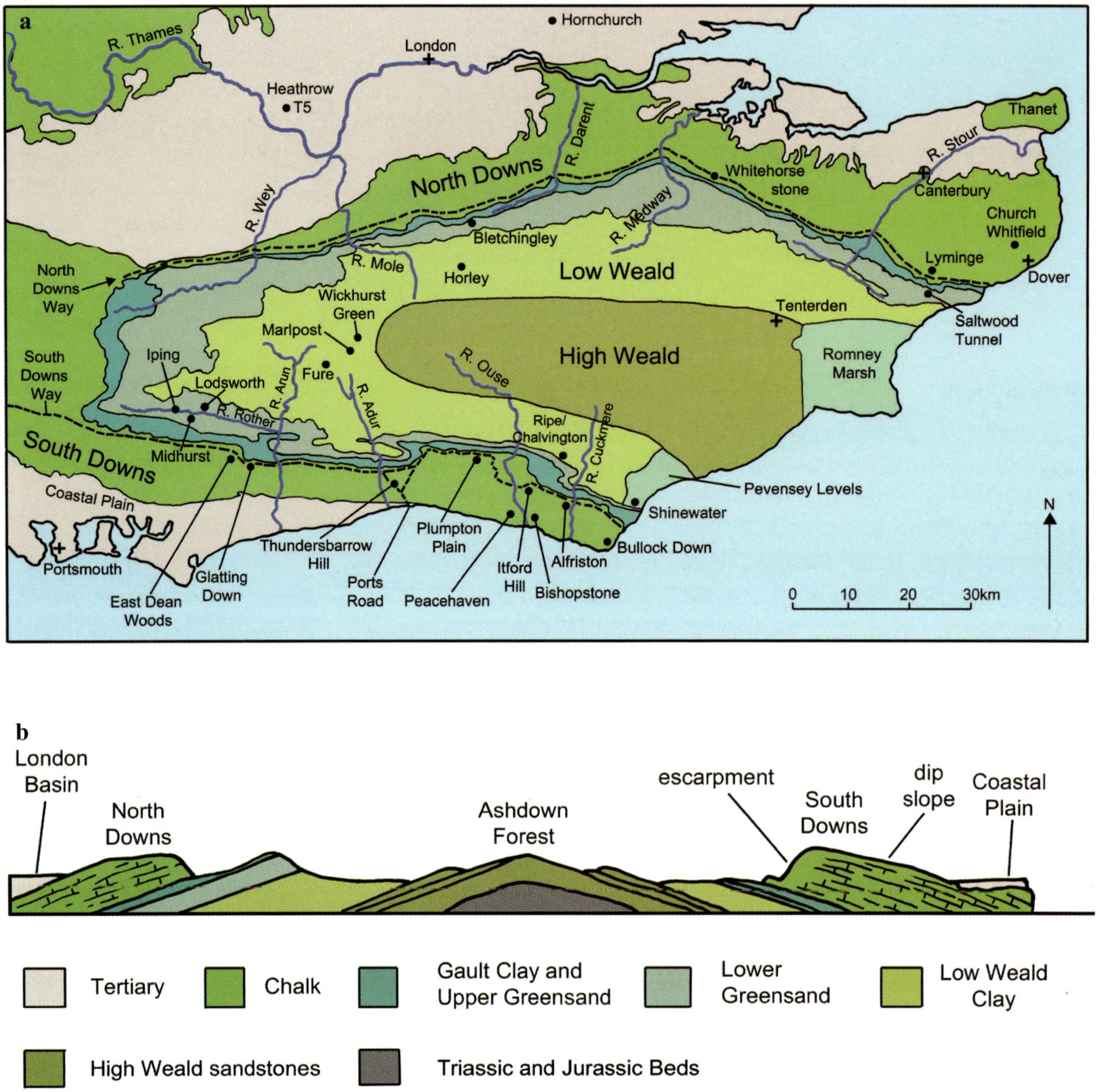

Figure 10.1 The Weald in south-east England showing: (a) the distribution of trackways sites discussed, the Ridgeways and the geology; (b) a north–south section showing the geological structure and topography (graphic: J. Foster after Wooldridge and Golding 1953).

settlements, which is evident in the medieval period, is partly obscured in *Domesday Book* (AD 1086) because daughter settlements are listed as part of the parent settlements, so that the Weald appears lacking in woodland and swine pasture, whereas that is where those resources were located (Darby and Campbell 1962). Historical sources and place-names enable the links between parent manors and Wealden outliers to be mapped in West Sussex (Chatwin and Gardiner 2005, fig. 6) and in Surrey (Blair 1991) and in both cases the pattern follows the axis of the drove roads. The general assumption in the past has been that the Weald was largely unoccupied woodland until clearance began in the later Saxon period. However, there was an extensive Romano-British iron industry and there is increasing evidence for prehistoric settlement and clearance (Drewett *et al.* 1988; Gardiner 1990; Margetts 2018a; Marples and Poulton 2019). The nature of that activity, and how it relates to the parallel drove ways, if indeed some of those are of pre-Saxon origin, is only now beginning to be appreciated.

The South Downs

The South Downs Way is a ridgeway, along the escarpment of the chalk from Winchester to Beachy Head. It was the first long distance trail in Britain designated in 1963 and today it is the backbone of the South Downs National Park, established in 2011. Although a footpath with magnificent views, it does not form such an obviously natural historic route as some other ridgeways which connect regions because the ridge is interrupted by four river valleys, the Arun, Adur, Ouse and Cuckmere which, in prehistory and until after *Domesday Boo*k, were estuaries. At the foot of the escarpment is an alternative low ground route sometimes called 'Underhill Road' which provides a less dramatically undulating route but would have involved the same river crossings. These routes were not included in the Ordnance Survey (1962) map of Iron Age Britain and they have not been subject to much archaeological discussion, or interpretation. The omission is significant since many pioneering observations on prehistoric trackways were made on the South Downs in the first two decades of the twentieth century, as noted in Chapter 1 (e.g. Curwen and Curwen 1923).

A feature of the South Downs, and ridge-way crests elsewhere, is that they are crossed by cross-ridge dykes, which are thought to be mainly of later Bronze and Iron Age date (Curwen 1951; Tilley 2010). Their number has recently been very dramatically increased by ongoing survey (Lea *et al.* 2017). They tend to occur in two main situations: one crossing over the ridge from the head of a dry valley to the steep escarpment crest, the second running across dip or escarpment slope spurs from the ridge crest. Examples crossing the ridge crest are so numerous in places, such as the South Downs west of the Arun, that they would seem to rule out any ridgeway route when they were in use. Just to complicate matters there are a few examples which could be interpreted as respecting a pre-existing ridgeway route. Curwen (1951) records that in three instances the dykes change direction where today's ridgeway crosses, in one case on Glatting Down a dyke is associated with a middle Bronze Age globular urn, and that dyke is broken by a possible entrance which would facilitate movement along the ridge (Curwen 1929, pl. xxvi). However, the most economical explanation for the cross-ridge dykes is that the ridges were long term pasture, and the dykes separated the pasture of different communities, analogous in a way to the longer linear ditches of Wessex (Bradley *et al.* 1994). The dykes crossing spurs may have served to separate open grazing on the ridge from the fields of particular communities on lower spurs.

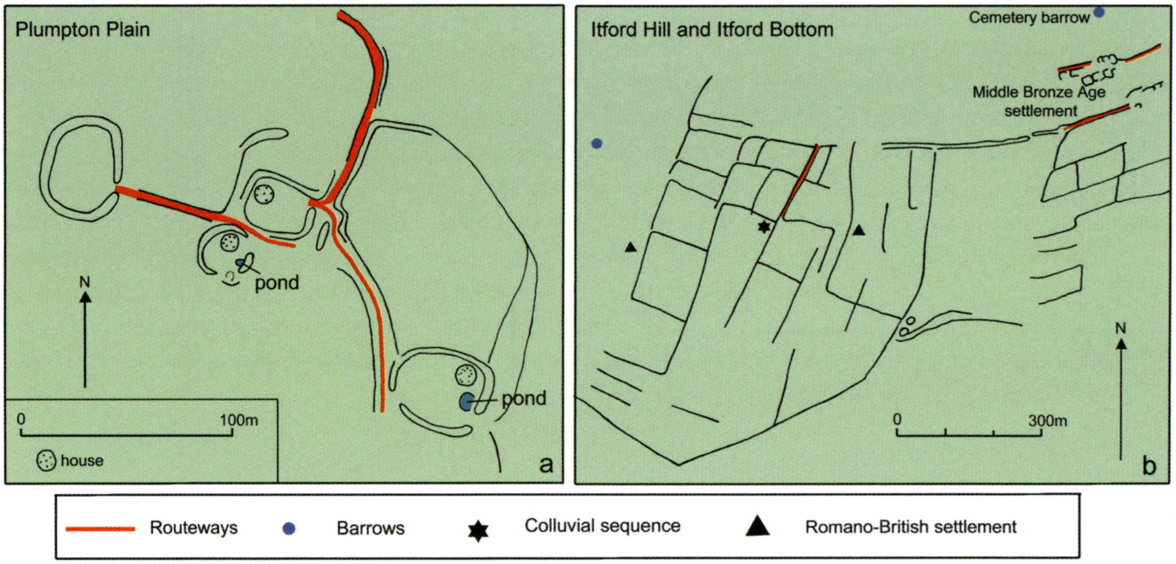

Figure 10.2. Fields and trackways at: (a) Plumpton Plain middle Bronze Age settlement (after Holleyman and Curwen 1935); (b) Itford Hill and Itford Bottom (after Bell 1983) (graphics: J. Foster).

There are many examples of trackways in this area directly associated with settlements and we will begin with these before considering wider patterns of landscape connectivity. In the middle Bronze Age, the settlement pattern is characterised by small farms with roundhouses and ponds with embanked enclosures linked to fields and double-lynchet trackways. An example is Plumpton Plain, Sussex (Fig. 10.2a) Here embankments around huts are linked by tracks, slight hollow ways, partly created by erosion and, where they pass through surrounding fields, by double lynchet trackways (Holleyman and Curwen 1935; Curwen 1954). At Plumpton, and most of the comparable sites, the tracks cannot be followed very far from the settlement, presumably either because they have been obliterated by the effects of later cultivation or ran into unenclosed land without defined routes. At Itford Hill, Sussex a well preserved, excavated middle Bronze Age settlement dated *c* 1150 cal BC has an associated trackway, partly defined by lynchets, but this also could not be traced far beyond the settlement (Fig. 10.2b; Burstow and Holleyman 1957). Even so this east–west route is just 250 m south of the present-day long distance ridgeway footpath and hints at the possibility of an early east–west route. There are 'Celtic' fields on the spur to the south and in Itford Bottom valley to the west. Investigation of the colluvial sediment sequence in this valley showed that the fields originated with clearance in the Beaker period *c* 2070 cal BC and there was middle Bronze Age colluvium containing pottery contemporary with the excavated settlement. Overlying this was a flinty lynchet deposit forming one side of a double lynchet trackway down the valley axis. Artefact distributions showed that the lynchet, and thus the trackway, formed in the Iron Age and Romano-British periods (Bell 1983). That trackway does not appear to be as early as the middle Bronze Age settlement but it does demonstrate how trackways associated with lynchets can be dated.

Case Studies: Bishopstone and Bullock Down 'ghost routes'

This brings us to the question of how wider patterns of landscape connectivity may be established. Sometimes the relationship of archaeological features of different dates can reveal the existence of 'ghost routeways' i.e. those for which no visible surface line survives, perhaps because it has been obliterated by later cultivation. I will begin with a case from one of my own excavations which, 44 years ago, first sparked my fascination for this topic. The site is at Bishopstone, East Sussex (Bell 1977), where on top of a downland promontory projecting into the estuary of the river Ouse was a multi-period occupation site (Fig. 10.3). The sequence begins with Neolithic pits which appear to form a discontinuous circle just below the hill crest; the area was partly wooded at the time and it may be speculated that the pit-defined area may have been a sort of proto-enclosure defining a clearing *c* 120 m across. Later, in the early Bronze Age, a line of six round barrows lay along the ridge crest. What makes these background observations significant is that in the late Bronze Age, or early Iron Age, a rectangular enclosure was laid out, again just south of the spur crest, with two entrances precisely aligned on the spur crest and the barrow alignment. Subsequently Romano-British enclosures were laid out with one edge along the same line. It is hard to avoid the conclusion that this indicates the existence of a ridge-top route which might have existed in the Neolithic but which is convincingly demonstrated from the early Bronze Age to the fourth century AD, a period of 2000 years. It had gone out of use by the fifth century when Anglo-Saxon buildings were laid out across its line. This example points to the existence of an early route, linking the chalk downs to the north, down a dip slope spur to the resources of the estuary and perhaps the opportunities for maritime connectivity that provided.

The Bullock Down project at the east end of the South Downs provides similar evidence for a longer route (Drewett 1982). Here a double lynchet trackway could be traced for 1.7 km along the side of a downland spur with Celtic fields on either side (Fig. 10.4). This spur runs west and at right-angles to the escarpment which here runs north from Beachy Head. Excavation of a trench across the trackway and recording of the artefact distributions indicated that the fields were in use in the late Bronze Age/early Iron Age and the Romano-British period. Activity during the latter period was confirmed by two associated Romano-British settlements; a marl pit along the edge of the double lynchet trackway also proved to be Romano-British and two coin hoards of the same period have been found along the trackway highlighting its significance. There are indications that the origins of this trackway pre-date the fields; on its line 1.1 km west it passes three Bronze Age barrows and points directly to the entrance of a very large earthwork enclosure at Belle Tout which has been variously dated between the later Neolithic and Saxon period. Within Belle Tout, on the same line, is an excavated Beaker settlement (Bradley 1970) and a well shaft which has recently been dated to the late Bronze Age (M.J. Allen, pers. comm.); this suggests the surrounding large earthwork, probably a stock enclosure, may be of that date. There is a strong suspicion that this, not particularly prominent, trackway was in use for some 2500 years; indeed part of its length remains a bridle-way today.

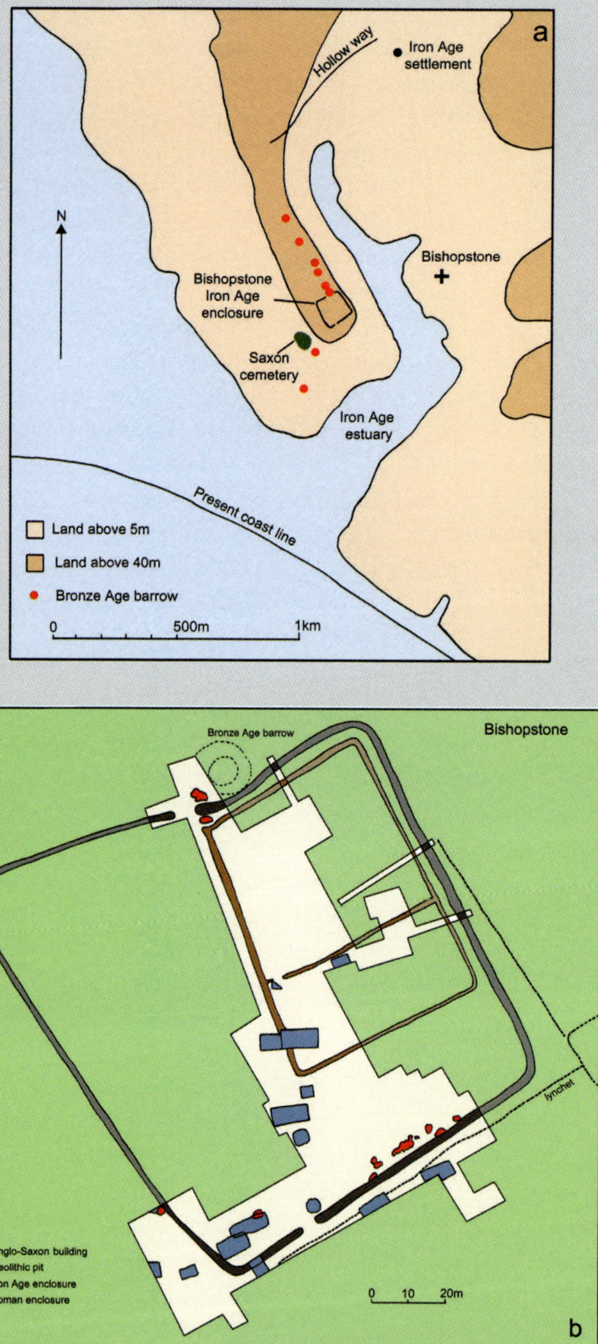

Figure 10.3 Bishopstone, Sussex, UK: the relationship between barrow alignments, Neolithic pits, an Iron Age enclosure and later features indicating a 'ghost' routeway (graphics: J. Foster after Bell 1977).

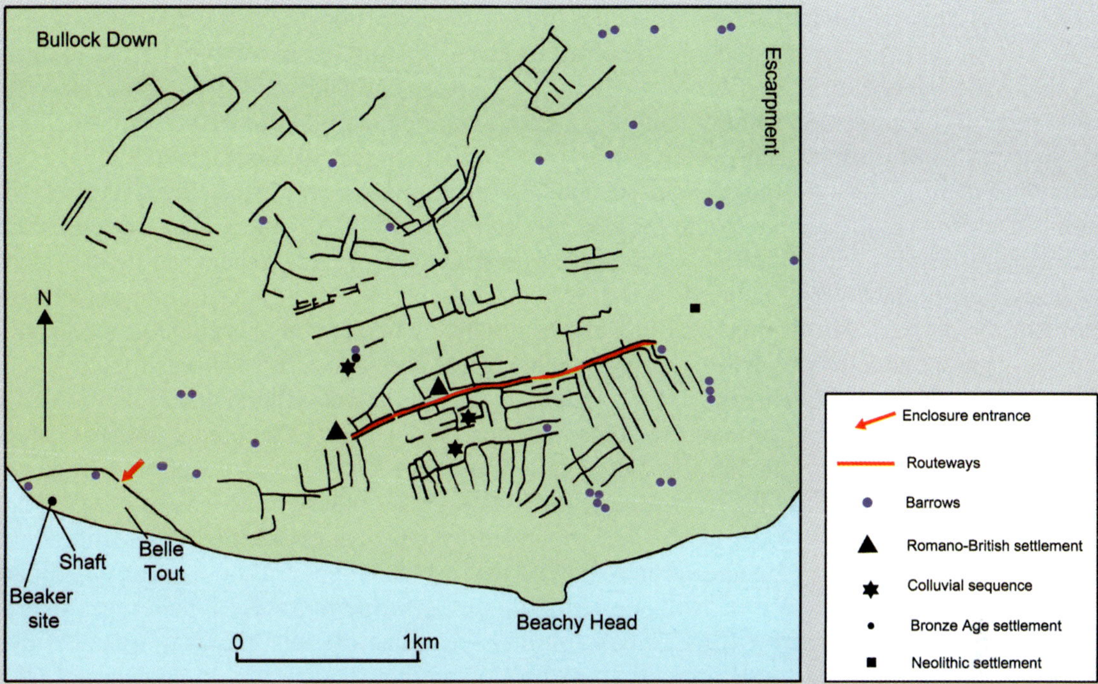

Figure 10.4 Fields and trackways at Bullock Down, East Sussex (graphic: J. Foster after Drewett 1982).

Other downland routes

Where really extensive landscape-scale excavations have taken place, the relationships between features can be especially revealing as an example at Peacehaven shows (Fig. 10.5). Here there is evidence for successive routeways following the line of a dry valley, part filled with later prehistoric colluvium (Hart 2015). The possibility of early origins for one route is suggested by a curvilinear band (Fig. 10.5A–B), formed by early Neolithic pits (*c* 3600 cal BC) and later Neolithic and Bronze Age barrows. Then at the end of the early Bronze Age a more clearly defined routeway between parallel ditches (C–D) was established at 35° to the earlier alignment. From this axis ditched field boundaries were laid out in the middle Bronze Age, *c* 1500 cal BC, with associated settlement and cemetery areas. There followed a period of abandonment after which, *c* 400 cal BC, middle Iron Age settlements, possibly of seasonal character, were established. Paddocks and tracks defined by parallel ditches were created in the southern part of the earlier Bronze Age field system but on a different, and less consistent, alignment. Thus, this landscape has three distinctive phases of trackways on different alignments, the first persisting perhaps for two millennia, the later two representing shorter periods of more intensive land use in the middle Bronze Age and middle to late Iron Age. All three routes trend in an east–west direction along the sides of the valley, perhaps connecting the grazing and arable of this plateau area with the valley of the River Ouse 2 km to the east.

The picture that is emerging is one of early routeways linking contrasting ecological zones: in the Peacehaven case the Tertiary plateau linked to the river valley, in the case of Bishopstone and

Bullock Down along downland spurs at right-angles to the escarpment. Some significant evidence for early routes at right-angles to the escarpment lies hidden in publications written in the early twentieth century before, as argued in Chapter 1, trackway research was nipped in the bud by the erroneous ideas of Alfred Watkins (1925). Two publications by Curwen and Curwen (1914; 1926) document what appears to be an 8.5 km route linking the Coastal Plain up along a downland south-facing dip slope spur to the Weald in the north (Fig. 10.6). On the northern part of this route at Saddlescombe multiple intercutting lines of trackways descended the escarpment towards the Weald as planned in Figure 10.7a. Presumably as one route became too gullied, or muddy, others were adopted. Some trackways are deeply incised hollow ways up to 5 m deep (Fig. 10.7c), others, on gentler slopes, take the form of terrace ways, their edges defined by lynchets (Fig. 10.7d). As Figure 10.7a shows, upslope of the tracks is a well-preserved field system which when examined on the ground is more extensive than shown on the Curwen's original plan. Three Romano-British pots, probably from a burial (Fig. 10.7e), were found in one of the lynchets of the field system (marked as H on Fig. 10.7a). Lynchets define the edges of some of the fields and thus the tracks appear to be at least as early as the Roman period. South of this the route they called 'the Port's Road' was traced by Curwen and Curwen (1926) to the mouth of the river Adur (Fig. 10.6a). This southern route was marked in places by terrace ways and hollow ways, with edges bounded by lynchets of 'Celtic' fields (Fig. 10.6b). The Curwens excavated on Round Hill and showed that a lynchet edging the trackway had partly accumulated prior to deposition of Romano-British pottery, furthermore the lynchet was overlain by a small undated round barrow (Fig. 10.6c). Another section across this trackway was later excavated in advance of making the Brighton bypass (A27) and this showed that the lynchets flanking the track were of at least as early as the Romano-British period (Rudling 2002). Part of this track was also excavated within the deserted medieval village at Hangleton which it pre-dated (Holden 1963). One stretch had the name 'The Drove'; it seems reasonable to infer that this was one of the drove roads linking the coastal plain to the seasonal pastures in the Weald.

Four kilometres west, another of the north–south routes down an escarpment spur on Thundersbarrow Hill (Fig. 10.6a) has a revealingly complex history. Thundersbarrow Camp originated in the middle Bronze Age as a sub-rectangular enclosure with two opposed entrances to east and west (Fig. 10.8 arrowed; Curwen 1933; Drewett *et al.* 1988, 131; Hamilton 2003). This was succeeded in the sixth–fifth centuries BC by a larger enclosure with opposed entrances in the opposite direction, north–south, this time on the line of the spur and leading out to a double lynchet trackway flanked by fields. This case illustrates not so much continuity, as a rather dramatic shift in the axis of communication, and it rather neatly indicates that this spur crest route is of about middle Iron Age date, which significantly accords with the Iron Age date which the Curwens suggested for the Ports Road. Several of the north–south routes on Figure 10.6 have been discussed by a succession of writers from 1818 onwards but in terms of possible alternative routes for a Roman road from London to the coast at Brighton, or the mouth of the Adur (Shields 2005). However, the preoccupation of these writers with a single Roman route, has served to obscure the existence of a series of earlier prehistoric routes.

Thirty kilometres west along the South Downs a recent project 'Secrets of the High Woods' has produced a new picture of the prehistoric landscape and its routeways. Here between Arundel and Petersfield this Heritage Lottery Funded project is investigating the archaeology of a landscape hidden by forestry (Thorne 2016a; 2016b; Manley 2016). Much of the area has been forested for centuries, but small patches of 'Celtic' fields had been recognised and protected in areas of grassland.

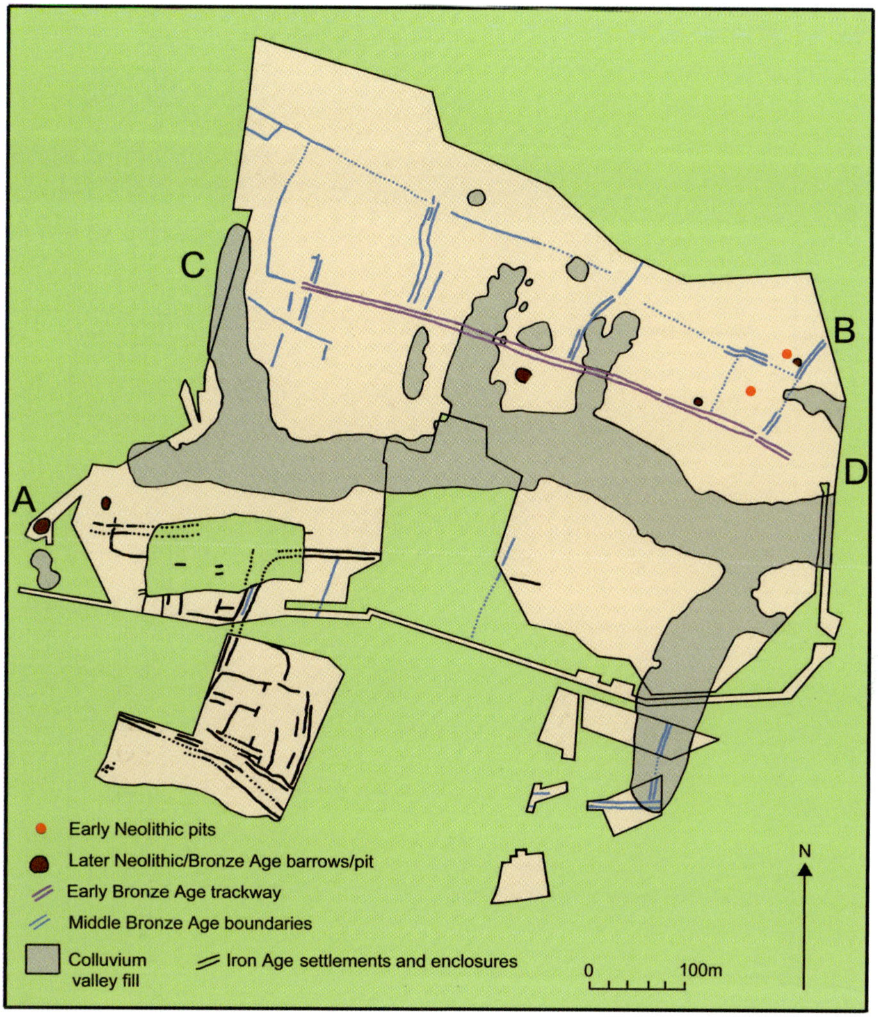

Figure 10.5 Peacehaven, East Sussex: extensive excavated area with Neolithic and early Bronze Age linear alignments, Bronze Age and Iron Age trackways, fields and settlements (graphic: J. Foster after Hart 2015).

Survey using LiDAR, has enabled the trees to be digitally stripped away revealing previously unimagined extents of 'Celtic' fields and trackways over much of the 305 km^2 area surveyed. Such results open up the possibility of understanding routeways on a greater scale than previously possible. The pattern is both complex and multiphase and presents significant challenges in terms of dating. Figure 10.9a shows a small part of the area in East Dean Woods. Within the 'Celtic' fields near the middle of the picture can be seen two double lynchet trackways running horizontally. To the left, marked by arrows, one broadens into a funnel as it approaches what at one stage would have been the open downland of the South Downs ridge crest. That funnel is suggestive of animals being driven from the open pasture on the crest, down the trackways to the lower dip slope and coastal plain pastures and settlements. The fields might date anywhere between the middle Bronze Age and the Romano-British period. Some are clearly earlier than the early Roman road Stane

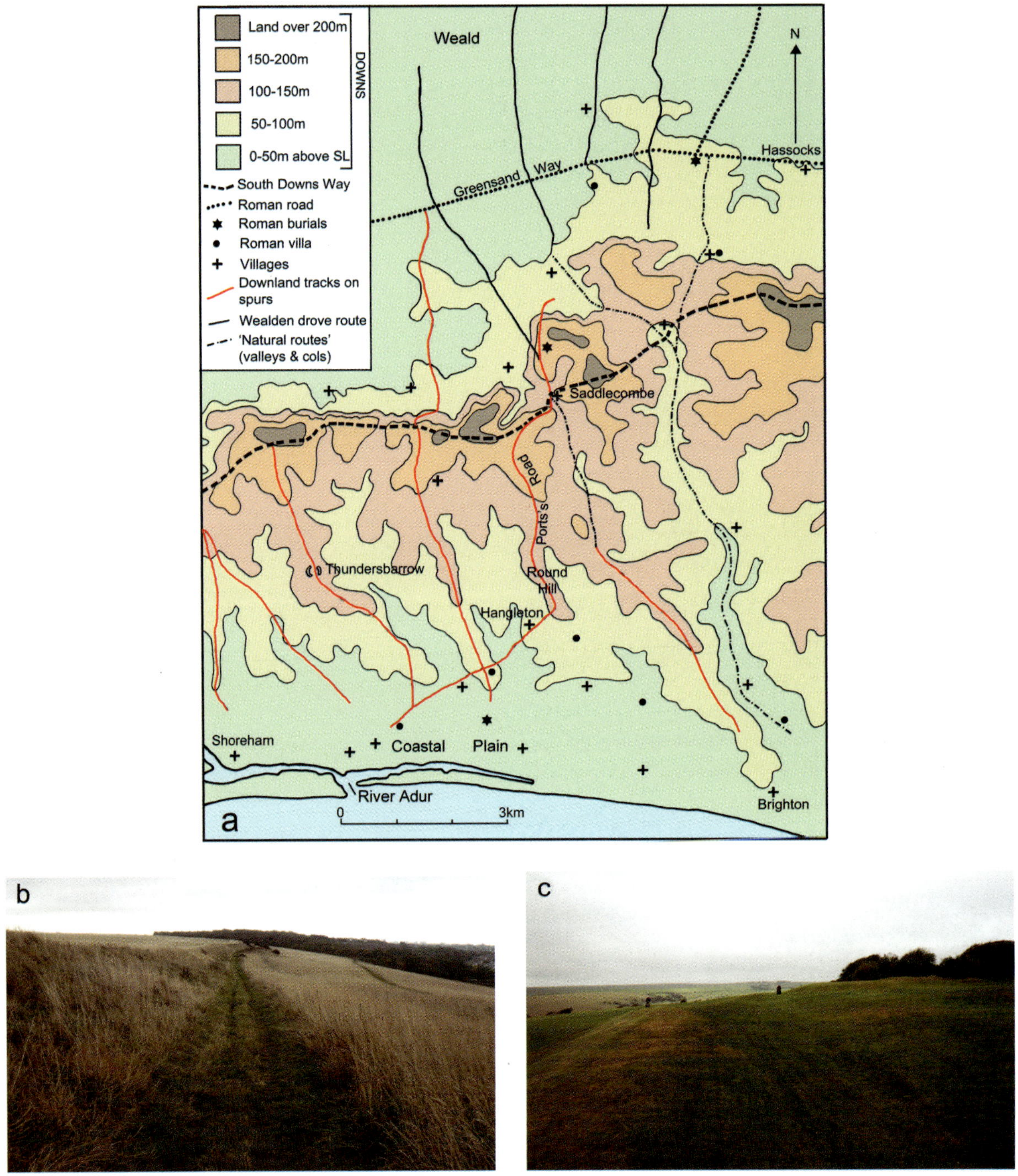

Figure 10.6 The South Downs, East Sussex in the Saddlescombe/Port's Road area showing: (a) trackways, mostly on dip slope spurs (graphic: J. Foster after Shields 2005); (b) double lynchet trackway on south side of Round Hill; (c) double lynchet trackway on the crest of Round Hill with barrow on lynchet to right (photos: M. Bell).

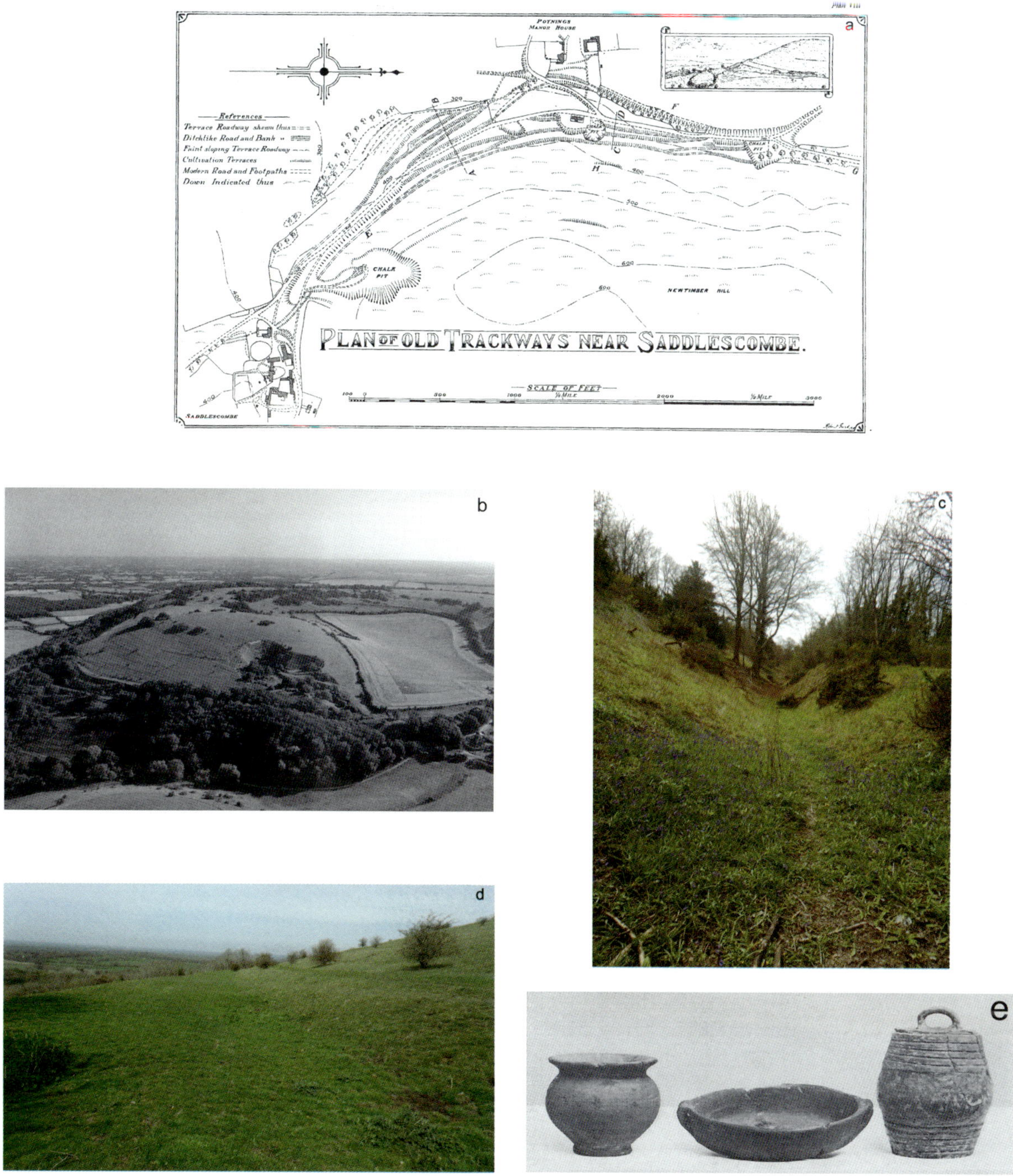

Figure 10.7 Trackways at Saddlescombe, East Sussex: (a) plan (after Curwen and Curwen 1914); (b) view of the trackways; (c) hollow way; (d) terrace way; (e) Romano-British pottery from lynchet (photos: c–d: M. Bell; e: Curwen and Curwen 1914).

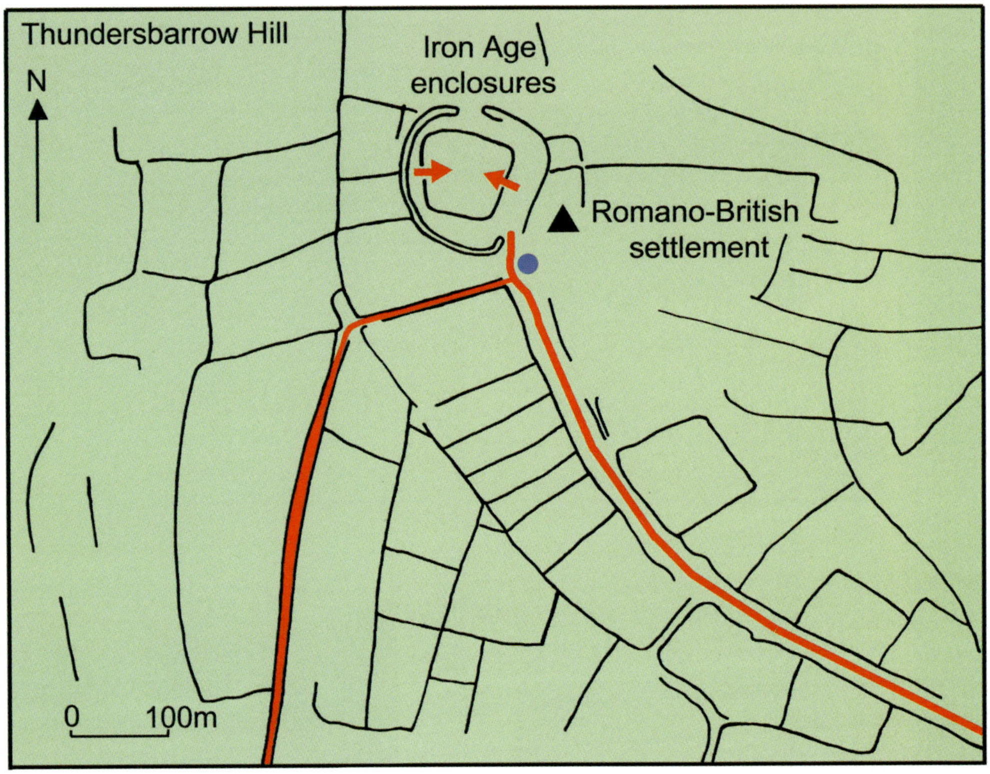

Figure 10.8 Thundersbarrow Hill, East Sussex showing the relationship between enclosures, routeways and fields (graphic: J. Foster after Curwen 1933).

Street from London to Chichester which cuts across them (Eve 2014). The scale of some lynchets, 3–4 m high in places, indicates a long period of use for those fields. A section was cut across one of the double lynchet trackways (Fig. 10.9b) and an OSL date obtained from the mid-point of the lynchet depth showed it was accumulating *c* AD 427±100 (GL16186; P. Toms, pers. comm.). Few artefacts were found in the lynchet and the sediment which formed it was decalcified, limiting the range of dating options, compared, for instance to the Lyminge hollow way case study below, where dating was achieved through the application of multiple techniques.

The Rother valley

The north–south routes identified on the downland spurs can be projected to connect with routes running into the Weald thus, in the medieval period, linking parent manors to the south with daughter pastures to the north. On the Lower Greensand in the Rother Valley these routes are marked by deeply incised hollow ways some 5 m deep (Fig. 10.10) which, during high rainfall events, act as temporary stream channels; they carry floodwaters and soil eroded from adjacent arable fields down to the floodplain of the Rother (Fig. 10.10b; Boardman 2013). This illustrates one of the ways that, in addition to traffic by people, animals and vehicles, these routes have become so deeply incised (Fig. 10.10c–d). The lanes in this area do not generally follow topographic depressions such as valleys but, to provide drier routes, tend to be on the slopes at interfluves between tributary

Figure 10.9 East Dean Woods: (a) LiDAR image of fields and trackways; (b) section across double lynchet trackway, marked by 2 m scale (images: courtesy of Secrets of the High Woods Project, South Downs National Park).

streams. The result is that development of the hollow ways has significantly extended the natural drainage network creating new channels for water and sediment transport. Boardman's perceptive study introduces some key principles with implications well beyond the Rother Valley. It illustrates the connectivity between human agency (the routeways) and geomorphic (i.e. landform) processes. Both have agency: the pattern of people's movement has changed the topography and drainage; the deep incision of hollow ways, through a combination of human actions and water erosion, has created landscape structures which influence future patterns of human activity. Both contribute in a marked way to the construction of this particular niche.

Land allotment, tracks and fields in the Low Weald

On the Wealden Clay of West Sussex the north–south axis of drove ways correlates with the historically attested, and place-name, links between parent manors to the south and outlying holdings in the Weald (Chatwin and Gardiner 2005). The coastal plain settlement of Poling is linked to the Wealden holding Palingas Schittas (the swine cott of the people of Poling) and to other outlier holdings at Pallingham and Pallinghurst. Earliest among the extant boundaries in this landscape are arc-shaped and oval areas which are thought to represent early clearing enclosures (Fig. 8.4b). Among the strongly linear north south axial boundaries of this coaxial landscape some curve round, and apparently post-date the earlier enclosures. Within the coaxial landscape early linear manors can also be identified, one up to 12 km long and only a few hundred metres wide, at Fure and Marlpost (Fig. 8.4b; Chatwin and Gardiner 2005). It is also notable that the main axial alignment of routeways is broadly parallel to the Roman roads in the western Weald (Chatwin and Gardiner 2005) but that leaves unresolved the question of which is the earlier. Understanding has recently been transformed by extensive landscape-scale excavations occasioned by housing developments at Winkhurst Green, Sussex and Horley, Surrey which have revealed early boundaries of middle to later Iron Age and Romano-British date, some on the same orientation as surviving boundaries and trackways. This suggests that parts of today's landscape, including droveways, may be of Iron Age or Romano-British origin (Margetts 2018a; 2018b). Further evidence comes from extensive excavations associated with quarrying at Bletchingly, Surrey on Lower Greensand where a Green Lane at right angles to the North Downs escarpment has medieval ditches on the same alignment and a scatter of Bronze Age features along its edge, hinting at prehistoric origins; it leads up the escarpment to War Coppice Iron Age hillfort (Marples and Poulton 2019).

The North Downs

A route leading along the North Downs escarpment from Dover to Winchester has long been proposed in the historic and antiquarian literature. The North Downs Way was described by Hilaire Belloc (1904) and is marked on the Ordnance Survey (1956 and 1962) maps of Iron Age and Roman Britain (Fig. 10.1). It was accepted as the most important prehistoric route in the south east by Margary (1965) in his work on Roman roads and more recently by Brookes (2007) in considering Anglo-Saxon routes. No doubt one reason it gained such acceptance was that it seems a logical 'natural' route from Dover, the closest crossing point from continental Europe, to Wessex. However, claims for its early origin were critiqued by Turner (1980) who showed that artefact distributions did not support the idea of a major east–west route in prehistory, rather they pointed to the importance of the Thames river corridor to the north. Turner makes the cogent point that

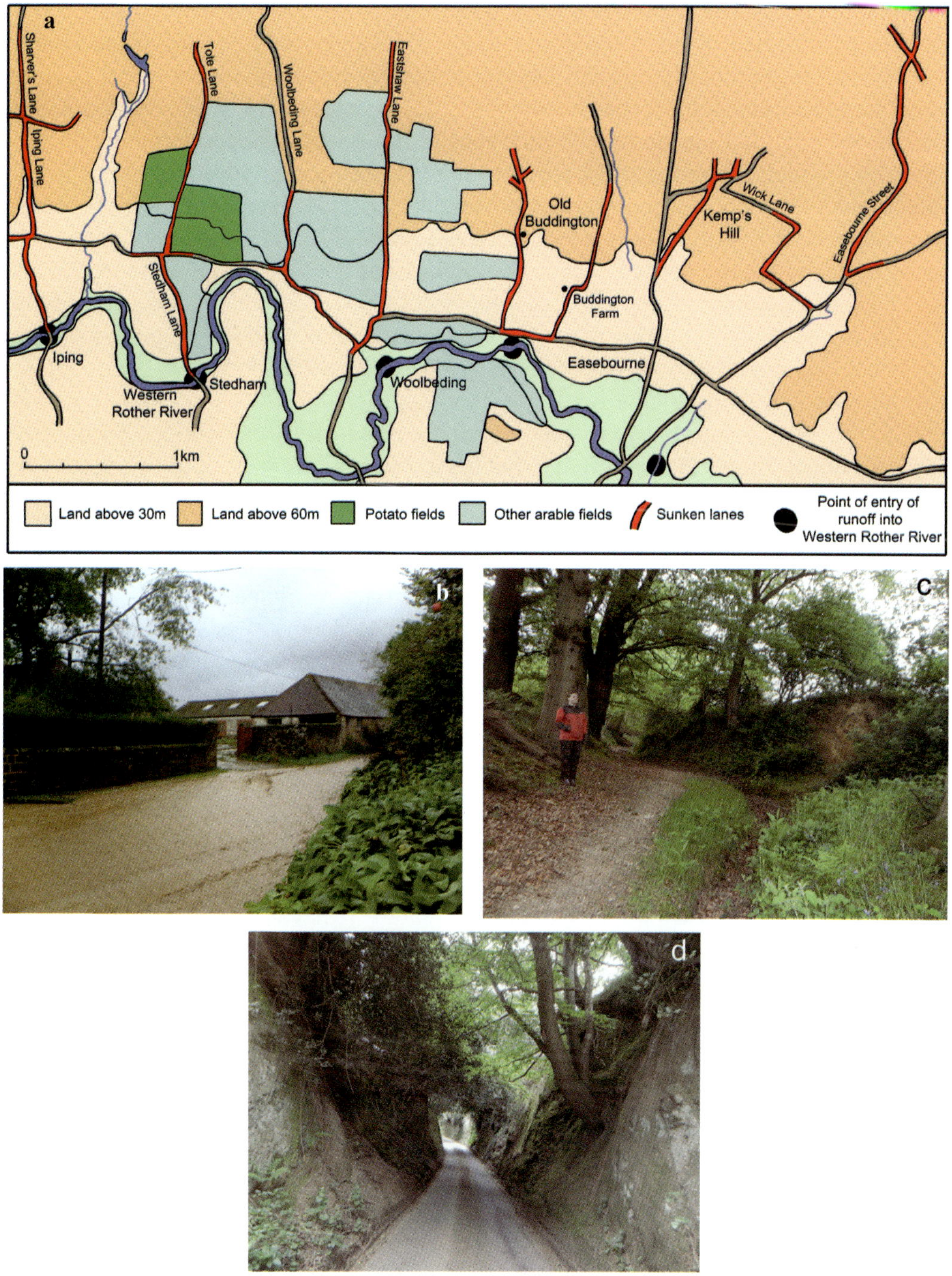

Figure 10.10 Sunken lanes in the valley of the River Rother, West Sussex, UK: (a) map (graphic: J. Foster after Boardman 2013); (b) water carrying sediment on the road at Stedham bridge close to the river (photo: J. Boardman); (c) sunken lane at Lodsworth; (d) the same lane deeply incised approaching the river at Ambersham Bridge (photos c–d: M. Bell).

there is a lack of prehistoric metalwork finds where the North Downs Way crosses rivers which might have been expected if this was a major route. The same could be said of the South Downs Way. The claimed route along the North Downs comprises two, sometimes more, parallel routes which now form national long-distance footpaths: the North Downs Way which generally runs along the Downs, and the Pilgrims Way, which runs near the base of the escarpment just above the spring line along which settlements of later Saxon origin are located. Figure 10.11a shows an area immediately east of the point where the River Medway dissects the escarpment where a series of tracks was mapped by Everitt (1986) at right angles to the ridge routes and the topography. These are more-or-less parallel and spaced at intervals of 1–2 km. Everitt interpreted these as the drove ways used for the movement of animals from the North Downs and the Thames Estuary lowlands to their north to the pastures of the Weald in the south. Everitt (1986, 39) makes the important observation that they cut directly across the Pilgrims Way and North Downs Way without deviation, dog-leg or termination, which might imply they were earlier.

Insights to the relationships between these routes are provided by excavations for the HS1 (High Speed 1: the Channel Tunnel Rail Link) at Whitehorse Stone at the west edge of Figure 10.11a and shown in more detail in Figure 10.11b (Booth *et al.* 2011). Possible early origins for routeways here are indicated by at least three, possibly more, early Neolithic tombs, including Kit's Coty House beside a hollow way which descends the escarpment; other sarsen stones may represent the remains of tombs, or wayside markers. The Neolithic tombs are part of a restricted group flanking the Medway valley and whilst they could indicate early routes, either north–south or east–west, they might equally point to the significance of early Neolithic riverine colonisation, the earliest Neolithic dates from Britain being round the Thames Estuary (Whittle *et al.* 2011). In the excavated area (Fig. 10.11b) Neolithic features, including a house site, were in a pronounced band 35 m wide just above the floor of a minor valley. This band was interpreted by the excavators as a possible routeway and it later became the line of a Roman road flanked by ditches. This would indicate a very early route down the escarpment which persisted from the Neolithic to the Roman period. An Anglo-Saxon burial of *c* sixth–seventh century date was found at the point where this route crosses the Pilgrims Way; burials of that period frequently occur at crossroads. In fact Brookes (2007) accepts the Pilgrims Way as a significant Saxon route because of its association with burials. Brookes and Harrington (2010) map the early Saxon estates which are distributed along Roman roads, but not along the Pilgrims Way, which may be further evidence that it only became a significant route in the mid to late Saxon period. The Pilgrims Way forms a parish and hundred boundary and these were established between the seventh and tenth centuries. A section across the Pilgrims Way revealed successive hollow ways, the earliest of which overlay a soil which was cultivated into the Roman period. Prior to the mid-Saxon period there is little evidence, except the rather equivocal testimony of the sarsen stones, to suggest earlier origins for the east–west ridgeway related lines of the North Downs Way and the Pilgrims Way. The position of the HS1 excavation is particularly fortuitous since, for reasons of topography near the Medway crossing, the North Downs Way and the Pilgrims Way share the same route at the point where they have been shown to be post-Roman. There being no obvious alternative route here, this casts significant doubt on the prehistoric origins of the escarpment route, whereas the route at right angles to the escarpment may have Neolithic origins.

Immediately west of the Medway in the Cobham area extensive excavations, again related to HS1 and road widening, exposed a landscape with hollow ways, ditched tracks and metalled tracks of later middle Bronze Age and mid Iron Age date, the latter with wheel ruts (Booth *et al.* 2011;

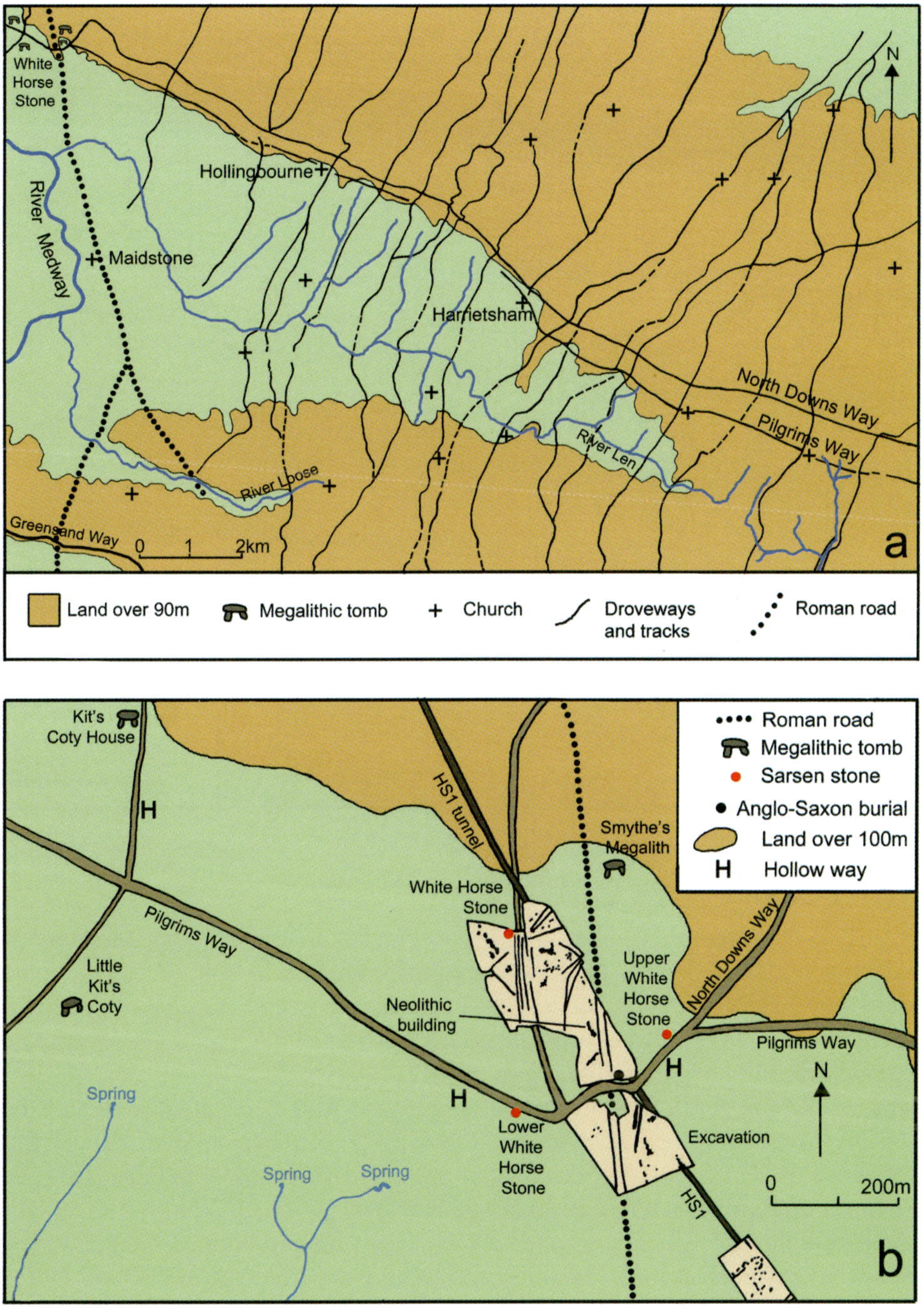

Figure 10.11 The North Downs east of the River Medway, Kent: (a) map showing North Downs Way and Pilgrims Way and drove ways at right angles to these and the escarpment; Whitehorse Stone is top left; (b) Whitehorse Stone excavations in relation to early routeways (graphics: J. Foster after Booth et al. *2011).*

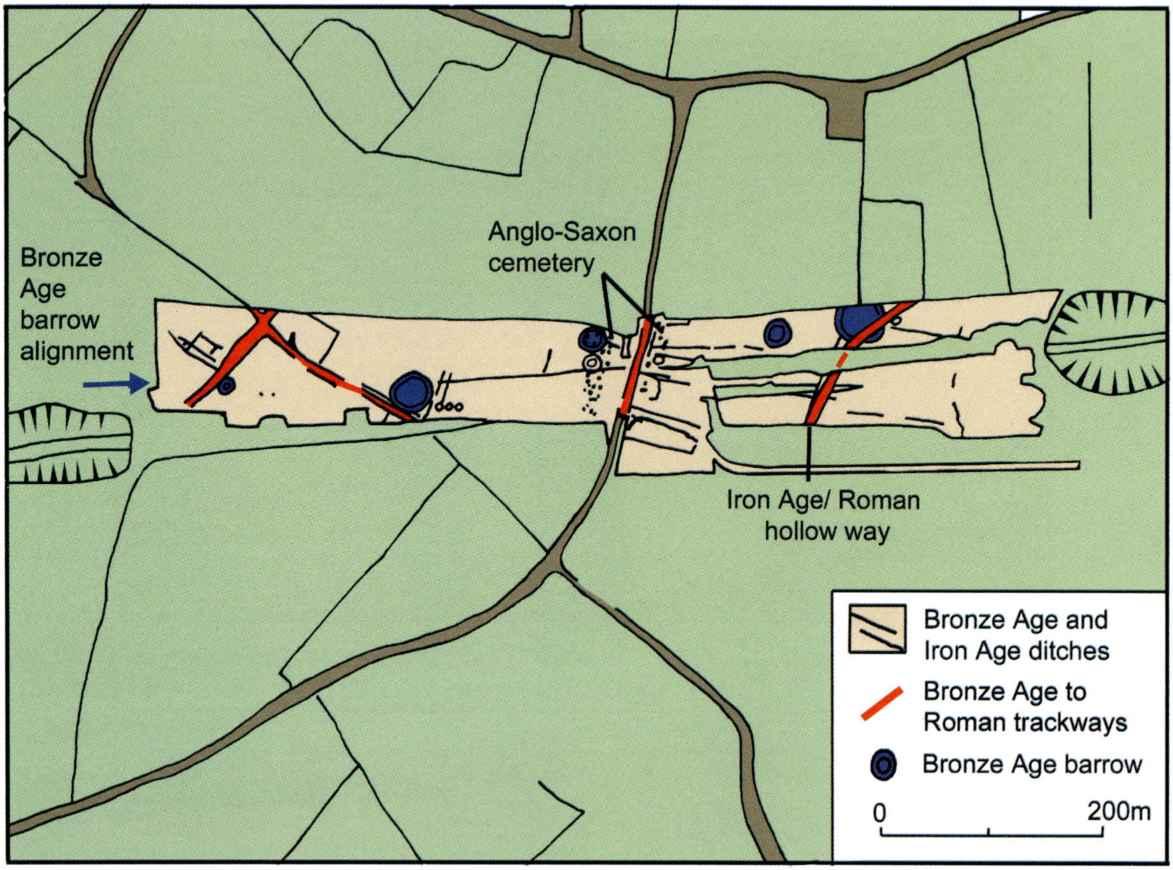

Figure 10.12 Saltwood Tunnel, Kent: barrow alignments, prehistoric to Roman ditches and trackways showing long term continuity into recent tracks (brown) and boundaries (graphic: J. Foster after Booth et al. *2011).*

Allen *et al.* 2012). The general orientation of these routes is from the Thames Estuary 3 km to the north up a dry valley to the downs. There are hints from the relationship of pits and other features that one route may have Neolithic origins. However, the landscape is quite a dynamic one and in later periods there are changes in trackway and orientation whilst maintaining an overall north–south axis of movement at right angles to the topography.

At the east end of the North Downs, but not directly on what is designated as the North Downs Way or Pilgrims Way, fascinating insights to the longevity of routes are provided by further excavations in advance of HS1 north of Folkestone at Saltwood Tunnel (Booth *et al.* 2011; Fig. 10.12). The earliest linear alignment is of five Bronze Age barrows, two dated 2290–1940 cal BC, on *either side* of a dead straight line which suggested a path. Then double ditched trackways with coaxial field systems were established in the middle Bronze Age and Iron Age at 30° and 110° to the earlier alignment. There was also a hollow way dated to the Iron Age and Roman period. Beside one of these roads, close to a crossroads, a cemetery was established in the Anglo-Saxon period and the significance of place perpetuated in the late Saxon period as a hundred meeting place (where local communities gathered to make legal decisions). What is especially striking about the plan revealed is that the prehistoric trackways are clearly perpetuated in roads, tracks and hedge lines in the recent landscape, demonstrating that some routeways and boundaries in today's landscape clearly originated in the middle Bronze Age.

A converse case comes from Church Whitfield, not far away (Parfitt 2014, 180). This excavated site is at a crossroads which was thought to be of some antiquity. An Iron Age enclosure ditch was astride a ridgeway with one entrance corresponding to the axis but there was no opposite entrance, and thus no through route in the Iron Age. Later an Anglo-Saxon village developed and by that date the crossroads seems to have been in existence.

A case study in multi-method dating: Lyminge, Kent

Also at the east end of the North Downs, inland of Folkestone, is Lyminge, which provides perhaps the clearest demonstration of how hollow ways may be dated (Bell *et al.* forthcoming). Here, in a valley on the dip slope, was a major Anglo-Saxon royal centre of the fifth–seventh centuries, succeeded by a monastery in the seventh century (Thomas, G. 2013). Running west from the village is a hollow way incised in places up to 4 m. It crosses a Roman road, Stone Street, and descends the chalk escarpment into the Weald (Fig. 10.13a). Collapse of a retaining wall along this route 1.3 km west of the Saxon site revealed that in one area the uphill side was formed of a very substantial positive lynchet which had built up against the routeway (Fig. 10.13b). As previously noted such colluvial accumulations on the upslope side of hollow ways are readily identified because the natural contour profile is broken by a roughly level terrace on top of the bank. It proved possible to date this lynchet accumulation using a range of techniques. The first phase of the lynchet contained pottery of late Bronze Age and Iron Age date. Samples from a second phase of the lynchet were analysed for land molluscs, which proved to be exceptionally abundant, indicating that the sediments had accumulated gradually (Fig. 10.13f). Three of the species present provide clues to date because they are species that were introduced to Britain at approximately known dates (Kerney 1999). Two were introduced in the Roman period: *Monacha cantiana* (Fig. 10.13c) which appears 0.5 m above the base; and *Cornu aspersum* which appears about half way up. *Candidula intersecta* (Fig. 10.13d) occurs near the top and was introduced in medieval or later times. Uranium Series dating on shells through the sequence by Dr Stuart Black shows that shells in the base of the sequence, which were *Pomatias elegans,* and almost certainly residual from an earlier woodland soil, are *c* 3000 years old and molluscs at the base of the second phase lynchet itself are *c* 2000 years old. The latest dated samples are about 500 years old. Optically Stimulated Luminescence dating by Dr P. Toms using quartz grains *c* 0.5 m from the base of the sequence dated to *c* 1700 years ago and 1 m from the base of the sequence about 1200 years ago. The fact that the OSL and Uranium Series dates fit the same straight-line graph (Fig. 10.13e) shows consistent results which are also in agreement with the artefactual and molluscan evidence. We can infer from this that the first phase lynchet may have originated in the late Bronze Age or Iron Age and the lynchet, and thus the trackway which it bounds, was certainly established by the Romano-British period.

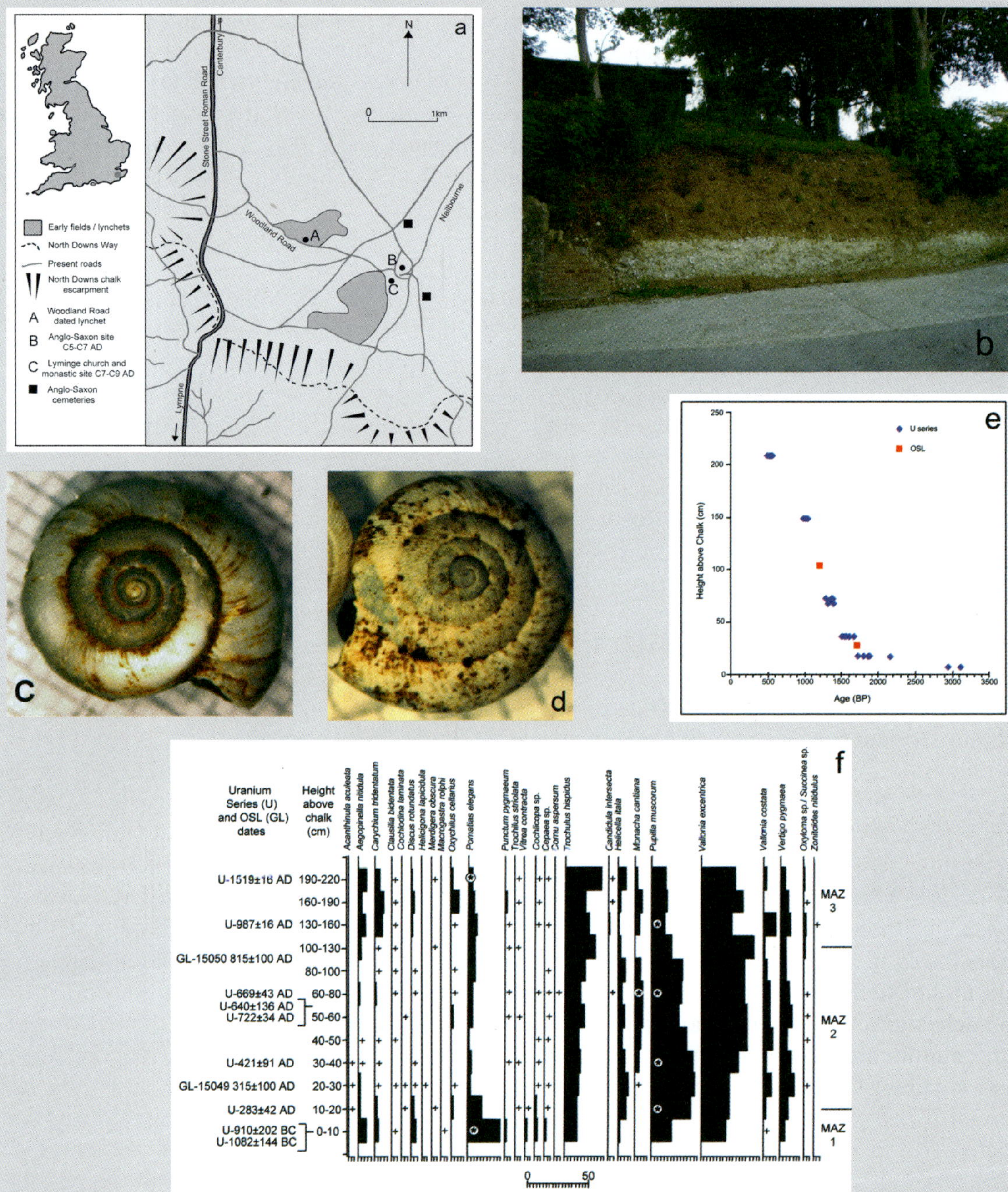

Figure 10.13 Lyminge, Kent: (a) map showing Lyminge area; (b) section of lynchet; (c) land mollusc Monarcha cantiana*; (d) land mollusc* Candidula intersecta*; (e) Uranium Series dates from molluscs and Optically Stimulated Luminescence dates from sediments in the lynchet; (f) land mollusc diagram with dates (GL =OSL, U = Uranium Series) (graphics: J. Foster; photos: M. Bell).*

Riverine and maritime connections

The question remains, how do the axes of connectivity identified in this chapter relate to the theme of Chapter 9, riverine and maritime connectivity? Continental connections via the straits of Dover are evident from the Dover boat, the probable Langdon Bay wreck and also from the distribution of continental metalwork in the later Bronze Age and Iron Age. The south-east English distribution of artefacts in both periods indicates that transport within the region was mainly riverine and coastal rather than overland (Cunliffe 1991, fig. 20.7; Lehoërff 2012, fig. 96). It has been noted that the coastal end of the Wealden rivers were substantial estuaries until the medieval period and these would have represented significant obstacles to the hypothetical ridgeway routes. There do seem to have been routes from the down top escarpments into the river valleys marked by deeply incised hollow ways, but these might indicate embarcation, rather than crossing, points. The now eroded Wealden dome has imposed a radial pattern of drainage which is comparable in orientation to the cross topographic drove way routes which have been a key theme in this chapter. It seems probable that they shared complementary roles. On the south coast, for instance, there are supposed Romano-British small trading river ports at Barcombe, 14 km upriver from the sea and at Bodiam 24 km up the River Rother. By the Norman period there were castles and presumably small ports at Lewes, 11 km up the Ouse, Bramber 7 km up the Adur and Arundel 6 km up the Arun.

Conclusions

This chapter has provided the opportunity to focus down on one region and to look beyond individual sites at ways in which different types of landscape were connected. By looking at the relationship between trackways which survive in today's landscape and archaeological evidence it has been possible to identify the early origins of some routes. Particularly notable is the survival of Bronze Age tracks into the present landscape at Saltwood Tunnel (Fig. 10.12). A theme of this chapter has been the origin of the cross-topographic routes. Of these the Ports Way and the Thundersbarrow Hill routes down spurs at right-angles to the escarpment (Fig. 10.6a.) seem to originate in the Iron Age. The Whitehorse Stone route down the escarpment has been argued to be Neolithic and was followed later by a Roman road (Fig. 10.11b). The Lyminge route is of probably Late Bronze Age or Iron Age origin, certainly it was well established by the Roman period well before the Saxon settlement to which it leads (Fig. 10.13). This example and the lessons drawn from Boardman's work in the Rother valley show that by combining landscape archaeology with geoarchaeological approaches to routeways it is possible to establish their date and wider significance. On the basis of the, admittedly slender, number of examples available it seems that the cross-topography routes may have prehistoric origins. It is these which have been previously assumed to originate in seasonal movement of stock attested in Saxon and medieval times. On present evidence, however, these seem to be essentially local routes of some 10–30 km relating to access to grazing and materials. It is not yet possible to single out specific cases from the many parallel drove ways which may represent more significant axes of longer distance travel, though these should be looked for in the future.

11. Conclusions: why paths matter

Bodily engagement, perception, anthropology and literature

The almost boundless nature of archaeology is both daunting and exciting. It is a discipline concerned with material evidence for the entire history of human activity. This is daunting, because so many lines of evidence contribute and are inevitably fragmentary, due to the destroying hand of time. It is exciting, because of the opportunities for interdisciplinary engagement which are much in evidence in the study of past patterns of movement. The daunting aspect of studying past routeways has too often been emphasised. Today, however, past patterns of mobility are the subject of active academic engagement, not just in archaeology, but in several cognate disciplines. Especially influential in this regard has been the anthropologically-developed ideas on mobility of Tim Ingold (e.g. Ingold and Vergunst 2008), who contends that one knows the world not purely by thinking but through the bodily engagement of movement, which is central to the way we structure space and understand our identify and place in the world. There is also a much wider and growing recognition of the cultural role and value of walking. Representative of this are the writings of Robert Macfarlane (e.g. 2012) who conveys empathy with landscape and enthusiasm for what can be perceived through movement. Such concepts are accessible to everybody because, at some level, this is something we all do all the time. The observations that we make are not a series of separate still frames, like photographs, but their significance is derived through movement, from sensory engagement creating perspectives, which change and are enriched as we move.

Prehistorians are increasingly aware of the need to develop an approach to mobility which can animate prehistory; this helps us to escape from a static picture of the past, with its focus on individual sites, to one which looks at the relationships between places as part of living landscapes. Many of the advances so far have been in terms of theoretical perspectives. Moving forward requires consideration of the full range of sources of evidence for past patterns of movement and the development of a practical toolkit of approaches for their application in the field and laboratory.

Steps forward

An obvious first step is getting out into the field, developing an understanding and empathy with existing routes and the landscape of which they form part. The approach draws on phenomenological perspectives but seeks to enhance this with other sources of evidence to facilitate evaluation of the strength and reliability of subjective interpretations. The same point applies to the identification of 'natural routes', those places where constricted topography, such as a watershed between two valleys, suggests a route once existed. They too can only be accepted if there is independent evidence demonstrating their use at a particular time, e.g. sites or rock art along the route.

Conceptual changes are also required, a shift of focus from the individual site to a far wider landscape perspective. Drawing on his work on prehistoric routeways in the Netherlands, Bakker (1976, 76–80) outlined some principles for the identification of ancient roads, particularly highlighting the need to draw on a combination of prehistory, protohistory and historical geographic sources and make full use of the evidence provided by topographic maps. As Bakker (1976, 79) acknowledges, there is a great deal of fragmentary information, e.g. in individual site reports, but there remains the considerable challenge of putting this together into an understanding of wider patterns of communication. Although, much significant work has been done recently, the essential challenge still remains 40+ years after Bakker wrote this. Other sage advice not yet adequately responded to is Coles' (1984) entreaty that we look at places where supposed ridgeways cross wetlands, and Bowen's (1961) emphasis on the need to establish ways of dating hollow ways and working out their relationship to the ridgeways, a call which has been responded to in Chapter 10.

A wider understanding of routeways and their significance also requires the identification of recurrent characteristics of routeways produced in different areas and by different cultural groups, thus developing a comparative approach to their study. A conference on *Landscapes of Movement* published by Snead *et al.* (2009) adopted this approach for cases mainly in middle America and the American South West. Variables included the amount of construction involved, technology of movement, sites at terminals, functions, scale and meaning. Many of those cases had the benefit of direct ethnographic evidence so that it was possible to establish that many routes had a primarily ideological, rather than practical purpose, such as linking settlements to ceremonial centres. A relationship was also demonstrated between the characteristics of routeways and the social complexity of the groups concerned. The semi-desert context, relatively late date and available anthropological perspectives contrasts with the European cases. However, as Chapter 2 on North America set out to show, perspectives from very different areas can stimulate new approaches and ways forward, in that case highlighting the contribution of niche construction and the manipulation of plant resources to the creation and marking of routes.

Multi-scalar and multi-disciplinary approaches

This book has progressed from the local scale of footpaths and footprints to longer routeways and riverine and marine movement. Evidence at a local scale has been deliberately foregrounded. This approach is encapsulated by two terms, often employed in a pejorative way, but amenable to more positive definition. One is *parochial*, highlighting the benefits of knowledge gained from an intimate understanding of an area. The other is *pedestrian*, highlighting the value of walking as a way of knowing. Our approach cannot, however, confine itself to a local scale; it is necessarily multi-scalar, from a footprint-track to a long-distance route.

Figure 11.1 Human footprints in estuarine silts at Goldcliff, Wales; two young people were represented with estimated ages 4–5 and 10–11 (photo: E. Sacre).

In the case of footprints, particularly important contributions concern the population structure and the axes of movement (Fig. 11.1). Prehistoric footprints, discussed in Chapter 4, highlighted the role of children in prehistoric communities. Footprints can also help us understand the relationship between animal and human movement and provide a new and independent perspective on human/animal relationships based on the animals actually present at the spot in question. This has, for instance, contributed to an understanding of the seasonality of early hominid activity at Laetoli and evidence for the role of saltmarsh exploitation in the later prehistory of the Severn Estuary. In considering the footprint evidence, a relatively new topic for most archaeologists, it has been important to identify the extent of present evidence worldwide and the range of sedimentary and topographic contexts in which it occurs. The preservation of footprints is inevitably restricted to certain geographic and sedimentary contexts and that will influence the conclusions we can draw. However, logic and observation suggests that this form of evidence is far more widespread than is generally appreciated. Wider recognition and recording could make a very significant contribution to work on patterns of past human movement.

Research is inhibited if we confine ourselves to one environment type or one class of evidence. As Chapter 6 showed, the connectivity implications of wetland tracks have been limited because so few wetland routes have successfully been related to those on dryland. Maybe many were so ephemeral and short-lived that there was no detectable continuation, but the question remains how often have such continuations been actively sought. A notable exception is the suggestion that Corlea Irish bog road formed part of a much longer route apparently linking two royal centres (Fig. 6.9).

Two chapters have focused on wetlands, Chapter 4 on footprints and Chapter 6 on constructed tracks, mostly of wood. It is striking how few trackway sites, beyond the Severn Estuary, have also produced footprints. This is puzzling, because in many ways wetlands provide ideal contexts for footprint preservation. Again, we must consider whether we have always looked hard enough. A common theme linking footprint-tracks and constructed trackways, particularly in wetlands, is the contribution they make to understanding patterns of connectivity at both a site and landscape scale. We have seen how, in the Severn Estuary, convergent patterns of both footprint-tracks (Fig. 4.10) and wooden trackways (Fig. 6.8) may enable us to predict the locations of occupation sites and activity areas and how, when, in the case of the Somerset Levels, the evidence for convergence can be combined with palaeoenvironmental evidence for foci of activity, the settlement pattern picture can be further refined (Fig. 6.7).

The value of a broad-scale and multi-period approach is demonstrated by excavations in advance of HS1, the high speed rail link between London and the Channel Tunnel (Booth *et al.* 2011). Because that work originated in a modern communication route, it was logical that the archaeological investigation included routeway themes and this resulted in particularly important case studies at Whitehorse Stone, where a cross topography route was apparently earlier than the ridgeway routes, and Saltwood Tunnel, where tracks established in the Bronze Age were shown to survive in the post-medieval landscape. Since major transport infrastructure inevitably changes and disrupts earlier patterns of communication, it is appropriate that environmental mitigation should include documentation, investigation and preservation of these earlier routes. The need to work across all periods has been highlighted by the Wealden case study in Chapter 10. Here there were three main types of routeway: the ridgeways which have been little investigated, the Roman roads which have been well studied, and the multiple parallel drove ways at right-angles to the topography. Paradoxically each has been considered in isolation respectively by prehistorians, Romanists and medieval historians, without any real focus on the opportunities which their spatial relationships offer for establishing chronology. Clearly this topic needs a great deal more investigation on the ground. However, on present evidence it appears that some of the droveways may be older than the ridgeways and the Roman roads. On the subject of Roman roads, the case of Sharpstone, Shropshire (Fig. 8.11) and several other cases have demonstrated there is clear evidence that some had pre-Roman origins.

The American North-west Coast has provided a rich seam of ethnographic analogy due to the lateness of European contact and the quality of the ethnohistorical record, particularly with regard to the wide range of plant resources utilised and the effect of people and movement on plant communities (Turner 2014). Evidence for the manipulation and encouragement of plant resources by hunter-gather communities, and the ways in which these were concentrated in particularly favoured places and along routeways, is illustrated by the Douglas map (Fig. 2.2). Many types of effect on landscape (e.g. hollow ways), vegetation, the creation of occupation sites and monuments give rise to spatial structures which influence subsequent patterns of movement. It has been argued that tracks, and especially their intersection points, provide a logical explanation for the frequently observed reoccupation of sites centuries, or even millennia, apart in date; examples include Megalithic tombs in Britain on sites which were previously occupied in the Mesolithic; or evidence of giant posts at Stonehenge millennia before it became enhanced by the great stone monument. The point about the intersection of routeways is especially illustrated by Avebury henge, the four opposed entrances of which make it clear that the grassy clearing which preceded henge construction was at the intersection of two routes.

Landscape structures and retrogressive analysis

Although the emphasis here has been on prehistory, there can be no apologies for straying at various points into historic periods, for instance in discussing the origins of Roman roads and, in the

Wealden case study, where drove ways are attested from early medieval times, but in some cases these have been shown to have prehistoric origins. The prehistoric focus has served a purpose in demonstrating that, contrary to many earlier opinions, there is much evidence for routeways in prehistory and ways of studying and dating them have been identified. However, when it comes to the identification of specific localised areas, it is acknowledged that a prehistoric focus is often not the most appropriate approach. More applicable is a uniformitarian approach which acknowledges that we have no other way of knowing the past except from the perspective of the world in which we live. Retrogressive landscape analysis starts from the landscape of today and works progressively back in a time-transgressive way. Use is made of air photographs, and old maps to identify patterns of landholding, settlement etc in historic periods (Rippon 2012), such that one arrives at a residue of trackways and fields which underlie the historic frameworks and may be prehistoric. LiDAR has great potential in this regard because it can detect the microtopographic features such as lynchets and hollow ways which, as we saw in earlier chapters, are important in tracing and dating past routeways. The case for LiDAR is well made by the success of the Secrets of the High Woods project on the South Downs, UK (Manley 2016) and the recent mapping of hollow ways in several areas of Europe enabling their prehistoric origins to be identified. It was argued in Chapter 8 that elements of continuity in co-axial landscapes may owe more to landscape structures imposed by multiple parallel droveways than to the continuity of individual fields or forms of land use. We have certainly seen some notable cases of routeway continuity, particularly at Saltwood, Kent (Fig. 10.12) and in the Kilen area of Denmark (Fig. 7.2). However, because so little research has been specifically focused on ancient routeways, and most discoveries were made during other investigations, we still have a very fragmentary picture of the extent to which continuities exist. It must also be acknowledged that many of the prehistoric routeways identified here have no apparent relationship to surviving, or historically recorded, routes. Indeed, many case studies demonstrate radical changes over time in the alignment of routes; these are just as culturally significant as the cases of continuity. Indeed many excavated Bronze Age and Iron Age drove ways have no obvious relationship to later routes. In Wessex during the late Bronze Age, *c* 850–750 cal BC, extensive linear ditches cut across the earlier fields and trackways alike. This created extensive ranch boundaries for cattle and sheep husbandry, with some boundaries radiating from sites which then developed as hillforts (Bradley *et al.* 1994; McOmish *et al.* 2002; Cunliffe 2004). At Danebury and on the Yorkshire Wolds (Fig. 8.7) trackways later became ranch boundaries. Again the relationship of this evidence to routeways requires further and wider investigation.

'Natural routes' and ridgeways

In some cases what generations of archaeologists have defined as natural routes are well supported by field evidence; examples are the axial route down the watershed of Jutland in Denmark (Fig. 7.1) and the Hoddersrug route in the Netherlands (Fig. 6.6). 'Natural routes' should, however, only be accepted when they are substantiated by other forms of evidence for use in the period in question. Other routes regarded as 'natural' by generations of archaeologists are more open to question and this is particularly so of the Ridgeways, seen as the main long-term routes of prehistoric movement in lowland England. In some cases the evidence is hopelessly weak, for instance the so-called Jurassic Way. Other cases may be seen as a little stronger, for instance the Wiltshire–Oxfordshire

Ridgeway. Even here, however, both the Uffington (Fig. 8.9) and Avebury/Overton (Fig. 5.5) studies showed that the present routes seem to be post-Roman, likewise the Pilgrim's Way at Whitehorse Stone (Fig. 10.11b).

Perhaps we have been misled by erroneous expectations as to how ridgetops were used, seeing them as a bounded trackway a few metres to tens of metres wide and used over long distances as they are today. Such expectations overlook two closely linked factors: the environmental evidence and the reasons prehistoric communities were using and moving through these landscapes. Burial monuments such as barrows of Neolithic and Bronze Age date tend to be concentrated along the ridges. Where we have environmental evidence from barrows of both periods, it generally indicates cleared landscapes characterised by grassland (Evans 1972). The maintenance of that environment indicates regular grazing; such areas are likely to have been grazed during the later spring and summer when it would have been necessary to keep animals away from growing crops and to allow the winter pastures to regrow for hay. It has been widely accepted that Neolithic and early Bronze Age communities in Britain remained quite mobile and moved around seasonally with their animals. The open landscape of the ridgetops are likely to have been used in this way. However, animals as they were moving around required grazing so a broad expanse of grassland a few kilometres wide is likely to have been utilised. Individual routes taken would not have been closely defined but meandered depending where grazing was best and the effects of the weather most favourable, or least harsh. What this envisages is not a closely defined route but a broad band, maybe hundreds of metres to, in places perhaps, kilometres wide. During the middle to late Bronze Age cross-ridge dykes may represent attempts to restrict movement, or perhaps to lay proprietary claim to some tracts of grazing where others could pass but not tarry. Later in about the Iron Age or Roman period, as pressure on land increased and fields encroached, broad ridge top belts may have become defined as multi-stranded trackways.

Droveways

Rather more tangible than the ridgeways has been the identification of multiple parallel routes often running down the crest of spurs at right-angles to the escarpments. Their significance lies in the fact that they link areas of contrasting topography, geology, vegetation and economic resource. They also connect areas on the ridgetops associated with burial, with lower areas more suitable for settlement. In the medieval period, these routes served as droveways connecting the resources of parent and daughter settlements. They are in places marked by very prominent hollow ways descending steep escarpments (Fig. 8.5) or crossing the Lower Greensand in the Rother valley (Fig. 10.10d). The Wealden case study has shown that some cross-topographic routes were of early origin. That at Whitehorse Stone (Fig. 10.11b) had possible Neolithic origins and, on the South Downs, routes down spurs at right-angles to the escarpment had Iron Age origins at Thundersbarrow Hill and Port's Road (Fig. 10.6). A route down an escarpment spur at Bishopstone (Fig. 10.3) may have Neolithic origins and have been in use for perhaps 4500 years until abandonment by the Anglo-Saxon period.

Where routeways were associated with the seasonal movement of animals it is probable that the requirements to provide grazing involved belts of grassland rather than narrowly constrained tracks. That might explain some curiously long thin landholdings in the Weald (Fig. 8.4; Chatwin and Gardiner 2012).

Many instances have been noted where there are multiple parallel routes, for instance in the Bronze Age at Fengate (Fig. 6.10) and Heathrow T5 (Fig. 8.8) and in the historical period, the cross topographic droveways; these have all been associated with the seasonal movements of animals, as particularly discussed in the Wealden case study in Chapter 10, but also reported in many other areas of the UK. It seems that individual communities each had their own droveway, which might at first seem like redundant duplication, but makes sense in terms of the need to provide adequate grazing, especially if animals were being moved from many settlements at the same time of year.

Ethnohistory of Lesser Transhumance

Around the uplands of western Britain trackways have been identified connecting lowland farms and upland seasonal pastures in a system of Lesser Transhumance which continued into historical times in Wales, Scotland and Ireland. In Wales the lowland settlement is known as the *hafod* or home farm and up in the hills is the *hendref* or summer farm. This system is attested in Cornwall by place-name evidence and, because the same Welsh terms are used, Herring (2012) was able to demonstrate that the system was in place before the seventh century AD when Anglo-Saxon influence extended to the area. In that area the home farm and upland pastures were at distances of 4–10 km. Similarly, writing of the upland moors beyond the enclosed landscape of Swaledale, Yorkshire, Fleming (1998) describes routeways down to Swaledale marked by pronounced hollow ways which can be documented at least as far back as the twelfth century. Fleming (1998) and Herring (2012) draw on the ethnohistorical evidence from Wales and Ireland to infer that seasonal settlement of the moorland was during the upland vegetation growing season between *Beltane* (early May) and *Samhain* (late October) (Kelly 1997). In recent ethnohistorical practice it was the teenage girls and young women who went to the upland pasture where they engaged in dairying, butter and cheese production; a spinning wheel was also invariably taken by them. Men and older women stayed at the home farm to work the fields. Although these Lesser Transhumance practices are generally thought of as involving the uplands, there is evidence from both Wales and Ireland of comparable seasonal movement between lowland farms and seasonally available saltmarsh grazing resources. This was the interpretation suggested in Chapter 6 for middle Bronze Age to Iron Age seasonal settlements in the Severn Estuary. Such practices may also be suggested by the drove ways associated with coaxial fields which lead to coastal and riverine lowlands. The Wealden case in Chapter 10 provided evidence that some of the multiple parallel long-distance droves, which have generally been associated with medieval lesser transhumance, may have their origins in prehistory. Thus it seems that parts of later prehistoric communities, at least in some areas, were a good deal more seasonally mobile than has generally been assumed.

Excavation

In resolving the many research questions concerning prehistoric routeways we cannot rely only on the sort of chance exposure which was exploited at Lyminge (Fig. 10.13), or assessments in advance of development. There is also a need for research-led excavation specifically to establish the date and character of past routeways. Some of the cases most in need of attention are: drove ways connecting uplands and lowlands, wetland and dry ground; those associated with surviving coaxial patterns; routes associated with prehistoric bridges; possible routeways of the late Iron Age *oppida*; and those connecting terrestrial, riverine or marine transport such as landing places.

Dating and sequencing palimpsests of field, settlement and trackway evidence can often not be resolved without excavation. However, sometimes this can be relatively small-scale excavations, as demonstrated, for instance, by the dating of double lynchet trackways at Itford Bottom and Bullock Down (Fig. 10.4). The value of excavation and dating programmes using a combination of radiocarbon, artefacts and OSL has been very effectively shown by the demonstration that the road at Sharpstone, Shropshire had its origins in the Iron Age rather than the Roman period (Fig. 8.11).

Linear environmental archaeology

The writer approached the topic of this book from the perspective of landscape archaeology and a background in environmental archaeology and geoarchaeology. Environmental archaeology has often had a site-based focus. Now, as the density of investigated sites increases, it has become possible, in some areas, to develop a more spatial picture derived from multiple environmental sequences in an area, as demonstrated for instance by reconstructions of environmental change round Stonehenge and Avebury (Allen, M. 1997; 2005). One of the messages of the review of hunter-gather landscapes in North America (Chapter 2) and Europe (Chapter 3), is that landscapes are likely to be at least partially structured by linear patterns of movement between sites, which was most clearly exemplified by the James Douglas map (Fig. 2.2). Linear heathland strips associated with barrow-marked routes on the Veluwe, Netherlands have also been identified by Doorenbosch (2013). This proposition, that environmental patterns in some areas may relate to routeways, is an hypothesis which is likely to repay testing elsewhere; it could make a significant contribution both to the identification of past routeways and recognising one of the ways in which humans contribute to niche construction. This constitutes the case for an approach to environmental archaeology which is less static and with a greater focus on mobility, i.e. one which has a linear focus. In Chapter 1 a wide range of palaeoenvironmental sources which can contribute to identification of routes used by people and animals was identified (Fig. 1.2). In subsequent chapters we have seen the contribution made by many of them, others such as non-pollen palynomorphs, stable isotopes, geochemical analysis and sedimentary DNA are set to make even greater contributions to the study of early routeways in the years to come.

Geoarchaeological approaches to human and landscape connectivity

It has been shown that dating can be achieved by developing a geoarchaeological approach focused on the sedimentary context of routeways and the composite landforms which they often represent, an approach which might be called archaeogeomorphology. It has been demonstrated that, in agricultural landscapes, hollow ways are often not simply negative erosive features, but have datable positive lynchets running along their upslope flanks, and negative lynchets along their downslope edges. It should be emphasised that these features were not deliberately constructed; they are the product of gradual soil movement downslope within cultivated fields. It was the realisation that a positive lynchet defined part of the route at Woodland Road, Lyminge, which facilitated its dating using a combination of techniques: artefacts, molluscan analysis, OSL and Uranium Series dating (Fig. 10.13).

The processes of water erosion which contribute to the formation of hollow ways have also been shown by Boardman's (2013) Rother valley study (Fig. 10.10) to be geomorphologically significant. These processes created new drainage channels, which in some cases went against the grain of the topography, e.g. by running down the axis of spurs. In this way, the density of drainage and amount of runoff was increased. The contribution of routeways to drainage has been quantified in wooded landscapes of the Polish Carpathians, where, on sedimentary flysch geology, 70–80% of suspended sediment entering drainage basins is derived from unmetalled road surfaces which, under modern conditions, are being incised at an average rate of 6.6 mm per year (Froehlich 1991). In the case of both agricultural fields in the Rother valley and wooded landscapes in the Carpathians, routeways are shown to be very significant sources of sediment entering river systems. We may infer that, in at least some river systems, hollow ways may be a significant contributor to the increased minerogenic alluvial sedimentation which is a widespread characteristic of later prehistory (Brown, A 1997). Identification of discharge fans of sediment from hollow ways to floodplains thus offers a further potential way of dating hollow ways by identifying and dating the sedimentary increments from eroded tracks.

The implications of this are not restricted to dating and geomorphology. Routeways are created by people and are thus reflective of social and economic relationships. By so doing, people have created additional channels for water and sediment movement. Such relationships exemplify the increasingly recognised interdependence of hydro-geomorphic, ecological and human processes and functions highlighted by Harden *et al.* (2014). They represent coupled relationships between human and natural systems with feedback loops (Chin *et al.* 2014). It is likewise illustrative of the interrelationships between multiple agents: human agents had a significant effect on geomorphic processes, which is not perhaps so surprising. More notable is the effect of geomorphic agents, literally making patterns of human movement entrenched and in that way influencing future patterns of movement, land use and the perspective from which perceptions are formed.

Movement as niche construction

The entrenchment of hollow ways is just one of the myriad contributions which movement makes to niche construction. This has emerged as a significant theme, linking the social connectivity represented by past routeways and the scientific aspects of environmental archaeology and geoarchaeology which contribute to their study. Niche construction was defined in Chapter 1 as the process whereby organisms modify their own and each other's niches, thus creating the potential for change and evolution. It recognises that all organisms can play a part in the construction of their own niches and those of others. The concept is particularly significant for environmental archaeology with its ecological emphasis. Early ecologists struggled to accommodate human agency into their models by attempting to define 'natural ecology'. This approach became increasingly untenable when it become apparent that there are few ecosystems on earth unaffected by humans. The attempt to isolate 'natural ecology' is also challenged by the realisation that many organisms, not just humans, contribute to niche construction, such that the difference between humans and other organisms becomes a matter of degree rather than fundamental difference. That resonates with one of the themes developed in Chapters 2 and 3 that some tracks which are influential in the patterns of human movement are likely to have originated as animal tracks. The influence exerted by tracks can be longstanding. When people move through a landscape on a route established centuries or millennia earlier, their approach to, and perception of, that landscape will be influenced by those who first took that route long before. That remains the case whether the original walkers were aurochs, Mesolithic hunter-gatherers, or prehistoric herders. This does not imply that their choices are conditioned by precedent, they are themselves active agents with free will, so in some cases the choice will be to deliberately avoid what has gone before and set off on a new path. To the landscape historian both continuity and radical change are of interest and can tell us much more than just the daily habits of those who used that route. Long continuities of use may serve to legitimise the landholdings of communities, or the authority of elites. Radical changes of routeway can help identify periods of social and ideological change and conflict.

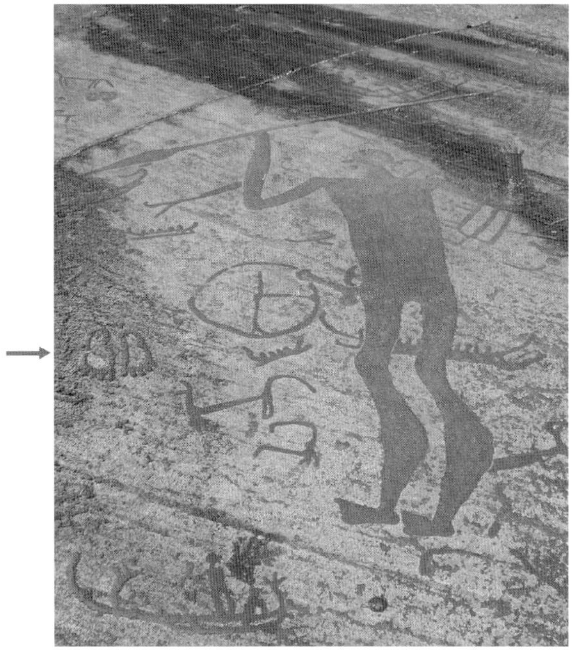

Figure 11.2 Bronze Age rock art Tanum, Bohuslan, Sweden: iconic images associated with movement. There is a large spear-carrying figure with feet in skewed perspective to show footprints, and two separate feet depicted on the left (arrowed). The scene also shows boats, animals, some probable horses, and a wheel or sun motif (photo: M. Bell).

Human geographers have demonstrated that it is not animals alone that exhibit agency in their capacity to engender change, but also plants, as Jones and Cloke (2002) showed in the context of the cultural significance of trees. Archaeologists have similarly argued, through concepts of the extended mind and material engagement theory (DeMarrais *et al.* 2004), that material culture, such as the objects made by people, can have agency and the capacity to lead to change. The material things that facilitated, or enhanced, travel clearly took on iconic significance. These range from human footprints, rock art depictions and models of boats, wheeled objects, deposits of wheels and depictions of the horse. Many such traits appear together in the Bronze Age rock art of Norway and Sweden (Fig. 11.2). Likewise significant are high status travel-related objects fashioned from exotic materials drawn from several places: the Chariot of the Sun, the Nebra Sky Disc and the Caergwle bowl (Chapter 9). These serve to highlight the social and cosmological significance of travel, which is especially apparent in the middle and late Bronze Age. This is further emphasised by Kaul's (1998) cosmological interpretations concerning the depictions of boats and horses on Danish bronze objects to which may now be added the apparent relevance of footprints to the cosmological scheme (Skogland *et al.* 2017). Communities acquired materials, esoteric knowledge and status through travel to distant and exotic places (Helms 1988).

Pathways are the material manifestation of the activities of human and animals in the landscape and as such they represent classic cases of both agency and niche construction. In the foregoing chapters we have seen many ways in which routeways of human movement became part of living landscapes and thus contributed to the construction of those niches. In hunter-gatherer landscapes there are the effects on vegetation of trampling, transplanting, burning, marks on trees from utilisation or trail marking blazes, plants propagated from faeces or those which flourished because of the chemical conditions created by faeces. In later prehistoric landscapes the traces become more tangible, and potentially more long lasting, with the construction of monuments and barrows which may be designed to channel movement and convey a particular message to travellers about ancestry, genealogy and ownership. With the development of agricultural landscapes in which routes run between fields, trackways can be more easily identified. Technology, engineering and the development of elite centres are likewise contributory factors, along with the development of bridges, constructed roads, wheeled transport and the horse.

Critical thresholds

From the various case studies reviewed we may begin to discern some key thresholds in terms of trackway development in North-west Europe. Some of these are perhaps predictable, because they correspond to significant boundaries between periods, others are perhaps more surprising. Some vary regionally. Some may be questionable, but still worth stating to stimulate debate and critical research.

One perhaps surprising conclusion has been that, on present evidence, there are no certain Mesolithic constructed wood trackways and the first clearly constructed wetland trackways appear with the earliest farmers in both north Germany and later in the UK. Barrow alignments indicating former routeways are also present from early in the Neolithic in Denmark, north Germany and the Netherlands. Such evidence from Britain has been argued on the grounds of antecedent activity episodes but is less securely attested. The first widespread, clearly defined dry ground tracks in Europe appear with extensive field systems in the middle Bronze Age. They are particularly clear in the coaxial field systems and associated drove ways of south-east England. The use of wheeled transport is a further critical threshold; in northern Europe this begins in the later Neolithic and, on present evidence, later in the British Isles, around the middle Bronze Age. Bridges appear in the UK in the later Bronze Age and one may speculate that they were associated with some form of road, so far unattested, which was at least partly constructed. A further threshold is represented by the greater speed and distance of communication facilitated by the horse from the later Bronze Age. The next threshold, which can be predicted, but again is not yet attested, may be associated with the emergence of *oppida* settlements in the later Iron Age. Finally, the most dramatic change of all, within the zone of Roman influence was the development of the Roman road network. In some ways the most intriguing aspect of this story is the evidence from Britain for chariots in the Iron Age and the pre-Roman origins of some Roman roads (Chapter 8). Such fragments of evidence seem to presage a network of routes in the final Iron Age which we have hardly begun to glimpse.

In terms of maritime and riverine communication thresholds are less clear; log boats are attested from the Mesolithic but, given evidence for Palaeolithic island colonisation, boats of some kind are clearly earlier. Advances in the use of skin boats are not archaeologically attested except in terms of their representation in Bronze Age and Iron Age boat models. However, with the identification of possible boats in burials near the coast and prehistoric coastal havens it is at least becoming clearer where we need to look. The first evidence of the sewn plank boat is early Bronze Age in

Britain, and may suggest introduction with Beakers and the use of metals. Iconographic evidence in Scandinavia begins *c* 1700 BC (Ling 2008). A further critical threshold is the advent of sail. In the Mediterranean this appeared during the Bronze Age and in North-west Europe apparently in the later Iron Age, with the substantially-built Romano-Celtic boat type which is described in historical sources such as Caesar and depicted on coins (McGrail 2001).

In the foregoing chapters a greater role has been proposed for maritime and riverine communication than is generally appreciated. This partly reflects a growing awareness of maritime archaeology but perhaps the most compelling evidence is negative, the surprisingly limited evidence for well-defined long distance routeways in prehistory, especially in Britain. This implies that the study of ancient routeways requires some reorientation towards integration of terrestrial, riverine and maritime routes and ways of identifying landing places. It would additionally be appropriate to look at those watershed locations where waterborne transport would have to be replaced by tracks, such as east to west across the Pennines, UK.

There is evidence of regional contrasts as between waterborne and terrestrial transport. The clearer evidence for longer routes and the earlier and more abundant use of the wheel in Denmark, north Germany and the Netherlands could reflect a greater terrestrial emphasis to communication, compared to the British Isles with radial drainage to an encircling sea. However, one might anticipate maritime communication being particularly important in Denmark, as in historic times. Regional contrasts may be as much cultural as environmental, and quite possibly cultural in two very different ways. First, during prehistory, particularly in Bronze Age northern Europe in terms of a link to metal producing areas to the south creating a need for overland transport, whereas Britain was connected by sea to the continent. Secondly, in terms of the emphasis of archaeologists who in continental Europe have perhaps given greater attention to patterns of prehistoric communication than in Britain where archaeologists can be argued to have been intimidated by the legacy of Alfred Watkins (1925) and mesmerised by the chimera of the ridgeways.

Route to sustainable heritage and nature conservation

It has hopefully been shown that past routeways are central to understanding landscapes and how they became structured through movement. It has also been observed that their study in many ways has been a neglected topic, especially in prehistory. Limited research has contributed to inadequate attention to routeways in Historic Environment Records in the UK. The paved paths or *Trods* of the North York Moors are a locally appreciated historic asset but only 20% of those recorded in the late nineteenth and early twentieth century survive (Evans 2008). In Britain some routes are protected because they form part of wider sites such, as Neolithic stone avenues, also some wooden trackways are protected, for instance in the Somerset Levels. The vast majority are unrecognised and without protection except tangentially through footpath and access legislation. In Denmark the situation is more positive, an inventory of early routes having been established for heritage protection purposes (Bang 2013). In Chapter 7 case studies were reviewed at Kilen, Denmark and Veluwe, Netherlands where walks along prehistoric routeways were marked by display boards or geocaches explaining the relationship between the barrow alignments, hollow ways and ancient routes. People informed in this way are more likely to support the preservation and enhancement of this significant evidence for past landscape connectivity.

Walking is increasingly recognised for the contribution it makes to quality of life and health. Political initiatives have increased access to the countryside, freedom to roam and in Britain to the creation of an almost complete coastal path. Some long-distance paths provide people with opportunities for encounters with the past which are stimulating and well attested, such as those along Hadrian's Wall and Offa's Dyke. Other initiatives aimed at walkers make much of the supposedly ancient origins of the Ridgeways, which as we have seen are open to question. In other cases attractive walks are promoted using quasi-historical tags of variable validity. Examples are the Monarch's Way, a serpentine 625 mile (1000 km) route approximating to that of Charles II on his escape from the battle of Worcester to the south coast, Robin Hood's Way in Nottinghamshire, or the Vangard Way in the Weald which follows the path of a 1960s rambling group. If such designations encourage people out in the countryside walking they are in many ways a good thing. Too often, however, those who go out into the landscape, are fascinated and moved by the obviously ancient nature of some of the routeways they are following, have only debatable assertions and quasi-historical associations to satisfy their curiosity. The case presented here is that if the origins of these route were the subject of focused scientific research then the encounters with landscape which they provide would be significantly more stimulating. The importance of learning about these things becomes especially apparent when we consider the imperative of getting children out into the landscape and the educational benefits derived from walking and thinking on one's feet.

Routeways are not just historic assets but contribute to wider agendas and inform broader debates concerning people, nature, connectivity, quality of life and migration. This is apparent from the strongly biographical flavour of much writing about landscape as outlined in Chapter 1. That literature demonstrates how walking is both a metaphor for, and a route to, self discovery. Giddens (1991) described thoughts engendered in such ways as contributing to a sense of ontological security, through which people negotiate their sense of self and place in the world. Increasingly the experience of heritage and landscape is seen as of therapeutic value because of its contribution to an individual's self-discovery (Reilly *et al.* 2018; Nolan 2019). When that occurs through the bodily engagement of walking the benefits become both physically and mentally embedded.

Nor should these issues be seen from a purely human perspective. Just as animals have contributed to the construction of routeway niches so the ancient routeways formed by human agency have created specific niches for plant and animal life which has significance in terms of sustainable nature conservation. In eastern Austria the deeply incised hollow ways in loess are recognised for their nature conservation significance because of the habitats they provide for rare plants, insects etc (Wiesbauer and Zettel 2004). In her book on rewilding the Knapp estate in Sussex Isabella Tree (2019) highlights the contribution which Wealden drove ways make to the connectivity of nature conservation sites. Having walked along many miles of hollow ways and other early routes in conducting this research the writer is aware from the footprint-tracks and signs of wild animals that they serve as significant corridors for animal movement, and with them seeds etc, today. This is important because in Britain a recent review of the condition of wildlife sites conducted for the government concluded that nature conservation sites are generally too small, fragmented and isolated from one another by tracts of intensively farmed landscape with significant risks of loss of species biodiversity and local or wider extinction (Lawton *et al.* 2010). A proposed solution is the creation of corridors of connectivity between nature reserves. It may not be sufficiently appreciated that in many areas such corridors already exist in the form of the long established routeways. Many are deeply incised, semi-abandoned and overgrown. Of special significance is the connectivity they often provide across topographic and environmental zones. Quite often these favourable circumstances have been created by benign neglect of a routeway that has gone out of active use, a form of unconscious rewilding. In other cases it may be that targeted management of the trackway habitat and enhancement of the connectivity they provide could significantly add to their conservation value. The proposition is that learning to appreciate the significance of past routeways makes a major contribution to understanding landscape origins and structures, contributes to access to the outdoors with the associated mental and physical benefits and also contributes to developing agendas for wildlife sustainability.

Bibliography

Aberet, O. and Jacomet, S. 1997 Analysis of plant macrofossils in goat/sheep faeces from the Neolithic lake shore settlement of Horgen Scheller – an indication of prehistoric transhumance, *Vegetation History and Archaeobotany* 6, 235–9.

Achterberg, I., Bauerochse, A., Giesecke, T., Metzler, A. and Leuschner, H. 2015 Contemporaneousness of trackway construction and environmental change: a dendrochronological study in Northwest German mires, *Interdisciplinaria Archaeologica* VI, 19–20.

Alcock, S., Talbert, R. and Bodel, J. 2012 *Highways, Byways and Road Systems in the Pre-Modern World.* New York: Wiley-Blackwell.

Aldhouse-Green, S.H.R. and Housley, R.A. 1993 The Uskmouth mattock: a radiocarbon date, *Archaeologia Cambrensis* 142, 340.

Aldhouse-Green, S.H.R., Whittle, A., Allen, J.R.L., Caseldine, A.E., Culver, S.J., Day, M.H., Lundqvist, J. and Upton, D. 1992 Prehistoric human footprints from the Severn Estuary at Uskmouth and Magor Pill, Gwent, Wales, *Archaeologia Cambrensis* 141, 14–55.

Aldred, O. 2014 Past movements, tomorrow's anchors. On the relational entanglements between archaeological mobilities. In Leary (ed.) 2014, 21–48.

Allen, C. 2009 *Exchange and Ritual at the Riverside: Late Bronze Age life in the lower Witham valley at Washingborough, Lincs.* Lincoln: Pre-Construct Archaeology Monograph 1.

Allen, J.R.L. 1987 Late Flandrian shoreline oscillations in the Severn Estuary: the Rumney Formation and its typesite, *Philosophical Transactions of the Royal Society* B 315, 157–74.

Allen, J.R.L. 1997 Subfossil mammalian track (Flandrian) in the Severn Estuary, S.W. Britain: mechanics of formation, preservation and distribution, *Philosophical Transactions of the Royal Society of London* B 352, 481–518.

Allen, J.R.L. 2001 Late Quaternary stratigraphy in the Gwent Levels (southeast Wales): the subsurface evidence, *Proceedings of the Geologists' Association* 112, 289–315.

Allen, J.R.L. 2004 Annual textural banding in Holocene estuarine silts, Severn Estuary Levels (SW Britain): patterns, causes and implications, *The Holocene* 14, 536–52.

Allen, J.R.L. 2017 *Geology for Archaeologists.* Oxford: Archaeopress.

Allen, J.R.L., Bell, M. and Scales, R. 2003 Animal and human footprint-tracks in archaeology: description and significance, *Archaeology in the Severn Estuary* 14, 55–68.

Allen, J.R.L. and Haslett, S.K. 2006 Granulometric characterisation and evaluation of annually banded mid-Holocene estuarine silts, Welsh Severn Estuary (UK): coastal change, sea level and climate, *Quaternary Science Reviews* 25, 1418–46.

Allen, M.J. 1992 Products of erosion and the prehistoric land-use of the Wessex Chalk. In Bell, M. and Boardman, J. (eds), *Past and Present Soil Erosion.* Oxford: Oxbow Monograph 22, 37–52.

Allen, M.J. 1995 Before Stonehenge. In Cleal and Walker 1995, 41–62.

Allen, M.J. 1997 Environment and land-use: the economic development of the communities who built Stonehenge (an economy to support the stones). In Cunliffe, B. and Renfrew, C. (eds), *Science and Stonehenge.* Oxford: Oxford University Press/Proceedings of the British Academy 92, 115–44.

Allen, M.J. 2005 Considering prehistoric environmental changes on the Marlborough Downs. In Brown, G., Field, D. and McOmish, D. (eds), *The Avebury Landscape.* Oxford: Oxbow Books, 77–86.

Allen, M.J. 2017 *Molluscs in Archaeology: methods, approaches and applications.* Oxford: Studying Scientific Archaeology 3.

Allen, M.J. and Gardiner, J. 2000 *Our Changing Coast: a survey of the intertidal archaeology of Langstone Harbour, Hampshire.* York: Council for British Archaeology Research Report 124.

Allen, M.J. and Gardiner, J. 2002 A sense of time: cultural markers in the Mesolithic of southern England. In David, B. and Wilson, M. (eds), *Inscribed Landscapes: marking and making place.* Honolulu: University of Hawai'i Press, 139–53.

Allen, M.J. and Maltby, M. 2012 Chalcolithic land-use, animals and economy – a chronological changing point? In Allen, M.J., Gardiner, J. and Sheridan, A. (eds), *Is there a British Chalcolithic? People, Place and Polity in the Later 3rd Millennium,* Oxford: Prehistoric Society Research Papers 4, 281–97.

Allen, T., Donnelly, M., Hardy, A. and Powell, K. 2012 *A Road Through the Past: Archaeological discoveries on the A2 Pepperhill to Cobham road-scheme in Kent.* Oxford: Oxford Archaeology Monograph 16.

Allen, T., Barclay, A., Cromarty, A.M., Anderson-Whymark, H., Parker, A., Robinson, M. and Jones, G. 2013 *Opening the Wood, Making the Land.* Oxford: Thames Valley Landscapes Monograph 38.

Ames, K.M. and Maschner, H.D.G. 1999 *Peoples of the Northwest Coast.* London: Thames and Hudson.

Andersen, S.H. 1993 History of vegetation and agriculture at Hassing House, Mose Thy, Northwest Denmark since the Ice Age, *Journal of Danish Archaeology* 11, 57–79.

Andersen, S.H. 2000 'Køkkenmøddinger' (Shell middens) in Denmark: a survey, *Proceedings of the Prehistoric Society* 66, 361–384.

Andersen, S.H. 2013 *Tybrind Vig: submerged Mesolithic setttlements in Denmark.* Aarhus: Jutland Archaeological Society.

Andersen, S.H. and Rasmussen, A.P. 1993 Radiocarbon wiggle-dating of elm declines in Northwest Denmark and their significance, *Vegetation History and Archaeobotany* 2, 125–35.

Anderson, M.K. 2009 *The Ozette Prairies of Olympic National Park: their former indigenous uses and management.* Washington DC: report to Olympia National Park.

Anderson-Whymark, H., Garrow, D. and Sturt, F. 2015a *Continental Connections: Exploring cross-Channel relationships from the Mesolithic to the Iron Age.* Oxford: Oxbow Books.

Anderson-Whymark, H., Garrow, D. and Sturt, F. 2015b Micoliths and maritime mobility: a continental European-style late Mesolithic flint assemblage from the Isles of Scilly, *Antiquity* 89, 954–71.

Anthony, D.W. 2007 *The Horse, the Wheel and Language: how Bronze-Age riders from the Eurasian steppes shaped the modern world.* Oxford: Princeton University Press.

Aramayo, S. 2009 A brief sketch of the Monte Hermoso human footprint sites, South Coast, Buenos Aires province, Argentina, *Ichnos* 16 (1–2), 49–54.

Arnold, B. 1995 *Pirogues Monoxyles d'Europe Centrale. Tome 1: construction, typologie, évolution.* Neuchâtel: Archéologie Neuchâteloise 20.

Arnold, B. 1996 *Pirogues Monoxyles d'Europe Centrale. Tome 2: construction, typologie, évolution.* Neuchâtel: Archéologie Neuchâteloise 21.

Arnold, T.G. 2002 Radiocarbon dates from the ice-free corridor, *Radiocarbon* 44, 437–54.

Ashley, G. 2003 Tracks, trails and trampling by large vertebrates in a Rift Valley palaeo-wetland, Lowermost Bed II, Olduvai Gorge, Tanzania, *Ichnos* 9 (1–2), 23–32.

Ashton, N. 2017 *Early Humans.* London: William Collins.

Ashton, N., Lewis, S.G., De Groote, I., Duffy, S.M., Bates, M., Bates, R., Hoare, P., Lewis, M.P., Parfitt, S.A., Peglar, S., Williams, C. and Stringer, C. 2014 Hominin footprints from early Pleistocene deposits at Happisburgh, UK, *PLOS ONE* 9: 2, e88329 doi:10.1371/journal.pone.

Aston, M. 1985 *Interpreting the Landscape.* London: Batsford.

Aston, M. and Gerrard, C. 2013 *Interpreting the English Village.* Oxford: Windgather Press.

Aston, M. and Rowley, T. 1974 *Landscape Archaeology: An introduction to fieldwork techniques on Post-Roman landscapes.* London: David and Charles.

Atkinson, R.J.C. 1955 The Dorset cursus, *Antiquity* 29, 4–9.

Avanzini, M., Mietto, P., Panarello, A., De Angelis, M. and Rolandi, G. 2008 The Devil's Trails: middle Pleistocene human footprints preserved in a volcanoclastic deposit of South Italy, *Ichnos* 15 (3–4), 179–89.

Bahn, P. and Pettitt, P. 2009 *Britain's Oldest Art: The Ice Age Cave Art of Creswell Crags.* Swindon: English Heritage.

Bahn, P. and Vertut, J. 1997 *Journey Through the Ice Age.* London: Weidenfeld and Nicolson.

Bakels, C. and Zeiler, J. 2005 Neolithic subsistence. In Louwe Kooijmans, L.P., van den Broeke, P.W., Fokkens, H. and van Gijn, A.L. (eds), *The Prehistory of the Netherlands* Volume 1. Amsterdam: Amsterdam University Press, 311–66.

Bakker, J.A. 1976 On the possibility of reconstructing roads from the TRB period, *Berichten van de Rijksdienst voor het Oudheidkundig Bodemonderzoek Jaargang* 26, 63–91.

Bakker, J.A. 1991 Prehistoric long-distance roads in North-West Europe. In Lichardus, J. (ed.), *Die Kupferzeit als historische Epoche.* Bonn: Habelt, 502–28.

Bakker, J.A. 2008 A note on prehistoric routes on the Veluwe and near Uelzen, *Analecta Praehistoria Leidensia* 40, 281–6.

Bakker, J.A., Kruk, J., Lanting, A.E. and Milisouskas, S. 1999 The earliest evidence of wheeled vehicles in Europe and the Near East, *Antiquity* 73, 778–90.

Balaam, N.D., Smith, K. and Wainwright, G.J. 1982 The Shaugh Moor Project: fourth report – environment, context and conclusion, *Proceedings of the Prehistoric Society* 48, 203–78.

Balaam, N.D., Bell, M., David, A., Levitan, B., Macphail, R., Robinson, M.A. and Scaife, R. 1987a Prehistoric and Romano-British sites at Westward Ho!, Devon: archaeological and palaeoenvironmental surveys 1983 and 1984. In Balaam *et al.* (eds) 1987b, 163–264.

Balaam, N.D., Levitan, B. and Straker, V. (ed.) 1987b *Studies in Palaeoeconomy and Environment in South West England.* Oxford: British Archaeological Report 181.

Banffy, E. 2004 *The Sixth Millennium BC Boundary in Western Transdanubia and its role in the Central European Transition.* Budapest: Institute of Archaeology Hungarian Academy.

Bang, J. 2013 The route to a history of the cultural landscape: a Danish record of prehistoric and historic roads, tracks and related structures. In Bergerbrant, S. and Sabatini, S. (ed.), *Counterpoint: essays in archaeology and heritage studies in honour of Professor Kristian Kristiansen.* Oxford: British Arcaheological Report S2508, 703–15.

Bang, P. and Dahlstrøm, P. 2001 *Animal Tracks and Signs.* Oxford: Oxford University Press.

Barber, K.E., Chambers, F.M. and Maddy, D. 2003 Holocene palaeoclimates from peat stratigraphy: macrofossil proxy climatic records from three oceanic raised bogs in England and Ireland, *Quaternary Science Reviews* 22, 521–39.

Barber, M. 2014 *Mount Pleasant, Dorset: a survey of the Neolithic 'henge enclosure' and associated features.* Swindon: English Heritage Research Report 70.

Barclay, A. and Bayliss, A. 1999 Cursus monuments and the radiocarbon problem. In Barclay and Harding (eds) 1999, 11–29.

Barclay, A. and Harding, A. 1999 *Pathways and Ceremonies: the cursus mouments of Britain and Ireland.* Oxford: Oxbow Books.

Barclay, A., Lambrick, G., Moore, J. and Robinson, M. 2003 *Lines in the Landscape: cursus monuments in the Upper Thames Valley.* Oxford: Thames Valley Landscape Monographs 15.

Barker, G. 2006 *The Agricultural Revolution in Prehistory.* Oxford: Oxford University Press.

Barnett, C. 2009 The chronology of early Mesolithic occupation and environmental impact at Thatcham reedbeds, southern England. In Combé, P., van Strydonck, M., Sergant, J., Boudin, M. and Bats, M. (eds), *Chronology and Evolution within the Mesolithic of North-West Europe.* Newcastle-upon-Tyne: Cambridge Scholars, 57–76.

Barr, K. 2018 Prehistoric avian, mammalian and *Homo sapiens* footprint-tracks from intertidal sediments as evidence of human palaeoecology. Reading University: Unpublished PhD thesis.

Barr, K. and Bell, M. 2016 Neolithic and Bronze Age ungulate footprint-tracks of the Severn Estuary: species, age, identification and the interpretation of husbandry practices, *Environmental Archaeology, DOI* 10.1080/14614103.2016.1222691.

Barrett, J. 1994 *Fragments from Antiquity.* Oxford: Blackwell.

Barrett, J., Bradley, R. and Green, M. 1991 *Landscape, Monuments and Society: the prehistory of Cranborne Chase.* Cambridge: Cambridge University Press.

Barriere, C. and Sahly, A. 1964 Les empreintes humaines de Lascaux. In Ripoll, E. (ed.), *Miscelanea en Homenaje a Abate H. Breuil* Volume 1. Barcelona: Diputación Provincial, 173–80.

Barton, R.N.E. and Roberts, A. 2015 Marine shell beads from three inland Later Mesolithic sites in western Britain. In Ashton, N. and Harris, C. (eds), *No Stone Unturned: papers in honour of Roger Jacobi.* London: Lithic Studies Society, 191–208.

Barton, R.N.E., Berridge, P.J., Walker, M.J.C. and Bevins, R.E. 1995 Persistent places in the Mesolithic landscape: an example from the Black Mountain uplands of South Wales, *Proceedings of the Prehistoric Society* 61, 81–116.

Basso, K.H. 1996 *Wisdom Sits in Places: Landscape and language among the Western Apache.* Albuquerque NM: University of New Mexico Press.

Bates, M. and Stafford, E. 2013 *Thames Holocene: a geoarchaeological approach to the investigation of the river floodplain for the Channel Tunnel Road Link.* Salisbury: Wessex Archaeology.

Bayliss, A. and Whittle, A. 2007 Histories of the dead: building chronologies for 5 south British long barrows, *Cambridge Archaeological Journal* 17, 1–147.

Bayliss, A., Benson, D., Galer, D., Humphrey, L., McFadyen, L. and Whittle, A. 2007 One thing after another: the date of the Ascott-under-Wychwood long barrow, *Cambridge Archaeological Journal* 17, 29–44.

BBC 2012 Ancient footprints found in peat at Borth Beach www., *bbc.c.uk*/news/uk_Wales-midwales-, 17353470.

Beattie, O. 2000 The Kwaday Dön Ts'inchi discovery from a glacier in British Columbia, *Canadian Journal of Archaeology* 24, 129–48.

Bech, J.-H. and Mikkelsen, M. 1999 Landscapes, settlement and subsistence in Bronze Age Thy, NW Denmark. In Fabech, C. and Ringtved, J. (eds), *Settlement and Landscape: proceedings of a conference in Århus, Denmark May 4–7 1998*. Århus: Jutland Archaeological Society, 69–77.

Beck, K.W. and Shennan, S. 1991 *Amber in Prehistoric Britain.* Oxford: Oxbow Books.

Beckett, S.C. 1978 The environmental setting of the Meare Heath Track, *Somerset Levels Papers* 4, 42–6.

Beckwith, B.R. 2004 'The queen root of this clime': ethnoecological investigations of blue camas (*Camassia leichtlinii* (Baker) Wats., *C. quamash* (Pursh) Greene; Liliaceae) and its landscapes on southern Vancouver Island, British Columbia. University of Victoria, Canada, BC: Unpublished PhD thesis.

Behre, K.-E. 1988 The role of man in European vegetation history. In Huntley, B. and Webb, T. (eds), *Vegetation History.* Dordrecht: Kluwer, 633–72.

Behre, K.-E. 2008 *Lanschafts-geschichte Norddeutschlands.* Neumünster: Wachholtz.

Behrensmeyer, A. and Laporte, L. 1981 Footprints of a Pleistocene hominid in northern Kenya, *Nature* 289, 167–9.

Bell, M. 1977 *Excavations at Bishopstone. Sussex Archaeological Collections* 115.

Bell, M. 1981 Valley sediments and environmental change. In Jones, M. and Dimbleby, G.W. (ed.), *The Environment of Man: the Iron Age to the Anglo-Saxon period.* Oxford: British Archaeological Report S87, 75–91.

Bell, M. 1983 Valley sediments as evidence of prehistoric land-use on the South Downs, *Proceedings of the Prehistoric Society* 49, 119–150.

Bell, M. 2005 Derek John Upton, 1941–2005: an appreciation, *Archaeology in the Severn Estuary* 16, 1–6.

Bell, M. 2007a *Prehistoric Coastal Communities: the Mesolithic in Western Britain.* York: Council for British Archaeology Research Report 149.

Bell, M. 2007b Wetland-dryland relationships in the Severn Estuary and surroundings during the Mesolithic and Neolithic. In Sidell, E.J. and Haughey, F. (eds), *Neolithic Archaeology in the Intertidal Zone of Rivers and Estuaries.* Oxford: Oxbow Books, 26–47.

Bell, M. 2013 *The Bronze Age in the Severn Estuary.* York: Council for British Archaeology Research Report 172.

Bell, M. 2015 Experimental archaeology at the cross roads: a contribution to interpretation or evidence of 'xeroxing'? In Wylie, A. and Chapman, R. (ed.), *Material Evidence: learning from archaeological practice.* Abingdon: Routledge, 42–58.

Bell, M. and Boardman, J. 1992 *Past and Present Soil Erosion.* Oxford: Oxbow Monograph 22.

Bell, M. and Walker, M.J.C. 2005 *Late Quaternary Environmental Change. Physical and Human Perspectives.* Second edition. Harlow: Pearson Education.

Bell, M., Caseldine, A.E. and Neumann, H. (ed.) 2000 *Prehistoric Intertidal Archaeology in the Welsh Severn Estuary.* York: Council for British Archaeology, Research Report 120.

Bell, M., Allen, M.J., Smith, R.W. and Johnson, S. 2008 Molluscan and sedimentary evidence for the palaeoenvironmental history of Hambledon Hill and its surroundings. In Mercer, R. and Healy, F. (eds), *Hambledon Hill, Dorset, England: excavation and survey of a Neolithic monument complex and its surrounding landscape.* Swindon: English Heritage, 412–53.

Bell, M., Black, S., Maslin, S. and Toms, P. forthcoming Multi-method solutions of the problems of dating early trackways and associated colluvial sequences, *Journal of Archaeological Science Reports.*

Bell, T.W. and Lock, G. 2000 Topographic and cultural influences on walking the Ridgeway in later prehistoric times. In Lock, G. (ed.), *Beyond the Map: archaeology and spatial technology. Nato Advanced Research Workshop, Netherlands.* Amsterdam: IOS Press, 85–100.

Belloc, H. 1904 *The Old Road.* London: Constable.

Belperio, A.P. and Fotheringham, D.G. 1990 Geological setting of two Quaternary footprint sites, western South Australia, *Australian Journal of Earth Sciences* 37, 37–42.

Bender, B. 1986 *The Archaeology of Brittany, Normandy and the Channel Islands.* London: Faber and Faber.

Bender, B., Hamilton, S. and Tilley, C. 2007 *Stone Worlds: narrative and reflexity in landscape archaeology.* Walnut Creek CA: Left Coast Press.

Bendrey, R. 2012 From wild horses to domestic horses: a European perspective, *World Archaeology* 44, 135–57.

Bendrey, R., Thorpe, N., Outram, A. and van Wijngaarden-Bakker, L.H. 2013 The origins of domestic horses in north-west Europe: new direct dates on the horses of Newgrange, Ireland, *Proceedings of the Prehistoric Society* 79, 91–103.

Bengtsson, B. 2017 *Sailing Rock Art Boats*. Oxford: British Archaeological Report S2865.

Bennett, K.D. 2015 Comment on 'Sedimentary DNA from a submerged site reveals wheat in the British Isles 8000 years ago', *Science* 349, 247.

Bennett, M.R. and Morse, S.A. 2014 *Human Footprints: fossilised locomotion?* London: Springer.

Bennett, M.R., Gonzalez, S., Huddart, D., Kirby, J. and Toole, E. 2010 Probable Neolithic footprints preserved in intertidal peat at Kenfig, South Wales (UK), *Proceedings of the Geologists' Association* 121, 66–76.

Bennett, M.R., Harris, J.W.K., Richmond, B.G., Braun, D.R., Mbua, E., Kiura, P., Olago, D., Kibunjia, M., Omuombo, C., Behrensmeyer, A., Huddart, D. and Gonzalez, S. 2009 Early hominin foot morphology based on 1.5-million-year-old footprints from Ileret, Kenya, *Science* 323, 1197–201.

Benson, D. and Whittle, A. 2006 *Building Memories: the Neolithic Cotswold Long Barrow at Ascott-under-Wychwood, Oxfordshire*. Oxford: Oxbow Books.

Bentley, R.A., Bickle, P., Fibiger, L., Nowell, G., Dale, C., Hedges, R., Hamilton, J., Wahl, J., Francken, M., Grupe, G., Lenneis, E., Teschler-Nicola, M., Arbogast, R.-M., Hofmann, D. and Whittle, A. 2012 Community differentiation and kinship among Europe's first farmers, *Proceedings of the National Academy of Sciences of the United States of America* 109, 9326–30.

Berglund, B.E., Digerfeldt, G., Engelmark, R., Gaillard, M.-J., Karlsson, S., Miller, U. and Risberg, J. 1996 Sweden. In Berglund, B.E., Birks, H.J.B., Ralska-Jasiewiczowa, M. and Wright, H.E. (eds), *Palaeoecological Events During the Last 15,000 years*. Chichester: Wiley, 233–80.

Bickle, P. 2016 Varied mobility in the Neolithic: the LBK on the move. In Leary, J. and Kadar, T. (ed.), *Moving on in Neolithic Studies*. Oxford: Neolithic Studies Group Seminar Papers 14, 14–27.

Binford, L. 1983 *In Pursuit of the Past*. London: Thames and Hudson.

Binford, L. 2001 *Constructing Frames of Reference*. Berkeley CA: University of California Press.

Binford, L. 2006 Bands as characteristic of 'mobile hunter-gatherers' may exist only in the history of anthropology. In Sellet *et al.* (eds) 2006, 3–22.

Bintliff, J. (ed.) 1999 *Structure and Contingency: evolutionary processes in life and human society*. Leicester: Leicester University Press.

Bird-David, N. 2005 Studying children in 'hunter-gatherer' societies: reflections from a Nayaka perspective. In Hewlett, B.S. and Lamb, M.E. (eds), *Hunter-Gatherer Childhoods: evolutionary, developmental and cultural perspectives*. Piscataway NJ: Transaction Publishers, 92–101.

Bishop, M.C. 2014 *The Secret History of the Roman Roads of Britain*. Barnsley: Pen and Sword.

Bishop, R.R., Church, M.J. and Rowley-Conwy, P. 2015 Firewood, food and human niche construction: the potential role of Mesolithic hunter-gatherers in actively structuring Scotland's woodlands, *Quaternary Science Reviews* 108, 51–75.

Blackburn, J. 2012 *Thin Paths*. London: Vintage Books.

Blackburn, J. 2019 *Timesong: searching for Doggerland*. London: Jonathan Cape.

Blackburn, T.C. and Anderson, K. (ed.) 1993 *Before the Wilderness*. Menlo Park CA: Balkema.

Blackstock, M. 2001 *Faces in the Forest: First Nations art created on living trees*. Montreal: McGill University Press.

Blair, J. 1991 *Early Medieval Surrey: landholding, church and settlement*. Stroud: Alan Sutton/Surrey Archaeological Society.

Bliege Bird, R., Taylor, N., Codding, B.F. and Bird, D.W. 2013 Niche construction and Dreamland logic: Aboriginal patch mosaic burning and varanid lizards (*Varanus gouldii*) in Australia, *Royal Society Proceedings Biological Sciences B* 280, 20132297.

Bliege Bird, R., Bird, D.W., Codding, B.F., Parker, C.H. and Jones, J.H. 2008 The 'fire-stick farming' hypothesis: Australian Aboriginal foraging strategies, biodiversity and anthropogenic fire mosaics, *Proceedings of the National Academy of Sciences of the United States of America* 105, 14796–801.

Boardman, J. 1993 The sensitivity of downland arable land to erosion by water. In Thomas, D.S.G. and Allison, R.J. (eds), *Landscape Sensitivity*. Chichester: Wiley, 211–28.

Boardman, J. 1999 *The Greeks Overseas*. London: Thames and Hudson.

Boardman, J. 2013 The hydrological role of 'sunken lanes' with respect to sediment mobilization and delivery to watercourses with particular reference to West Sussex, southern England, *Journal of Soils and Sediments* 13, 1636–44.

Boardman, J. 2014 Sunken lanes. Historical landmark or flood risk?, *Geography Review* April 2014, 28–30.

Boardman, J. 2016 The value of Google Earth™ erosion mapping, *Catena* 143, 123–7.

Boas, F. 1921 *Ethnology of the Kwakuitl*. Washington DC: Smithsonian Institution.

Bogaard, A. 2004 *Neolithic Farming in Central Europe*. London: Routledge.

Bogaard, A. and Jones, G. 2007 Neolithic farming in Britain and central Europe: contrast or continuity. In Whittle, A. and Cummings, V. (eds), *Going Over: the Mesolithic–Neolithic transition in North-West Europe.* Oxford: Proceedings of the British Academy 144, 357–75.

Bond, C.J. 2004 The Sweet Track, Somerset: a place mediating culture and spirituality? In Insoll, T. (ed.), *Belief in the Past.* Oxford: British Archaeological Report S1212, 37–50.

Bonsall, C. and Smith, C. 1989 Late Palaeolithic and Mesolithic bone and antler artifacts from Britain: first reactions to accelerator dates, *Mesolithic Miscellany* 10, 33–8.

Booth, P. 2011 Romano-British trackways in the Upper Thames Valley, *Oxoniensia* 76, 1–14.

Booth, P., Champion, T., Foreman, S., Garwood, P., Glass, H., Munby, J. and Reynolds, A. 2011 *On Track: the archaeology of High Speed I Section I in Kent.* Oxford: Oxford-Wessex Archaeology.

Bork, H.-R., Bork, H., Dalchow, C., Faust, B., Piorr, H.-P. and Schatz, T. 1998 *Landschafts-entwicklung in Mitteleuropa.* Stuttgart: Klett-Perthes.

Borrow, G. 1862 *Wild Wales.* London: John Murray.

Bortenschlager, S. and Oeggl, K. (eds) 2000 *The Iceman and his Natural Environment.* Vienna: Springer.

Bos, J.A.A. and Janssen, C.R. 1996 Local impact of Palaeolithic man on the environment during the end of the last glacial in The Netherlands, *Journal of Archaeological Science* 23, 731–9.

Bos, J.A.A. and Urz, R. 2003 Late Glacial and early Holocene environment in the middle Lahn river valley (Hessen, central-west Germany) and the local impact of early Mesolithic people – pollen and macrofossil evidence, *Vegetation History and Archaeobotany* 12, 19–36.

Both, F. and Fansa, M. 2011a Die Moorwege im Wesr-Ems-Gebiet. In Fansa, M. and Both, F. (eds), *"O, schaurig ist's, übers Moor zu gehn": 220 Jahre Moorarchäologie.* Mainz: Philipp von Zabern, 61–188.

Both, F. and Fansa, M. 2011b Geschichte der Moorwegforschung zwischen Weser und Ems. In Fansa, M. and Both, F. (ed.), *"O, schaurig ist's, übers Moor zu gehn": 220 Jahre Moorarchäologie.* Mainz: Philipp von Zabern, 43–60.

Bourdieu, P. 1977 *Outline of a Theory of Practice.* Cambridge: Cambridge University Press.

Bourgeois, Q. 2013 *Monuments on the Horizon: the formation of the barrow landscape throughout the 3rd and 2nd millennium BC.* Leiden: Sidestone Press.

Bouzek, J. 1993 The shifts of the amber route. In Beck, K.W. and Bouzek, J. (ed.), *Amber in Archaeology: proceedings of conference in Liblice, 1990.* Prague: Czech Academy of Science, 141–6.

Bowden, M., Soutar, S., Field, D. and Barber, M. 2015 *The Stonehenge Landscape.* Swindon: Historic England.

Bowen, H.C. 1961 *Ancient Fields.* Wakefield: S.R. Publishers.

Boyd, R. 1999 *Indians, Fire and the Land in the Pacific Northwest.* Corvallis OR: Oregon State University Press.

Bradley, R. 1970 The excavation of a Beaker settlement at Belle Tout, East Sussex, England, *Proceedings of the Prehistoric Society* 36, 312–79.

Bradley, R. 1975 Maumbury Rings, Dorchester: The excavations of 1908–13, *Archaeologia* 105, 1–97.

Bradley, R. 1984 *The Social Foundations of Prehistoric Britain.* London: Longman.

Bradley, R. 1990 *The Passage of Arms.* Cambridge: Cambridge University Press.

Bradley, R. 1993 *Altering the Earth.* Edinburgh: Society of Antiquaries of Scotland.

Bradley, R. 1997 *Rock Art and the Prehistory of Atlantic Europe.* London: Routledge.

Bradley, R. 2000 *An Archaeology of Natural Places.* London: Routledge.

Bradley, R. 2007 *The Prehistory of Britain and Ireland.* Cambridge: Cambridge University Press.

Bradley, R. 2009 *Image and Audience: rethinking prehistoric art.* Oxford: Oxford University Press.

Bradley, R. 2012 *The Idea of Order: the circular archetype in prehistoric Britain.* Oxford: Oxford University Press.

Bradley, R. 2016 *A Geography of Offerings: deposits of valuables in the landscape of ancient Europe.* Oxford: Oxbow Insights in Archaeology 3.

Bradley, R 2018 Early Bronze Age boat graves in the British Isles. In Bakels, C., Bourgeois, Q., Fontijn, D. and Jansen, R (eds), *Local Communities in the Big World of Prehistoric Europe.* Leiden: Anelecta Praehistorica Leidensia 49, 23–28.

Bradley, R. and Chambers, R. 1988 A new study of the cursus complex at Dorchester-on-Thames, *Oxford Journal of Archaeology* 7 (3), 271–89.

Bradley, R. and Fulford, M.G. 2008 The chronology of co-axial field systems. In Rainbird, P. (ed.), *Monuments in the Landscape.* Stroud: Tempus, 114–22.

Bradley, R., Entwistle, R. and Raymond, F. 1994 *Prehistoric Land Divisions on Salisbury Plain.* London: English Heritage.

Bradley, R., Haselgrove, C., Vander Linden, M. and Webley, L. 2015 *The Later Prehistory of North-West Europe: the evidence of development-led fieldwork.* Oxford: Oxford University Press.

Bradley, R., Rogers, A., Sturt, F. and Watson, A. 2016 Maritime havens in earlier prehistoric Britain, *Proceedings of the Prehistoric Society* 82, 125–59.

Brady, J. and Stone, A. 1986 Naj Tunich: Entrance to the Maya underworld, *Archaeology* 39 (6), 18–25.

Brady, K. 2014 The logboats in the lake, *Current Archaeology* 292, 10–15.

Brandon, P. 1974 *The Sussex Landscape.* London: Hodder and Stoughton.

Brandon, P. and Short, B. 1990 *The South East from AD 1000.* London: Longman.

Brennand, M. and Taylor, M. 2003 The survey and excavation of a Bronze Age timber circle at Holme-next-the-Sea, Norfolk, 1998–9, *Proceedings of the Prehistoric Society* 69, 1–84.

Brindley, A.L. and Lanting, J.N. 1998 Radiocarbon dates for Irish trackways, *Journal of Irish Archaeology* 8, 5–8.

British Columbia Ministry of Forests 2001 *Culturally Modified Trees of British Columbia: handbook of identification and records.* Vancouver: British Columbia Ministry of Forests.

Britnell, W.J. and Savory, H.N. 1984 *Gernvale and Penwrlod: two Neolithic long cairns in the Black Mountains of Brecknock.* Cardiff: Cambrian Archaeological Monograph 2.

Britton, K. and Müldner, G. 2013 Stable isotope evidence of saltmarsh economies. In Bell (ed.) 2013, 263–73.

Britton, K., Müldner, G. and Bell, M. 2008 Stable isotope evidence for salt-marsh grazing in the Bronze Age Severn Estuary, UK: implications for palaeodietary analysis at coastal sites, *Journal of Archaeological Science* 35, 2111–18.

Broadbent, N.D. 2010 *Lapps and Labyrinths.* Washington DC: Arctic Studies Center.

Brody, H. 2002 *Maps and Dreams: Indians and the British Columbia frontier.* London: Faber and Faber.

Bromley, R.G. 1996 *Trace Fossils: biology, taphonomy and applications.* London: Chapman and Hall.

Brongers, J.A. 1974 *Vaassen: een akkercomplex vit de Ijzertijd.* Bussum/Amersfoort: Fibula-Van Dishoeck.

Brongers, J.A. 1976 *Air Photography and Celtic Field Research in The Netherlands.* Amersfoort: ROB: Nederlandse Oudheden 6.

Brongers, J.A. and Woltering, P.J. 1973 Prehistory in the Netherlands; an economic-technological approach, *Berichten van de Rijksdienst voor het Oudheidkundig Bodermonderzoek* 23, 7–47.

Broodbank, C. 2013 *The Making of the Middle Sea.* London: Thames and Hudson.

Brookes, S. 2007 Walking with Anglo-Saxons: landscapes of the dead in Early Anglo-Saxon Kent, *Anglo-Saxon Studies in Archaeology and History* 14, 143–53.

Brookes, S. and Harrington, S. 2010 *The Kingdom and People of Kent AD 400–1066.* Stroud: History Press.

Brown, A.G. 1986 Flint and chert small finds from the Somerset Levels. Part 1: the Brue Valley, *Somerset Levels Papers* 12, 12–26.

Brown, A.G. 1997 *Alluvial Geoarchaeology.* Cambridge: Cambridge University Press.

Brown, K.A. 2014 Women on the move: the DNA evidence for female mobility and exogamy in prehistory. In Leary (ed.) 2014, 155–74.

Brown, K.J. and Hebda, R.J. 2002 Ancient fires on southern Vancouver Island, British Columbia, Canada: a change in causal mechanisms at about 2,000 ybp, *Environmental Archaeology* 7, 1–12.

Brown, K.J. and Hebda, R.J. 2003 Coastal rainforest connections disclosed through a Late Quaternary vegetation, climate, and fire history investigation from the Mountain Hemlock Zone on southern Vancouver Island, British Columbia, Canada, *Review of Palaeobotany and Palynology* 123, 247–69.

Brown, R., Ferguson, J., Lawrence, M. and Lees, D. 2003 *Tracks and Signs of the Birds of Britain and Europe.* London: Christopher Helm.

Brown, T. 1997 Clearances and clearings: deforestation in Mesolithic/Neolithic Britain, *Oxford Journal of Archaeology* 16, 133–46.

Brunning, R. 1997 Two Bronze Age wooden structures in the Somerset Moors. *Archaeology in the Severn Estuary* 8, 5–8.

Brunning, R. 2007 Structural Wood Remains from Prehistoric England and Wales. University of Exeter: Dept of Archaeology, unpublished PhD thesis.

Brunning, R. 2013 *Somerset's Peatland Archaeology: Managing and investigating a fragile resource.* Oxford: Oxbow Books.

Brunning, R. and McDermott, C. 2013 Trackways and roads across the wetlands. In Menotti and O'Sullivan (eds) 2013, 359–85.

Bryan, A.L. 1973 New light on ancient Nicaraguan footprints, *Archaeology* 26, 146–7.

Bryant, S. and Burleigh, G. 1995 Later prehistoric dykes of the eastern Chilterns. In Holgate, R. (ed.), *Chiltern Archaeology: recent work. A handbook for the new decade.* Dunstable: Book Castle, 17–27.

Bryant, S., Perry, B. and Williamson, T. 2005 A 'relict' landscape in south-east Hertfordshire: archaeological and topographic investigations in the Wormley area, *Landscape History* 27, 5–16.
Buchwald, E. and Svenning, J.-C. 2006 Premature rejection of a controversial hypothesis? Primeval forest openess revisited, *Journal of Ecology* 93, 168–77.
Buckland, P.C. and Edwards, K.J. 1984 The longevity of pastoral episodes of clearance activity in pollen diagrams: the role of post-occupation grazing, *Journal of Biogeography* 11, 243–9.
Budd, P., Chenery, C., Montgomery, J. and Evans, J. 2003 You are where you ate: isotopic analysis in the reconstruction of prehistoric residency. In Parker Pearson, M. (ed.), *Food, Culture and Identity in the Neolithic and early Bronze Age.* Oxford: British Archaeological Report S1117, 69–78.
Bull, E.J. 1993 The bi-axial landscape of prehistoric Buckinghamshire, *Records of Bucks* 35, 11–18.
Bunting, M.J. and Farrell, M. 2017 Seeing the wood for the trees: recent advances in the reconstructon of woodland in archaeological landscapes using pollen data, *Environmental Archaeology* http://www.tandfonline.com/10.1080/14614103.2017.1377405.
Bunyan, J. 1678 *The Pilgrim's Progress From this World to That Which is to Come.* London: Nathaniel Ponder.
Burl, A. 2000 *The Stone Circles of Britain, Ireland and Brittany.* London: Yale University Press.
Burstow, G.P. and Holleyman, G.A. 1957 Late Bronze Age settlement on Itford Hill, Sussex, *Proceedings of the Prehistoric Society* 23, 167–212.
Bush, M.B. 1988 Early Mesolithic disturbance: a force in the landscape, *Journal of Archaeological Science* 15, 453–62.
Buteux, S. and Chapman, H. 2009 *Where Rivers Meet: the archaeology of Catholme and the Trent-Tame confluence.* York: Council for British Archaeology Research Report 161.
Canti, M. 1999 The production and preservation of faecal sperulites: animals, environment and taphonomy, *Journal of Archaeological Science* 26 (3), 251–8.
Canti, M. and Brochier, J.E. 2017 Faecal spherulites. In Nicosia and Stoops (eds) 2017, 51–4.
Carlstein, T. 1982 *Time, Resources, Society and Ecology Vol 1: Pre-Industrial Societies.* Lund: Allen and Unwin.
Carter, S. 1990 The stratification and taphonomy of shells in calcareous soils: implications for land snail analysis in archaeology, *Journal of Archaeological Science* 17, 495–507.
Carter, S., Hunter, F. and Smith, A. 2010 A 5th century BC Iron Age chariot burial from Newbridge, Edinburgh, *Proceedings of the Prehistoric Society* 76, 31–74.
Caseldine, C., Hatton, J. and Caseldine, A. 1996 Palaeoecological studies at Corlea (1988–1992). In Raftery (ed.) 1996, 379–92.
Caseldine, A., Griffiths, C. and Druce, D. 2013 The botanical evidence from Redwick. In Bell (ed.) 2013, 95–129.
Casparie, W.A. 1982 The Neolithic wooden trackway XXI (Bou) in the raised bog at Niew-Dordrecht (The Netherlands), *Palaeohistoria* 24, 115–64.
Casparie, W.A. 1984 The three Bronze Age footpaths XVI (Bou), XVII (Bou) and XVIII (Bou) in the raised bog of Southeast Drenthe (The Netherlands), *Palaeohistoria* 26, 41–94.
Casparie, W.A. 1986 The two Iron Age wooden trackways XIV (Bou) and XV (Bou) in the raised bog of Southeast Drenthe (The Netherlands), *Palaeohistoria* 28, 169–210.
Casparie, W.A. 1987 Bog trackways in the Netherlands, *Palaeohistoria* 29, 35–65.
Casparie, W.A. 2001 Prehistoric building disasters in Derryville Bog, Ireland: trackways, floodings and erosion. In Raftery, B. and Hickey, J. (eds), *Recent Developments in Wetland Research.* Dublin: Seandálaíocht, 115–28.
Casparie, W.A. 2005 Opening up the peat bogs: the timber trackways of Drenthe. In Louwe-Kooijmans, L.P., van den Broeke, P.W., Fokkens, H. and van Gijn, A.L. (eds), *The Prehistory of the Netherlands* Volume 1. Amsterdam: Amsterdam University Press, 401–5.
Casparie, W.A. and Moloney, A. 1996 Corlea 1: Palaeo-environmental aspects of the trackway. In Raftery (ed.) 1996, 3670–77.
Chadwick, A.M. 2007 Trackways, hooves and memory-days: human and animal movements and memories around Iron Age and Romano-British rural landscapes of the English north Midlands. In Cummings and Johnston (eds) 2007, 131–52.
Chadwick, A.M. 2016 Foot-fall and hoof-hit agencies. Movements, materialities and identities; and later prehistoric and Romano-British trackways, *Cambridge Archaeological Journal* 26, 93–120.
Chamberlain, A.T. 1997 Missing stages of life: towards the perception of children in archaeology. In Moore and Scott (eds) 1997, 248–50.
Champion, T. 2014 Food, technology and culture in the late Bronze Age of southern Britain: perforated clay plates of the lower Thames Valley, *Proceedings of the Prehistoric Society* 80, 279–98.

Chan, B., Viner, S., Parker Pearson, M., Albrella, U. and Ixer, R. 2016 Resourcing Stonehenge: paths of human, animal and goods mobility in the late Neolithic. In Leary and Kadar (eds) 2016, 28–44.
Channel 4 2016 *Britain's Ancient Tracks*. 3 part TV series: first broadcast in UK Autumn 2016.
Chapman, H. and Gearey, B.R. 2000 Palaeoecology and the perception of prehistoric landscapes: some comments on visual approaches to phenomenology, *Antiquity* 74, 316–31.
Chapman, H. and Gearey, B.R. 2006 The henge that went straight, *British Archaeology* Sept/Oct 2006, 43–7.
Chapple, R. 2016 https//sites.google.com/site/Chapple archaeology.
Chatterton, R. 2006 Ritual. In Conneller, C. and Warren, G. (eds), *Mesolithic Britain and Ireland: new approaches*. Stroud: Tempus, 101–20.
Chatwin, B. 1989 *The Songlines*. London: Jonathan Cape.
Chatwin, D. and Gardiner, J. 2005 Rethinking the early medieval settlement of woodlands: evidence from the western Sussex Weald, *Landscape History* 27 (1), 31–49.
Chaucer, G. 1400 (1951 edition: translation by Nevill Coghill) *The Canterbury Tales*. Harmondsworth: Penguin.
Chin, A., Galvin, K.A., Gerlak, A.K., Harden, C.P. and Wohl, E. 2014 The future of human landscape introductions: Drawing on the past, anticipating the future, *Environmental Management* 53, 1–3.
Chisham, C. 2004 Early Mesolithic Human activity and Environmental Change: a case study of the Kennet Valley. University of Reading: Dept of Archaeology, Unpublished PhD thesis.
Christensen, K. 1997a Oak trunks from the Halsskov Fjord. In Pedersen *et al.* (eds) 1997, 37–44.
Christensen, K. 1997b Wood from fish weirs – forestry in the Stone Age. In Pedersen *et al.* (eds) 1997, 147–56.
Claassen, C.P. 1991 Gender: shellfishing and the Shell Mound Archaic. In Gero, J.M. and Conkey, M.W. (ed.), *Engendering Archaeology: women and prehistory*. Oxford: Blackwell, 276–97.
Clark, J.D.G. 1954 *Excavations at Star Carr*. Cambridge: Cambridge University Press.
Clark, J.S. and Royall, P.D. 1995 Transformation of a northern hardwood forest by aboriginal (Iroquois) fire: charcoal evidence from Crawford Lake, Ontario, Canada, *The Holocene* 5, 1–9.
Clark, J.S., Merkt, J. and Müller, H. 1989 Post-glacial fire, vegetation and human history on the Northern Alpine Forelands, south-western Germany, *Journal of Ecology* 77, 897–925.
Clark, P. 2004a *The Dover Bronze Age Boat*. Swindon: English Heritage.
Clark, P. 2004b *The Dover Bronze Age Boat in Context: society and water transport in prehistoric Europe*. Oxford: Oxbow Books.
Clarke, D. 1976 Mesolithic Europe: the economic basis. In Sieveking, G.d.G., Longworth, I.H. and Wilson, K.E. (eds), *Problems in Economic and Social Archaeology*. London: Duckworth, 449–81.
Cleal, R.M.J. and Walker, K.E. with Montague, R. 1995 *Stonehenge in its Landscape*. London: English Heritage.
Close, A.E. 2006 *Finding the People who Flaked the Stone at English Camp (San Juan Island)*. Salt Lake City UT: University of Utah Press.
Clottes, J. 2003 *Return to Chauvet Cave*. London: Thames and Hudson.
Clough, T. and Cummins, W. 1988 *Stone Axe Studies*. London: Council for British Archaeology Research Report 23.
Cohen, N. and Wragg, E. 2017 *The River's Tale: archaeology on the Thames foreshore in Greater London*. London: Museum of London Archaeology.
Coles, B. 1994 *Trisantona* rivers: a landscape approach to the interpretation of river names, *Oxford Journal of Archaeology* 13, 295–312.
Coles, B. 2000 Doggerland: the cultural dynamics of a shifting coastline. In Pye, K. and Allen, J.R.L. (ed.), *Coastal and Estuarine Environments: sedimentology, geomorphology and geoarchaeology*. London: Geological Society Special Publication 175, 393–402.
Coles, B. 2006 *Beavers in Britain's Past*. Oxford: WARP Occasional Paper 19.
Coles, B. and Coles, J. 1986 *Sweet Track to Glastonbury*. London: Thames and Hudson.
Coles, B. and Coles, J. 1989 *People of the Wetlands*. London: Guild Publishing.
Coles, B. and Coles, J. 1998 Passages of time, *Archaeology in the Severn Estuary* 9, 3–16.
Coles, B. and Brunning, R. 2009 Following the Sweet Track. In Cooney, G., Becker, K., Coles, J. and Sievers, S. (eds), *Relics of Old Decency: archaeological studies in later prehistory*. Dublin: Wordwell, 25–37.
Coles, J. 1984 Prehistoric roads and trackways in Britain: problems and possibilities. In Fenton and Stell (eds) 1984, 1–21.
Coles, J. 1989 Prehistoric settlement in the Somerset Levels, *Somerset Levels Papers* 15, 14–32.
Coles, J. 1990 *Images of the Past*. Vitlycke: Hallristningsmuseet.
Coles, J. 2005 *Shadows of a Northern Past: rock carvings of Bohuslän and Østfold*. Oxford: Oxbow Books.

Coles, J. and Orme, B. 1978a The Meare Heath Track, *Somerset Levels Papers* 4, 11–39.

Coles, J. and Orme, B. 1978b Multiple trackways from Tinney's Ground, *Somerset Levels Papers* 4, 47–81.

Coles, J. and Orme, B. 1983 *Homo sapiens* or *Castor fiber*?, *Antiquity* 57, 95–102.

Coles, J. and Orme, B. 1984 Ten excavations along the Sweet Track (3200 bc), *Somerset Levels Papers* 10, 5–45.

Connah, G. 1965 Excavations at Knap Hill, Alton Priors, 1961, *Wiltshire Archaeological and Natural History Magazine* 60, 1–23.

Conneller, C. 2006 Death. In Conneller, C. and Warren, G. (eds), *Mesolithic Britain and Ireland: new approaches.* Stroud: Tempus, 139–64.

Conneller, C. and Schadla-Hall, T. 2003 Beyond Star Carr: Vale of Pickering in the 10th millennium BP, *Proceedings of the Prehistoric Society* 69, 85–105.

Conolly, J. and Lake, M. 2006 *Geographical Information Systems in Archaeology.* Cambridge: Cambridge University Press.

Cook, J. 1784 (1967 edition) *The Journals of Captain James Cook Volume III.1: The Voyage of the Resolution and Discovery 1776–1780.* Cambridge: Hakluyt Society.

Cooney, G. 2000 *Landscapes of Neolithic Ireland.* London: Routledge.

Cooney, G. 2009 Tracing lines across landscapes: corporality and history in later prehistory. In Cooney, G., Becker, K., Coles, J., Ryan, M. and Sievers, S. (eds), *Relics of Old Decency: archaeological studies in Later Prehistory.* Dublin: Wordwell, 375–88.

Copeland, T. 2009 *Akeman Street: moving through Iron Age and Roman landscapes.* Stroud: History Press.

Cowell, R.W. 2008 Coastal Sefton in the prehistoric period. In Lewis, J.M. and Stanistreet, J.E. (ed.), *Sand and Sea: Sefton's Coastal Heritage.* Bootle: Sefton Council, 20–7.

Cowell, R.W. and Innes, J.B. 1994 *The Wetlands of Merseyside.* Lancaster: Lancaster Imprints.

Crane, N. 1997 *Clear Waters Rising: a mountain walk across Europe.* London: Penguin.

Crane, N. 2010 *Coast: our island story.* London: Random House.

Crawford, O.G.S. 1922 *The Andover District: an account of Sheet 283 of the one-inch Ordnance Survey Map.* Oxford: Oxford University Press.

Creighton, J. 2005 *Britannia: the creation of a Roman province.* London: Routledge.

Cresnar, M. 2016 Lubljana marshes, landscape and movement, Slovenia. UK: Lecture to Wetland Archaeology Research Project Conference, 28/6/16 and pers. comm.

Crockett, A.D., Allen, M.J. and Scaife, R.G. 2002 A Neolithic trackway within peat deposits at Silvertown, London, *Proceedings of the Prehistoric Society* 68, 185–213.

Croes, D.R. 1995 *The Hoko River Archaeological Site Complex, the Wet/Dry Site (45CA213), 3,000–2,600 B.P.* Pullman, WA: Washington State University Press.

Croes, D.R. 2015 The Salish Sea: using wet and dry site archaeology to explore the defining temporal characteristics of this inland sea, *Journal of Wetland Archaeology* 15 (1), 72–108.

Croes, D.R., Fagan, J.L. and Zehender, M.N. 2009 Sunken village, Sauvie Island, Oregon, USA: a report on the 2006–7 investigations of National Historic Landmark wet site 35MU4, *Journal of Wetland Archaeology* 9, 1–216.

Crombé, P. 1993 Tree-fall features on Final Palaeolithic and Mesolithic sites situated on sandy soils: how to deal with it, *Helinium* 33.1, 50–66.

Cronon, W.M. 1983 *Changes in the Land: Indians, colonists and the ecology of New England.* New York: Hill and Wang.

Cross, S., Murray, C., Ó Néill, J. and Stevens, P. 2001 Derryville Bog: a vernacular landscape in the Irish Midlands. In Raftery, B. and Hickey, J. (eds), *Recent Developments in Wetland Research.* Dublin: Seandálaíocht, 87–97.

Crumley, C.L. and Marquardt, W.H. 1987 *Regional Dynamics: Burgundian landscapes in historial perspective.* London: Academic Press.

Crumlin-Pedersen, O. 2010 *Archaeology and the Sea in Scandinavia and Britain.* Roskilde: Viking Ship Museum.

Crumlin-Pedersen, O. and Trakadas, A. 2003 *Hjortspring: A Pre-Roman Iron-Age Warship in Context.* Roskilde: National Museum.

Crutchley, S. and Crow, P. 2010 *The Light Fantastic: using airborne laser scanning in archaeological survey.* Swindon: English Heritage.

Cummings, V. 2002 Landscapes in motion: interactive computer imagery and Neolithic landscapes of the Outer Hebrides. In Buck, C., Cummings, V., Henley, C., Mills, S. and Trick, S. (ed.), *UK Chapter of Computer Applications and Quantitative Methods in Archaeology.* Oxford: British Archaeological Report S844, 11–20.

Cummings, V. and Harris, O. 2011 Animals, people and places: the continuity of hunting and gathering practices across the Mesolithic-Neolithic transition in Britain, *European Journal of Archaeology* 14, 361–82.

Cummings, V. and Johnston, R. 2007 *Prehistoric Journeys.* Oxford: Oxbow Books.

Cummings, V. and Whittle, A. 2004 *Places of Special Virtue: megaliths in the Neolithic landscapes of Wales.* Oxford: Oxbow Books.

Cunliffe, B. 1965 *Winchester Excavations Vol 1, 1949–60.* Winchester: Winchester Museum.

Cunliffe, B. 1973 Chalton, Hampshire: the evolution of a landscape, *Antiquaries Journal* 53, 173–90.

Cunliffe, B. 1991 *Iron Age Communities in Britain.* London: Routledge.

Cunliffe, B. 2001a *Facing the Ocean: the Atlantic and its peoples.* Oxford: Oxford University Press.

Cunliffe, B. 2001b *The Extraordinary Voyage of Pytheas the Greek.* New York: Walker.

Cunliffe, B. 2004 Wessex cowboys, *Oxford Journal of Archaeology* 23 (1), 61–81.

Cunliffe, B. 2015 *By Steppe, Desert and Ocean: the birth of Eurasia.* Oxford: Oxford University Press.

Cunliffe, B. and Poole, C. 1991 *Danebury: An Iron Age fort in Hampshire. Vol 4: The excavations 1979–1988: the site.* York: Council for British Archaeology Research Report 73.

Curwen, E. 1933 Exavations on Thundersbarrow Hill, Sussex, *Antiquaries Journal* 13, 109–33.

Curwen, E. 1951 Cross-ridge dykes in Sussex. In Grimes, W.F. (ed.), *Aspects of Archaeology in Britain and Beyond.* London: HW Edwards, 93–107.

Curwen, E. and Curwen, E.C. 1914 Ancient trackways near Saddlecombe, *Brighton and Hove Archaeologist* 1914, 36–42.

Curwen, E. and Curwen, E.C. 1923 Sussex lynchets and their associated field ways, *Sussex Archaeological Collections* 64, 1–65.

Curwen, E. and Curwen, E.C. 1926 Port's Road, the ancient road of Portslade, *Brighton and Hove Archaeologist* 3, 28–41.

Curwen, E.C. 1929 *Prehistoric Sussex.* London: Homeland Association.

Curwen, E.C. 1954 *The Archaeology of Sussex.* London: Methuen.

Darby, H.C. and Campbell, E.M.J. 1962 *The Domesday Geography of South-East England.* Cambridge: Cambridge University Press.

Dark, P. 2000 Revised absolute dating of the early Mesolithic site of Starr Carr, North Yorkshire in the light of changes in the early Holocene tree ring chronology, *Antiquity* 74, 304–8.

Dark, P. 2004 New evidence for the antiquity of the intestinal parasite *Trichuris* in Europe, *Antiquity* 78, 676–81.

Dark, P. 2007 Plant communities and human activity in the Lower Submerged Forest and on Mesolithic occupation sites. In Bell, M. (ed.) 2007, 169–87.

Dark, P. and Allen, J.R.L. 2005 Seasonal deposition of Holocene banded sediments in the Severn Estuary Levels, southwest Britain: palynological and sedimentological evidence, *Quaternary Science Reviews* 24, 11–33.

Darvill, T. 1986 Neolithic Avon 3500–1650 BC. In Aston, M. and Iles, R. (ed.), *The Archaeology of Avon.* Bristol: Avon County Council, 13–28.

Darwin, C. 1881 *The Formation of Vegetable Mould through the Action of Worms with Observations on their Habits.* London: Faber and Faber.

David, A., Cole, M., Horsley, T., Linford, N., Linford, P. and Martin, L. 2004 A rival to Stonehenge? Geophysical survey at Stanton Drew, England, *Antiquity* 78, 341–58.

David, B. and Thomas, J. (ed.) 2008 *Handbook of Landscape Archaeology.* Walnut Creek CA: Left Coast Press.

Davies, H. 2006 *From Trackways to Motorways: 5000 years of highway history.* Stroud: Tempus.

Davies, P. 2010 Land and Freshwater molluscs. In O'Connor, T. and Sykes, N. (ed.), *Extinctions and Invasions: a social history of British fauna.* Oxford: Windgather Press, 175–80.

Davies, T.G., Pomeroy, E., Shaw, C.N. and Stock, J.T. 2014 Mobility and the skeleton: a biomedical view. In Leary (ed.) 2014, 129–53.

Davis, P. 1996 The footprint surfaces at the Barnfield Pit, Swanscombe. In Conway, B., McNabb, J. and Ashton, N. (eds), *Excavations at Barnfield Pit, Swanscombe 1968–72.* London: British Museum Press, 169–85.

Davis, S. 1987 *The Archaeology of Animals.* London: Batsford.

De Navarro, J.M. 1972 *The Finds from the Site of La Tène Volume I: scabbards and the swords found in them.* London: Oxford University Press.

De Reu, J., Plets, G., Verhoeven, G., De Smedt, P., Bats, M., Cherretté, B., De Maeyer, W., Deconynck, J., Herremans, D., Laloo, P., Van Meirvenne, M. and De Clerq, W. 2013 Towards a three-dimensional cost-effective registration of the archaeological heritage, *Journal of Archaeological Science* 40, 1108–21.

Deakin, R. 2000 *Waterlog: a swimmer's journey through Britain.* London: Vintage.

Deakin, R. 2007 *Wildwood: a journey through trees.* London: Hamish Hamilton.

Deakin, R. 2008 *Notes from Walnut Tree Farm.* London: Hamish Hamilton.

DeLumley, H. 1969 A Palaeolithic camp site at Nice, *Scientific American* 220, 42–50.

De Marrais, E., Gosden, C. and Renfrew, C. (eds) 2004 *Rethinking Materiality: the engagement of mind with the material world.* Cambridge: McDonald Institute.

Dent, J.S. 1982 Cemeteries and settlement patterns of the Iron Age on the Yorkshire Wolds, *Proceedings of the Prehistoric Society* 48, 437–57.

Dent, J.S. 1985 Three cart burials from Wetwang Slack, *Antiquity* 59, 85–92.

Deur, D. 1999 Salmon, sedentism and cultivation: toward an environmental prehistory of the Northwest Coast. In Goble and Hirt (eds) 1999, 129–55.

di Vitro, M.A., Zanella, E., Gurioli, L., Lanza, R., Sulpizio, R., Bidshop, J., Tema, E., Boenzi, G. and Laforgia, E. 2009 The Afragola settlement near Versuvius, Italy: the destruction and abandonment of a Bronze Age village revealed by archaeology, volcanology and rock-magnetism, *Earth and Planetry Science Letters* 277, 408–21.

Dickson, J.H., Bortenschlager, S., Oeggl, K., Porley, R. and McMullen, A. 1996 Mosses and the Tyrolean Iceman's southern provenance, *Proceedings of the Royal Society of London B* 263, 567–71.

Dickson, J.H., Richards, M.P., Hebda, R.J., Mudie, P.J., Beattie, O., Ramsay, S., Turner, N.J., Leighton, B.J., Webster, J.M., Hobischak, N.R., Anderson, G.S., Troffe, P.M. and Wigen, R.J. 2004 Kwäday Dän Ts'ìnchí, the first ancient body of a man from a North American glacier: reconstructing his last days by intestinal and biomolecular analyses, *The Holocene* 14, 481–6.

Diers, S., Jonsen, D., Alsleben, A., Dörfler, W., Müller, J. and Mischka, D. 2014 The western Altmark versus Flintbek: palaeoecological research on two megalithic regions, *Journal of Archaeological Science* 41, 185–98.

Dietler, M. and Herbich, I. 1993 Living on Luo time: reckoning sequence, duration, history and biography in a rural society, *World Archaeology* 25, 248–60.

Dillehay, T.D. 1989 *Monte Verde: a late Pleistoene settlement in Chile: palaeoenvironment and site context,* Volume 1. Washington DC: Smithsonian Institutional Press.

Dillehay, T.D., Ramirez, C., Pino, M., Collins, M.B., Rossen, J. and Pino-Navarro, J.D. 2008 Monte Verde: seaweed, food, medicine and the peopling of South America, *Science* 320, 784–6.

Dimbleby, G.W. 1978 *Plants and Archaeology.* London: Baker.

Dincauze, D.F. 2000 *Environmental Archaeology: principles and practice.* Cambridge: Cambridge University Press.

Doneus, M., Janner, M. and Fera, M. 2011 Flugzeuggetragenes Laserscanning im Leithagebirge, *Archäeologie Österreichs Spezial* 3, 1–12.

Doorenbosch, M. 2013 A history of open space: barrow landscapes and the significance of heaths: the core of the Echoput barrows. In Fontijn, D., Louwen, A.J., van der Vaart, S. and Wentink, K. (eds), *Beyond Barrows.* Amsterdam: Sidestone Press, 197–223.

Doughty, C., Wolf, A. and Malhi, Y. 2013 The legacy of the Pleistocene mega fauna extinctions on nutrient availability in Amazonia, *Nature Geoscience* 6, 761–4.

Douka, K. 2017 Radiocarbon dating of marine and terrestrial shell. In Allen (ed.) 2017, 381–99.

Drewett, P. 1982 *The Archaeology of Bullock Down, Eastbourne, East Sussex: the development of a landscape.* Lewes: Sussex Archaeological Society.

Drewett, P., Rudling, D. and Gardiner, M. 1988 *The South-east to AD 1000: a regional history of England.* London: Longman.

Duller, G.A.T. 2008 *Luminescence Dating.* Swindon: English Heritage.

Dunkley, M. 2014 Travelling by water: a chronology of prehistoric boat archaeology/mobility in England. In Leary (ed.) 2014, 187–200.

Dunning, C. 1991 La Tène. In Moscati, S., Frey, O.H., Kruta, V., Raftery, B. and Szabó, M. (eds), *The Celts.* Milan: Bompiani, 366–8.

Earnshaw, J. 2017 Cultural forests in cross-section: the exposure and destruction of CMT chronologies on Vancouver Island's West Coast. Lecture to Society of American Archaeology Conference, Vancouver: 30.3.17.

Edgeworth, M. 2011 *Fluid Pasts: archaeology of flow.* London: Bristol Classical Press.

Edgeworth, M. 2014 Enmeshments of shifting landscapes and embodied movements of people and animals. In Leary (ed.) 2014, 49–62.

Edmonds, M. 1995 *Stone Tools and Society.* London: Batsford.

Edmonds, M. 1997 Taskscape, technology and tradition, *Analecta Praehistoria Leidensia* 29, 99–110.

Edwards, K.J. 1988 The hunter-gatherer/agricultural transition and the pollen record in the British Isles. In Birks, H., Birks, H.J.B., Kaland, P.E. and Moe, D. (ed.), *The Cultural Landscape – Past, Present and Future*. Cambridge: Cambridge University Press, 255–66.

Edwards, K.J., Whittington, G. and Tipping, R. 2000 The incidence of microscopic charcoal in late glacial deposits, *Palaeogeography, Palaeoclimatology and Palaeoecology* 164, 247–62.

Ehser, A., Borg, G. and Pernicka, E. 2011 Provenance of the gold of the early Bronze Age Nebra sky disc, central Germany: geochemical characterisation of natural gold from Cornwall, *European Journal of Mineralogy* 23 (6), 895–910.

Ellenberger, P., Mossman, D.J., Mossman, A.D. and Lockley, M.G. 2007 Bushmen cave paintings of ornithropoid dinosaurs. Palaeolithic trackers interpret early Jurassic footprints, *Ichnos* 12 (3), 223–6.

Ellis, C., Allen, M.J., Gardiner, J., Harding, P., Ingrem, C., Powell, A. and Scaife, R. 2003 An Early Mesolithic seasonal hunting site in the Kennet Valley, southern England, *Proceedings of the Prehistoric Society* 69, 107–35.

Eogan, G. 1994 *The Accomplished Art: gold and gold working in Britain and Ireland during the Bronze Age*. Oxford: Oxbow Books.

Evans, C. 1987 Nomads in "waterland"? Prehistoric transhumance and fenland archaeology, *Proceedings of the Cambridge Antiquarian Society* 76, 27–40.

Evans, C. 1988 Acts of enclosure: a consideration of concentrically organised causewayed enclosures. In Barrett, J.C. and Kinnes, I. (eds), *The Archaeology of Context in the Neolithic and Bronze Age: recent trends*. Sheffield: Department of Archaeology and Prehistory, University of Sheffield, 85–96.

Evans, C. 2002 Metalwork and cold 'clay-lands': pre-Iron Age occupation on the Isle of Ely. In Lane, T. and Coles, J. (eds), *Through Wet and Dry: essays in honour of David Hall*. Exeter: WARP, 33–53.

Evans, C. 2009 *Grounding Knowledge Walking Land: archaeological research and ethnohistory identity in Central Nepal*. Cambridge: McDonald Institute.

Evans, C., Pollard, J. and Knight, M. 1999 Life in woods: tree throws, 'settlement' and forest cognition, *Oxford Journal of Archaeology* 18 (3), 241–54.

Evans, C.P. 2008 *Trods of the North York Moors: a gazeteer of flagged paths*. Scarborough: Scarborough Archaeological and Historical Society Research Report 13.

Evans, J. 1972 *Land Snails in Archaeology*. London: Seminar Press.

Evans, J., Pitts, M. and Williams, D. 1985 An excavation at Avebury, Wiltshire, 1982, *Proceedings of the Prehistoric Society* 51, 305–20.

Evans, J., Limbrey, S., Máté, I. and Mount, R. 1993 An environmental history of the Upper Kennett Valley, Wiltshire, for the last 10,000 years, *Proceedings of the Prehistoric Society* 59, 139–95.

Evans, J.A., Chenery, C. and Fitzpatrick, A.P. 2006 Bronze Age childhood migration of individuals near Stonehenge, revealed by strontium and oxygen isotope tooth enamel analysis, *Archaeometry* 48, 309–21.

Evans, J.G. 1975 *The Environment of Early Man in the British Isles*. London: Paul Elek.

Evans, J.G. and O'Connor, T. 1999 *Environmental Archaeology*. Stroud: Sutton.

Eve, S. 2014 An undiscovered county? New field systems found on the South Downs, *PAST* 78, 3–5.

Everitt, A. 1986 *Continuity and Colonization: the evaluation of Kentish settlement*. Leicester: Leicester University Press.

Ewert, E.C. 1999 Setting the Pacific Northwest stage: the influence of the natural environment. In Goble and Hirt (eds) 1999, 3–25.

Fagan, B.M. 1991 *Ancient North America*. London: Thames and Hudson.

Fairbairn, A.S. 2000 *Plants in Neolithic Britain and Beyond*. Oxford: Neolithic Studies Group Seminar Papers 5.

Fansa, M. and Burmeister, S. (eds) 2004 *Rad und Wagen: der Ursprung einer Innovation Wagen im Vorderen Orient und Europa*. Mainz: Philipp von Zabern.

Fedje, D.W. and Mathewes, R.W. 2005 *Haida Gwaii*. Vancouver: UBC Press.

Felstead, N.J., Gonzalez, S., Huddart, D., Noble, S.R., Hoffmann, D.L., Metcalfe, S.E., Leng, M.J., Albert, B.M., Pike, A.W.G., Gonzalez-Gonzalez, A. and Jiminez-Lopez, J.C. 2014 Holocene-aged human footprints, *Journal of Archaeological Science* 42, 250–59.

Fenton, A. and Stell, G. 1984 *Loads and Roads in Scotland and Beyond: road transport over 6000 years*. Edinburgh: John Donald.

Fenton-Thomas, C. 2005 *The Forgotten Landscapes of the Yorkshire Wolds*. Stroud: Tempus.

Field, D., Anderson-Whymark, H., Linford, N., Barber, M., Bowden, M., Linford, P. and Topping, P. 2015 Analytical surveys of Stonehenge and its environs, 2009–2013: Part 2 – the Stones, *Proceedings of the Prehistoric Society* 81, 125–48.

Field, N. and Parker Pearson, M. 2003 *Fiskerton: an Iron Age timber causeway with Iron Age and Roman votive offerings.* Oxford: Oxbow Books.

Finlayson, B. and Warren, G.M. 2010 *Changing Nations: hunter-gatherers, first farmers and the modern world.* London: Duckworth.

Fischer, A. 1991 Pioneers in deglaciated landscapes: the expansion and adaptation of late Palaeolithic societies in South Sandinavia. In Barton, R.N.E., Roberts, A.J. and Roe, D.A. (ed.), *The Lateglacial in North-west Europe.* York: Council for British Archaeology Research Report 71, 100–21.

Fischer, A. 2002 Food for feasting? An evaluation of explanations of the neolithicisation of Denmark and southern Sweden. In Fischer, A. and Kristiansen, K. (eds), *The Neolithicisation of Denmark.* Sheffield: J.R.Collis Publications, 343–93.

Fischer, A. 2003 Trapping up the rivers and trading across the sea-steps towards the neolithicisation of Denmark. In Larsson, L. (ed.), *Mesolithic on the Move: papers presented at the sixth International Conference on the Mesolithic in Europe, Stockholm 2000.* Oxford: Oxbow Books, 404–13.

Fitzgerald, M. 2007 Catch of the day at Clowanstown, Co. Meath, *Archaeology Ireland* 21, 12–15.

Fitzpatrick, A.P. 2011 *The Amesbury Archer and the Boscombe Bowmen: Bell Beaker burials at Boscombe Down, Amesbury, Wiltshire.* Salisbury: Wessex Archaeological Report 27.

Fitzpatrick, A.P., Ellis, C. and Allen, M.J. 1996 Bronze Age "jetties" or causeways at Testwood lakes, Hampshire, Great Britain, *NewsWARP* 20, 19–22.

Flannery, K.V. 1969 Origins and ecological effects of early domestication in Iran and the Near East. In Ucko, P.J. and Dimbleby, G.W. (ed.), *The Domestication and Exploitation of Plants and Animals.* London: Duckworth, 73–100.

Flannery, K.V. and Marcus, J. 1993 Cognitive archaeology, *Cambridge Archaeological Journal* 3 (2), 260–7.

Fleckinger, A. and Steiner, H. 1999 *The Fascination of the Neolithic Age: the Ice Man.* Bolzano: Folio.

Fleming, A. 1971 Territorial patterns in Bronze Age Wessex, *Proceedings of the Prehistoric Society* 37, 138–66.

Fleming, A. 1988 *The Dartmoor Reaves.* London: Batsford.

Fleming, A. 1998 *Swaldedale: valley of the wild river.* Edinburgh: Edinburgh University Press.

Fleming, A. 1999 Phenomenology and the megaliths of Wales: a dreaming too far, *Oxford Journal of Archaeology* 18 (2), 119–25.

Fleming, A. 2005a *St Kilda and the Wider World.* Macclesfield: Windgather Press.

Fleming, A. 2005b Megaliths and Post-Modernism: the case of Wales, *Antiquity* 79, 921–32.

Fleming, A. 2006 Post-Processual landscape archaeology: a critique, *Cambridge Archaeological Journal* 16, 267–80.

Fleming, A. 2009a The making of a medieval road: the Monk's Trod routeway, mid Wales, *Landscapes* 1, 77–100.

Fleming, A. 2009b The Monk's Trod: Britain's best medieval road, *British Archaeology* Nov/Dec 2009, 22–27.

Fleming, A. 2010 Horses, elites … and long-distance roads, *Landscapes* 11 (2), 1–20.

Fleming, A. 2011 The Lich Way: a path for all seasons, *Transactions of the Devonshire Association* 143, 91–103.

Fleming, A. 2012 Devon's early roads: Devon Archaeological Society Presidential Lecture 2011, *Proceedings of the Devon Archaeological Society* 70, 1–23.

Fokkens, H. 2005 Late Neolithic, Early and Middle Bronze Age: introduction. In Louwe-Kooijmans, L.P., van den Broeke, P.W., Fokkens, H. and van Gijn, A.L. (eds), *The Prehistory of the Netherlands* Volume 1. Amsterdam: Amsterdam University Press, 357–69.

Ford, B. and Teague, S. 2011 *Winchester: a city in the making.* Oxford: Oxford Archaeology Monograph 12.

Forman, R.T.T. 1995 *Land Mosaics: the ecology of landscapes and regions.* Cambridge: Cambridge University Press.

Foster, J. 1980 *The Iron Age moulds from Gussage All Saints.* London: British Museum Occasional Paper 12.

Fowler, C.S. 1996 Historical perspectives on Timbisha Shoshone land management practices, Death Valley, California. In Reitz, E.J., Newsom, L.E. and Scudder, S.J. (eds), *Case Studies in Environmental Archaeology.* London: Plenum Press, 87–102.

Fowler, C.S. 2009 Reconstructing Southern Paiute-Chemehuevi trails in the Mojave Desert of southern Nevada and California: ethnographic perspectives from the 1930s. In Snead *et al.* (eds) 2009, 84–105.

Fowler, P. 1998 Moving through the landscape. In Everson, P. and Williamson, T. (eds), *The Archaeology of Landscape: studies presented to Christopher Taylor.* Manchester: Manchester University Press, 25–41.

Fowler, P. 2000 *Landscape Plotted and Pieced: landscape history and local archaeology in Fyfield and Overton, Wiltshire.* London: Society of Antiquaries.

Fowler, P. and Blackwell, I. 1998 *The Land of Lettice Sweetapple.* Stroud: Tempus.

Fox, C. 1923 *The Archaeology of the Cambridge Region.* Cambridge: Cambridge University Press.

Fox, C. 1932 *The Personality of Britain.* Cardiff: National Museum of Wales.

Fox, H. 2012 *Dartmoor's Alluring Uplands.* Exeter: Exeter University Press.

Frankopan, P. 2015 *The Silk Roads.* London: Bloomsbury.

Fredengren, C. 2010 *Lough Kinale: studies of an Irish lake.* Bray: Wordwell.

French, C. 2015 *A Handbook of Geoarchaeological Approaches for investigating Landscapes and Settlement Sites.* Oxford: Studying Scientific Archaeology 2.

French, C., Scaife, R.G. and Allen, M.J. 2012 Durrington Walls to West Amesbury by way of Stonehenge: a major transformation of the Holocene landscape, *Antiquaries Journal* 92, 1–36.

French, C., Lewis, H., Allen, M.J., Scaife, R.J. and Gardiner, J. 2007 *Prehistoric Landscape Development in the Upper Allen Valley, Cranborne Chase, Dorset.* Cambridge: McDonald Institute.

Froehlich, W. 1991 Sediment production from unmetalled road surfaces. In Peters, N.E. and Walling, D.E. (ed.), *Sediment and Stream Water Quality in a Changing Environment: trends and explanation.* Wallingford: IAHS Press, 21–9.

Froom, R. 2012 *The Mesolithic of the Kennet Valley.* Newbury: privately printed.

Froom, R. and Cook, J. 2005 *Late Glacial Long Blade Sites in the Kennet Valley.* London: British Museum.

Fuchs, M. and Lang, A. 2009 Luminescence dating of hillslope deposits – a review, *Geomorphology* 109, 17–26.

Fulford, M.G. 2012 *Silchester and the Study of Romano-British Urbanism.* Portsmouth, RI: *Journal of Roman Archaeology* Supplement 90.

Fulford, M.G. and Creighton, J. 1998 A Late Iron Age mirror burial from Latchmere Green, near Silchester, Hampshire, *Proceedings of the Prehistoric Society* 64, 331–42.

Fyfe, R. 2006 GIS and the application of a model of pollen deposition and dispersal: a new approach to testing landscape hypotheses using the POLLANDCAL models, *Journal of Archaeological Science* 33 (4), 483–93.

Gallois, R.W. and Edmunds, F.H. 1965 *British Regional Geology: the Wealden District.* London: HMSO.

Gamble, C. 1993 *Timewalkers: the prehistory of global colonisation.* Stroud: Alan Sutton.

Gardiner, M. 1990 The archaeology of the Weald – a survey and a review, *Sussex Archaeological Collections* 128, 33–53.

Garner, M. 2007 *The Neolithic and Bronze Age Settlement at Oversley Farm, Styal, Cheshire: excavation in advance of Manchester airport's second runway 1997–8.* Oxford: British Archaeological Report 435.

Garrow, D. and Sturt, F. 2011 Grey waters bright with Neolithic argonauts? Maritime connections and the Mesolithic-Neolithic transition within the 'western seaways' of Britain, *c* 5000–3500 BC, *Antiquity* 85, 59–72.

Garrow, D. and Sturt, F. 2017 *Neolithic Stepping Stones: excavation and survey within the western seaways of Britain.* Oxford: Oxbow Books.

Garton, G., Elliot, L. and Salisbury, C. 2001 Ashton-upon-Trent, Argosy Washolme, 196–200 in Guilbert, G. and Garton, D. Some recent fieldwork in Derbyshire by Trent Peak Archaeological Unit, *Derbyshire Archaeological Journal* 121, 196–231.

Garwood, P. 2012 The present dead: the making of past and future landscapes in the British Chalcolithic. In Allen, M.J., Gardiner, J. and Sheridan, A. (eds), *Is there a British Chalcolithic? People, Place and Polity in the Later 3rd Millennium*, Oxford: Oxbow: Prehistoric Society Research Papers 4, 298–316.

Gasser, A. 2003 World's oldest wheel found in Slovenia, www.ukom.gov.sl/en/media, oldest wheel.

Gatesy, S.M. and Falkingham, P.L. 2017 Neither bones nor feet: track morphological variation and 'preservation quality', *Journal of Vertebrate Paleontology* DOI, 10.1080/02724634.2017.1314 298.

Gerrard, C. and Aston, M. 2007 *The Shapwick Project, Somerset: a rural landscape explored.* Leeds: Society for Medieval Archaeology Monograph 25.

Gibson, A., MacPherson-Grant, N. and Stewart, I. 1997 A Cornish vessel from farthest Kent, *Antiquity* 71, 438–41.

Gibson, D., Knight, M. and Murrell, K. 2012 Waterworld: Must Farm's Bronze Age boats, *Current Archaeology* 263, 13–19.

Gibson, D., Knight, M., Davenport, S. and Wakefield, C. 2016 The Must Farm inferno, *Current Archaeology* 312, 12–18.

Giddens, A. 1984 *The Constitution of Society.* Oxford: Polity Press.

Giddens, A. 1991 *Modernity and Self-identity: self and society in the late modern age.* Cambridge: Polity Press.

Giguet-Covex, C., Pansu, J., Arnaud, F., Rey, P.-J., Criggo, C., Gielly, L., Domaizon, I., Coissac, E., David, F., Choler, P., Poulenard, J. and Taberlet, P. 2014 Long livestock farming history and human landscape shaping revealed by lake sediment DNA, *Nature Communications* 5:3211 DOI: 10.1038, 1–7.

Giles, M. 2012 *A Forged Glamour: landscape, identity and material culture in the Iron Age.* Oxford: Windgather Press.

Gillings, M. and Pollard, J. 2016 Making megaliths: shifting and unstable stones in the Neolithic of the Avebury landscape, *Cambridge Archaeological Journal* 26, 1–32.

Gillings, M., Pollard, J., Wheatley, D. and Peterson, R. 2008 *Landscape of the Megaliths: excavation and fieldwork on the Avebury monuments 1997–2003.* Oxford: Oxbow Books.

Gillis, J.R. 2012 *The Human Shore.* Chicago IL: University of Chicago Press.

Giraldus Cambrensis 1191 (1908 edition) *The Itinerary through Wales: description of Wales.* London: Dent, Everyman.

Girling, M.A. 1984 Investigations of a second insect assemblage from the Sweet Track, *Somerset Levels Papers* 10, 79–91.

Goble, D.D. and Hirt, P.W. (eds) 1999 *Northwest Lands, Northwest Peoples.* Seattle WA and London: University of Washington Press.

Godwin, H. 1978 *Fenland: its ancient past and uncertain future.* Cambridge: Cambridge University Press.

Goldberg, P. and Macphail, R.I. 2006 *Practical and Theoretical Geoarchaeology.* Oxford: Blackwell.

Goldfarb, B. 2018 *Eager: the surprising secret life of beavers and why they matter.* London: Chelsea Green.

Gonzalez, S., Huddart, D. and Roberts, G. 1997 Holocene development of the Sefton coast: a multidisciplinary approach to understanding the archaeology. In Sinclair, A., Slater, E. and Gowlett, J. (ed.), *Archaeological Sciences 1995.* Oxford: Oxbow Monograph 64, 289–99.

Gooley, T. 2010 *The Natural Navigator.* London: Virgin Books.

Gooley, T. 2012 *The Natural Explorer: understanding your landscape.* London: Hodder and Stoughton.

Gooley, T. 2018 *Wild Signs and Star Paths: the keys to our lost sixth sense.* London: Sceptre.

Gosden, C. 1994 *Social Being and Time.* Oxford: Blackwell.

Gosden, C. 2010 The death of the mind. In Malafouris, L. and Renfrew, C. (ed.), *The Cognitive Life of Things.* Cambridge: McDonald Institute, 39–46.

Gould, R.A. 2011 *Archaeology and the Social History of Ships.* Cambridge: Cambridge University Press.

Gowan, M., O'Neill, J. and Phillips, M. 2005 *The Lisheen Mine Archaeological Project 1996–8.* Bray: Wordwell.

Gowlett, J. 2016 The discovery of fire by humans: a long and convoluted process, *Philosophical Transactions of the Royal Society* B 371, 20150164.

Greatorex, C. 2003 Living on the margins? The Late Bronze Age landscape of the Willingdon Levels. In Rudling, D. (ed.), *The Archaeology of Sussex to AD 2000.* Kings Lynn: Heritage Marketing, 89–100.

Green, C. 1985 The Caergwrle Bowl, not oak but shale, *Antiquity* 59, 116–17.

Green, C. 2004 Evidence of a marine environment associated with the Dover boat. In Clark (ed.) 2004b, 13–16.

Green, M. 1992 *Dictionary of Celtic Myth and Legend.* London: Thames and Hudson.

Greer, S. 2002 *Ice Patch.* Yukon: South Yukon First Nations.

Gregory, N.T.N. 1997 A Comparative Study of Irish and Scottish Logboats. Edinburgh University: Unpublished PhD thesis.

Gregory, R.A., Murphy, E.M., Church, M.J., Edwards, K.J., Guttmann, E.B. and Simpson, D.A. 2005 Archaeological evidence for the first Mesolithic occupation of the Western Isles of Scotland, *The Holocene* 15 (7), 944–50.

Griffiths, W.E. 1960 The excavation of stone circles near Penmaenmawr, north Wales, *Proceedings of the Prehistoric Society* 26, 303–39.

Grimes, W.F. 1951 The Jurassic Way across England. In Grimes, W.F. (ed.), *Aspects of Archaeology in Britain and Beyond: Essays presented to OGS Crawford.* London: H.W. Edwards, 144–71.

Groenewoudt, B.J., Deeben, J., van Geel, B. and Lauwerier, C.G.M. 2001 An early Mesolithic assemblage with faunal remains in a stream valley near Zutphen, The Netherlands, *Archäologisches Korrespondenzblatt* 31, 329–48.

Groenmann-van Waateringe, W. 1993 The effects of grazing on the pollen production of grasses, *Vegetation History and Archaeobotany* 2, 157–62.

Grove, R.H. 1995 *Green Imperialism.* Cambridge: Cambridge University Press.

Guttmann, E.B. and Last, J. 2000 A Late Bronze Age landscape at South Hornchurch, Essex, *Proceedings of the Prehistoric Society* 66, 319–59.

Haak, W., Forster, P., Bramanti, B., Matsumura, S., Brandt, G., Tänzer, M., Villems, R., Renfrew, C., Gronenborn, D., Werner Alt, K. and Burger, J. 2005 Ancient DNA from the first European farmers in 7500-year-old Neolithic sites, *Science* 310 (5750), 1016–18.

Hägerstrand, T. 1975 Spacetime and human conditions. In Karlqvist, A., Lundqvist, J. and Snickars, F. (ed.), *Dynamic Allocation of Urban Space.* Farnborough: Saxon House, 3–12.

Halfpenny, J.C. 2007 *Scats and Tracks.* Guilford, CO: Globe Pequot Press.

Hall, D. and Coles, J. 1994 *Fenland Survey: an essay in landscape and persistence.* London: English Heritage.

Hallam, S.J. 1989 Plant usage and management in Southwest Australian Aboriginal societies. In Harris, D.R. and Hillman, G.C. (eds), *Foraging and Farming.* London: Unwin Hyman, 136–51.

Hamilton, S. 2003 Sussex not Wessex: a regional perspective on Southern Britain c 12–200 BC. In Rudling, D. (ed.), *The Archaeology of Sussex to AD 2000.* Great Dunham: Heritage Marketing and Publications, 69–88.

Hansen, K.M. and Pedersen, K.B. (eds) 2006a *Across the Western Baltic.* Vordingborg: Sydsjællands Museum.

Hansen, K.M. and Pedersen, K.B. 2006b With or without bones – Late Palaeolithic hunters in South Zealand. In Hansen. and Pedersen (eds) 2006a, 93–110.

Harden, C.P., Chin, A., English, M., Fu, R., Galvin, K.A., Gerlak, A.K., McDowell, P., McNamara, D., Peterson, J., Puff, N.L., Rosa, E., Solecki, W. and Wohl, E. 2014 Understanding human-landscape interactions in the 'Anthropocene', *Environmental Management* 53, 4–13.

Harding, A. 1981 Excavations in the prehistoric ritual complex near Milfield, Northumberland, *Proceedings of the Prehistoric Society* 47, 87–135.

Harding, J. 2012 Conformity, routeways and religious experience – the henges of central Yorkshire. In Gibson, A. (ed.), *Enclosing the Neolithic: recent studies in Britain and Europe.* Oxford: British Archaeological Report S2440, 67–80.

Harding, J. 2013 *Cult, Religion and Pilgrimage.* York: Council for British Archaeology Research Report 174.

Harding, J. and Healy, F. 2007 *A Neolithic and Bronze Age Landscape in Northamptonshire.* Swindon: English Heritage.

Harlan, J.R. 1989 Wild-grass seed harvesting in the Sahara and Sub-Sahara of Africa. In Harris, D. and Hillman, G. (eds), *Foraging and Farming: the evolution of plant exploitation.* London: Unwin Hyman, 79–97.

Harris, D. 1969 Agricultural systems, ecosystems and the origins of agriculture. In Ucko, P.J. and Dimbleby, G.W. (eds), *The Domestication and Exploitation of Plants and Animals.* London: Duckworth, 3–16.

Harrison, S. 2003 The Icknield Way: some queries, *Archaeological Journal* 160, 1–22.

Harrison, S.P. and Sanchez Goni, M.F. 2010 Global patterns of vegetation response to millennial-scale variability and rapid climate change during the last glacial period, *Quaternary Science Reviews* 29, 2957–80.

Hart, D. 2010 Excavations at Belmarsh West, Woolwich, *London Archaeologist* 12, 203–7.

Hart, D. 2015 *Around the Ancient Track.* Portslade: SpoilHeap Publications Monograph 10.

Hartz, S., Heinrich, D. and Lübke, H. 2002 Coastal farmers – the neolithicisation of northernmost Germany. In Fischer, A. and Kristiansen, K. (eds), *The Neolithicisation of Denmark: 150 years of debate.* Sheffield: J.R. Collis Publications, 321–40.

Harvey, D. 1989 *The Condition of Postmodernity.* Oxford: Blackwell.

Hastorf, C.A. 1999 Cultural implications of crop introductions in Andean prehistory. In Gosden, C. and Hather, J. (eds), *The Prehistory of Food: appetites for change.* London: Routledge One World 32, 35–58.

Hatala, K.G., Richmond, B.G., Liutkus, C.M., Harcourt-Smith, W.E.H., Pobiner, B.L., Rossi, V., Metallo, A., Dingwall, H., Zimmer, B., OlleMoita, G. and Brett, J. 2011 Early modern human footprints from Engare Sero, Tanzania, *American Journal of Physical Anthropology* 144, 158.

Haughey, F.M. 1999 The archaeology of the Thames: prehistory within a dynamic landscape, *London Archaeologist* 9 (1), 16–21.

Haughey, F.M. 2003 From prediction to prospection: finding prehistory on London's river. In Howard, A.J., Macklin, M.G. and Passmore, D.G. (eds), *Alluvial Archaeology in Europe.* Abingdon: Balkema, 61–79.

Hauser, K. 2008 *Bloody Old Britain: O.G.S. Crawford and the archaeology of modern life.* London: Granta.

Hawkes, J. 1951 *A Land.* London: Cresset Press.

Hawkes, K., O'Connell, J.F. and Blurton Jones, N. 1995 Hardworking Hadza grandmothers. In Standen, V. and Goley, R. (eds), *Comparative Socioecology: the behavioural ecology of humans and other mammals.* London: Blackwell, 341–66.

Hay, R.L. 1987 The geology of the Laetoli area. In Leakey and Harris (eds) 1987, 23–47.

Hayden, B. and Cousins, S.M. 2004 The social divisions of roasting pits in a winter village site. In Prentiss, W.C. and Kuijt, I. (eds), *Complex Hunter-gatherers.* Salt Lake City UT: Utah University Press, 140–54.

Hayen, H. 1963 Zwei hölzerne Moorwege aus dem Fundgebiet Ipwegermoor B, Kreis Ammerland, *Neue Ausgrabungen und Forschungen in Niedersachsen* 1, 113–31.

Hayen, H. 1987 Peatbog archaeology in Lower Saxony, west Germany. In Coles, J. and Lawson, A.J. (ed.), *European Wetlands in Prehistory.* Oxford: Oxford University Press, 117–36.

Healy, F., Heaton, M. and Lobb, S.J. 1992 Excavations of a Mesolithic site at Thatcham, Berkshire, *Proceedings of the Prehistoric Society* 58, 41–76.

Healy, P.F. 1980 *Archaeology of the Rivas Region, Nicaragua.* Waterloo: Wilfrid Laurier Press.

Hebda, R.J., Greer, S. and Mackie, A.P. (eds) 2017 *Kwäday Dän Ts'ìnchi: teachings from Long Ago Dead Person Found.* Victoria: Royal British Columbia Museum.

Heidegger, M. 1962 *Being and Time.* Oxford: Blackwell.

Hellstrom, J. and Pickering, R. 2015 Recent advances and future prospects of the U-Th and U-Pb chronometers applicable to archaeology, *Journal of Archaeological Science* 56, 32–40.

Helm, J. 1972 The Dogrib Indians. In Bicchieri, M.G. (ed.), *Hunters and Gatherers Today.* New York: Holt, Rinehart and Winston, 51–89.

Helms, M.W. 1988 *Ulysses' Sail: an ethnographic odyssey of power, knowledge and geographical distance.* Princeton NJ: Princeton University Press.

Hencken, H.O. 1932 *The Archaeology of Cornwall and Scilly.* London: Methuen.

Hensey, R. 2015 *First Light: The Origins of Newgrange.* Oxford: Oxbow Insights in Archaeology 1.

Herring, P. 2009 Early medieval transhumance in Cornwall, GB. In Klapsle, J. and Somner, P. (eds), *Medieval Rural Settlements in Marginal Landscapes.* Cardiff: Ruralia VII 2007, 47–56.

Herring, P. 2012 Shadows of ghosts: early medieval transhumants in Cornwall. In Turner, S. and Silvester, B. (ed.), *Life in Medieval Landscapes: people and places in the Middle Ages.* Oxford: Windgather, 89–105.

Herring, P., Johnson, N., Jones, A.M., Nowakowski, J., Sharpe, A. and Young, A. 2017 *Archaeology and Landscape at the Land's End, Cornwall: the West Penwith Surveys 1980–2010.* Truro: Cornwall Heritage Trust.

Heumüller, M. 2000 Die Bohlenwege des Federseemoor, *Nachrichtenblatt Arbeitskreis Unterwasserarchäologie* 7, 75–80.

Hey, G., Garwood, P., Robinson, M., Barclay, A. and Bradley, P. 2011 *The Thames Through Time: The Archaeology of the Gravel Terraces of the Upper and Middle Thames: Early Prehistory to 1500 BC. Part 2 – the Mesolithic, Neolithic and early Bronze Age and the establishment of permanent human occupation in the valley.* Oxford: Thames Valley Landscapes Monograph 32.

Higgs, E.S. and Vita-Finzi, C. 1972 Prehistoric economies: a territorial approach. In Higgs, E.S. (ed.), *Papers in Economic Prehistory.* Cambridge: Cambridge University Press, 27–36.

Hill, G. 2015 Reconstructing Holocene Landscape Cover using Subfossil Coleoptera. Queen's University, Belfast: Unpublished PhD thesis.

Hillam, J., Groves, C.M., Brown, D.M., Baillie, M.G.L., Coles, J.M. and Coles, B.J. 1990 Dendrochronology of the English Neolithic, *Antiquity* 64, 210–20.

Hindle, B.P. 1993 *Roads, Tracks and their Interpretation.* London: Batsford.

Hindle, B.P. 2001 *Roads and Tracks for Historians.* Chichester: Phillimore.

Hinton, D.A. 1997 The 'Scole-Dickleburgh field system' examined, *Landscape History* 19, 5–12.

Historic England 2015 *Where on Earth are We? The role of global navigation satellite systems in archaeological field survey.* Swindon: Historic England.

Hodder, I. 1990 *The Domestication of Europe.* Oxford: Blackwell.

Hodder, I. 1999 *The Archaeological Process: an introduction.* Oxford: Blackwell.

Hodder, I. 2012 *Entangled: an archaeology of the relationship between humans and things.* Chichester: Wiley Blackwell.

Holden, E.W. 1963 Excavations at the deserted medieval village of Hangleton, *Sussex Archaeological Collections* 101, 54–181.

Holleyman, G. and Curwen, E.C. 1935 Late Bronze Age lynchet-settlements on Plumpton Plain, Sussex, *Proceedings of the Prehistoric Society* 1, 16–38.

Holst, D. 2010 Hazelnut economy of Early Holocene hunter-gatherers: a case study from Mesolithic Duvensee, northern Germany, *Journal of Archaeological Science* 37, 2871–80.

Homer nd *The Odyssey.* (English translation by Alexander Pope, published 1978), Norwalk CO: Easton Press.

Horton, D. 1994 *The Encyclopedia of Aboriginal Australia.* Canberra: Australian Institute of Aboriginal and Torres Strait Islander Studies.

Hoskins, W.G. 1955 *The Making of the English Landscape.* London: Hodder and Stoughton.

Hoskins, W.G. 1974 *The Making of the English Landscape* (Paperback Edition). Harmondsworth: Penguin.

Household, G. 1939 *Rogue Male.* London: Chatto and Windus.

Housley, R.A., Gamble, C., Street, M. and Pettitt, P. 1997 Radiocarbon evidence for the late glacial recolonisation of north Europe, *Proceedings of the Prehistoric Society* 63, 25–54.

Huck, B. 2006 *In Search of Ancient British Columbia.* Winipeg: Heartland.

Huddart, D. 1992 Coastal environmental changes and morphostratigraphy in southwest Lancashire, England, *Proceedings of the Geologists' Association* 103, 217–36.

Huddart, D., González, S. and Roberts, G. 1999a The archaeological record and mid-Holocene marginal coastal palaeoenvironments around Liverpool Bay. In Edwards, K.J. and Sadler, D. (ed.), *Holocene Environments of Prehistoric Britain.* London: Quaternary Proceedings 7, 563–74.

Huddart, D., Roberts, G. and González, S. 1999b Holocene human and animal footprints and their relationship with coastal environmental change, Formby Point, NW England, *Quaternary International* 55, 29–41.

Huddart, D., Bennett, M.R., González, S. and Velay, X. 2008 Analysis and preservation of Pleistocene human and animal footprints: an example from Toluquilla, Vasequillo Basin (Central Mexico), *Ichnos* 15, 232–45.

Huet, T. 2017 New perspectives on the chronology and meaning of Mont Bégo Rock Art (Alpes-Maritimes, France), *Cambridge Archaeological Journal* 272, 199–221.

Hughes, P.D.M., Mauquoy, D., Barber, K.E. and Langholm, P.G. 2000 Mire-development pathways and palaeoclimatic records from a full Holocene peat archive at Walton Moss, Cumbria, England, *The Holocene* 10, 465–79.

Hunn, E.S. 1999 Mobility as a factor limiting resource use on the Columbia Plateau. In Goble and Hirt (eds) 1999, 156–72.

Hunter, J. 1994 'Maritime culture': notes from the land, *International Journal of Nautical Archaeology and Underwater Exploration* 23, 261–4.

Hygen, A.-S. and Bengtsson, L. 2000 *Rock Carvings in the Borderlands.* Gothenburg: Warne.

Ingold, T. 1980 *Hunters, Pastoralists and Ranchers.* Cambridge: Cambridge University Press.

Ingold, T. 1986 *The Appropriation of Nature.* Manchester: Manchester University Press.

Ingold, T. 1993 The temporality of the landscape, *World Archaeology* 25, 152–74.

Ingold, T. 1996 Growing plants and raising animals: an anthropological perspective on domestication. In Harris, D. (ed.), *The Origins and Spread of Agriculture and Pastoralism in Eurasia.* London: UCL Press, 12–24.

Ingold, T. 2000 *The Perception of the Environment: essays in livelihood, dwelling and skill.* London: Routledge.

Ingold, T. 2007 *Lines: a brief history.* London: Routledge.

Ingold, T. 2011 *Being Alive: essays on movement, knowledge and description.* Abingdon: Routledge.

Ingold, T. and Vergunst, L.J. (eds) 2008 *Ways of Walking: ethnography and practice on foot.* Aldershot: Ashgate.

Ings, M. and Murphy, F. 2011 *The Lost Lands of our Ancestors.* Llandeilo: Dyfed Archaeological Trust.

Innes, J.B. and Simmons, A.H. 1988 Disturbance and diversity: floristic changes associated with pre-elm decline woodland recession in north east Yorkshire. In Jones, M. (ed.), *Archaeology and the Flora of the British Isles.* Oxford: University Committee for Archaeology, 7–20.

Innes, J.B., Blackford, J.J. and Simmons, I.G. 2010 Forest disturbance and possible land-use regimes during the Late Mesolithic in the English uplands: pollen, charcoal and non-pollen palynomorph evidence from Bluewath Beck, North York Moors, UK, *Vegetation History and Archaeobotany* 19, 439–52.

Innes, J.B., Blackford, J.J. and Rowley-Conwy, P. 2013 Late Mesolithic and early Neolithic forest disturbance: a high resolution palaeoecological test of human impact hypotheses, *Quaternary Science Reviews* 77, 80–100.

Jacobi, R.M. and Currant, A. 2011 The Late Pleistocene mammalian palaeontology and archaeology of Mendip. In Lewis, J. (ed.), *The Archaeology of Mendip.* Oxford: Heritage Publications, 45–84.

Jacques, D. 2014 New dates for Vespasian's camp, *Current Archaeology* 292, 8–9.

Jacques, D. and Phillips, T. 2014 Mesolithic settlement near Stonehenge: excavations at Blick Mead and Vespasian's Camp, Amesbury, *Wiltshire Archaeological and Natural History Magazine* 107, 7–27.

Jacques, D., Phillips, T. and Lyons, T. 2018 *Blick Mead: exploring the 'first place' in the Stonehenge landscape.* Oxford: JP Lang.

Jaeger, E. 1948 *Tracks and Trailcraft.* New York: Macmillan.

Jager 1984 A prehistoric route and ancient cart-tracks in the Gemeente, Antloo, Province of Drenthe, *Palaeohistoria* 27, 185–245.

James, C.S. 1939 Footprints and feet of natives in the Soloman Islands, *The Lancet* 1939 (2), 1390–3.

Janssen, S. 1994 A Hjortspring boat from North Sweden?, *Maritime Newsletter from Roskilde* 2, 16–17.

Jay, M., Parker Pearson, M., Richards, M.P., Nehlich, O., Montgomery, J., Chamberlain, A.T. and Sheridan, A. 2011 The Beaker People Project: an interim report on the progress of the isotopic analysis of the organic skeletal material. In Allen, M.J., Gardiner, J., Sheridan, A. and McOmish, D. (eds), *Is there a British Chalcolithic? People, Place and Polity in the Later 3rd Millennium.* Oxford: Oxbow: Prehistoric Society Research Papers 4, 226–36.

Jenkins, J.G. 1988 *The Coracle.* Carmarthen: Golden Grove.

Jennings, S., Greatorex, C., Smyth, C. and Spurr, G. 2003 The environmental archaeology of the Late Bronze Age occupation platform at Shinewater, near Eastbourne, UK. In Howard, A.J., Macklin, M.G. and Passmore, D.G. (eds), *Alluvial Archaeology in Europe.* Abingdon: Balkema, 93–110.

Johannsen, N. and Laursen, S. 2010 Routes and wheeled transport in late 4th-early 3rd millennium funerary customs of the Jutland peninsula: regional evidence and European context, *Praehistorische Zeitschrift* 88, 5–58.

Johnson, D.L. 2002 Darwin would be proud: bioturbation, dynamic denudation and the power of theory in science, *Geoarchaeology* 17, 7–40.

John, M. 2010 *Archaeological Theory: an introduction.* Oxford: Wiley Blackwell.

Johnson, L.M. 2010 *Trail of Story, Travellers' Path: reflections on ethnoecology and landscape.* Edmonton: Athabasca University Press.

Johnson, M. 2007 *Ideas of Landscape.* Oxford: Blackwell.

Johnson, N. and Rose, P. 1994 *Bodmin Moor: an archaeological survey:* volume 1*: human landscape to* c 1800. London: English Heritage.

Johnston, R. 1999 An empty path? Processions, memories and the Dorset cursus. In Barclay and Harding (eds) 1999, 39–48.

Jones, B. and Mattingly, D. 1990 *An Atlas of Roman Britain.* Oxford: Blackwell.

Jones, M. 2016 Food globalisation in prehistory: the agrarian foundations of an interconnected continent, *Journal of the British Academy* 4, 73–87.

Jones, O. and Cloke, P. 2002 *Tree Cultures: the place of trees and trees in their place.* Oxford: Berg.

Jones, P. 2013 *A Mesolithic 'Persistent Place' at North Park Farm, Bletchingley, Surrey.* Woking: Spoil Heap Publications.

Jones, R. 1969 Fire-stick farming, *Australian Natural History* 16, 224–8.

Jones, R. and Meehan, B. 1989 Food plants of the Gidjingali: ethnographic and archaeological perspectives from northern Australia on tuber and seed exploitation. In Harris, D. and Hillman, G. (eds), *Foraging and Farming.* London: Unwin Hyman, 120–35.

Jordan, P. 2003a *Material Culture and Sacred Landscape: the anthropology of the Siberian Khanty.* Oxford: Altamira Press.

Jordan, P. 2003b Peopling the Mesolithic: insights from ethnographies of landscape and material culture. In Bevan, L. and Moore, J. (ed.), *Peopling the Mesolithic in a Northern Environment.* Oxford: British Archaeological Report 1157, 27–34.

Jørgensen, L., Storgaard, B. and Gebauer Thomsen, L. 2003 *The Spoils of Victory: the North in the shadow of the Roman Empire.* Copenhagen: Nationalmuseet.

Kador, T. 2013 Taking on the final frontier: movement, mobility and social change in early prehistoric Ireland. In Preston and Schörle (eds) 2013, 17–28.

Kaul, P. 1998 *Ships on Bronzes: a study in Bronze Age religion and iconography.* Copenhagen: National Museum.

Kelly, E.P. 2006 Secrets of the bog bodies: the enigma of the Iron Age explained, *Archaeology Ireland* 20, 26–30.

Kelly, F. 1997 *Early Irish Farming.* Dublin: Institute of Irish Studies.

Kennett, D.J. 2005 *The Island Chumash.* Berkeley CA: University of California Press.

Kerney, M. 1999 *Atlas of the Land and Freshwater Molluscs of Britain and Ireland.* Colchester: Harley Books.

Kim, K.S., Kim, J.Y. and Kim, S.H. 2009 Preliminary report on hominid and other vertebrate footprints from the Late Quaternary strata of Jeju Island, Korea, *Ichnos* 16, 1–11.

Kirk, R. 2015 *Ozette: excavating a Makah whaling village.* Seattle WA: University of Washington Press.

Kjaerum, P. and Olsen, R.A. (ed.) 1990 *Oldtidens Ansigt: faces of the past.* Aarhus: Jysk Arkæologisk Selskab.

Klassen, L. 2014 *Along the Road: aspects of causewayed enclosures in South Scandinavia and beyond.* Aarhus: Aarhus University Press.

Klindt-Jensen, O. 1979 *Gundestrup Kedelen.* Copenhagen: Nationalmuseet.

Knapp, A.B. 2013 *The Archaeology of Cyprus.* Cambridge: Cambridge University Press.

Knight, M. and Murrell, K. 2012 Must Farm and Bradley Fen, *British Archaeology* 123, 15–21.

Knight, M., Harris, S. and Appleby, G. 2016 Must Farm: an extraordinary tale of the everyday, *Current Archaeology* 319, 12–18.

Knight, M., Ballantyne, R., Robinson Zeki, I. and Gibson, D. 2019 The Must Farm pile-dwelling settlement, *Antiquity* 93, 645–63.

Kohen, J. 1995 *Aboriginal Environmental Impacts.* Sydney: University of South Wales Press.

Kramer, P. 1999 *Totem Poles.* Vancouver: Altitude Publishing Canada.

Kristiansen, K. and Larsson, T.B. 2005 *The Rise of Bronze Age Society: travels, transmissions and transportations.* Cambridge: Cambridge University Press.

Kruckeberg, A.R. 1999 *The Natural History of Puget Sound Country.* Seattle WA: Weyerhaeuser Environmental Books.

Kuhnlein, K.V. and Turner, N.J. 1991 *Traditional Plant Foods of Canadian Indigenous Peoples: nutrition, botany and use.* Philadelphia PA: Garden and Breach.

Kunwald, G. 1962 Broskovvejene, *Nationalmuseets arbejdsmark* 1962, 149–67.

Kunwald, G. 1984 Ældre Vej, *Skalk* 4, 12–15.

Lake, M. 2000 MAGICAL computer simulation of Mesolithic foraging on Islay. In Mithen, S. (ed.) 2000, Vol. 2, 465–95.

Laland, K.N. and O'Brien, M.J. 2010 Niche construction theory and archaeology, *Journal of Archaeological Method and Theory* 17, 303–22.

Lambrick, G. and Robinson, M. 2009 *Thames through Time: the Archaeology of the gravel terraces of the Upper and Middle Thames: late prehistory 1500 BC–AD 50.* Oxford: Thames Valley Monograph 29.

Lang, A. and Wagner, G.A. 1996 Infrared stimulated luminescence dating of archaeosediments, *Archaeometry* 38, 129–41.

Lang, A. and Hönscheidt, S. 1999 Age and source of colluvial sediments at Vaihingen-Enz, Germany, *Catena* 38, 89–107.

Lang, A., Rieser, U., Habermann, J. and Wagner, G.A. 1998 Luminescence dating of sediments, *Naturwissenschaften* 85, 515–23.

Lang, W.L. 1999 The sense of place and environmental history. In Goble and Hirt (eds) 1999, 79–94.

Langohr, R. 1993 Types of tree windthrow, their impact on the environment and their importance for the understanding of archaeological excavation data, *Helinium* 31–2 (1), 36–49.

Lanting, J.N. 1997–8 Dates for origin and diffusion of the European logboat, *Palaeohistoria* 39/40, 627–50.

Larsson, L. 1983 *Agerod V: an Atlantic bog site in central Scania.* Lund: Acta Archaeologica Lundensia Series 8: 12.

Last, J. 1999 Out of line: cursuses and monument typology in eastern England. In Barclay and Harding (eds) 1999, 86–91.

Lawton, J.H., Brotherton, P.N.M., Brown, V.K., Elphick, C., Fitter, A.H., Forshaw, J., Haddow, R.W., Hilborne, S., Leafe, R.N., Mace, G.M., Southgate, M.P., Sutherland, W.J., Tew, T.E., Varley, J. and Wynne, G.R. 2010 *Making Space for Nature: a review of England's wildlife sites and ecological network.* Report to Defra https://www.researchgate.net/publication/268279426_Making_Space_for_Nature_A_Review_of_England's_Wildlife_Sites_and_Ecological_Network.

Lea, D.L., English, J. and Tapper, R.Q. 2017 *South Downs Cross-Ridge Dyke Project: Part 2: River Cuckmere to River Ouse,* https://www.worthingarchaeological.org/uploads/8/6/3/9/86390614/report_lea_english_tapper_arun_to_adur_cross_ridge_dyke_project_interim_5.pdf.

Leach, A.L. 1918 Flint-working sites on the submerged land (submerged forest) bordering the Pembrokeshire Coast, *Proceedings of the Geologists Association* 29, 46–7.

Leakey, M. 1984 *Disclosing the Past.* London: Doubleday.

Leakey, M.D. and Harris, J.M. 1987 *Laetoli: a Pliocene site in North Tanzania.* Oxford: Clarendon Press.

Leary, J. 2014 *Past Mobilities.* Farnham: Ashgate.

Leary, J. and Field, D. 2012 Journeys and juxtapositons: Marden henge and the view from the Vale. In Gibson, A. (ed.), *Enclosing the Neolithic: recent studies in Britain and Europe.* Oxford: British Archaeological Report S2440, 55–65.

Leary, J. and Kador, T. 2016 *Moving on in Neolithic Studies.* Oxford: Oxbow Neolithic Studies Group Seminar Papers 14.

Leary, J., Field, D. and Campbell, G. 2013 *Silbury Hill.* Swindon: English Heritage.

Legat, A. 2008 Walking stories; leaving footprints. In Ingold and Vergunst (eds) 2008, 37–49

Lehoërff, A., Bourgeois, J., Clark, P. and Talon, M. 2012 *Beyond the Horizon.* Paris: Somogy Art Publishers.

Leopold, E.B. and Boyd, R. 1999 An ecological history of old prairie areas in Southwestern Washington. In Boyd (ed.) 1999, 139–63.

Lepofsky, D. and Peacock, D. 2004 A question of intensity: exploring the role of plant foods in North Plateau prehistory. In Prentiss, W.C. and Kuijt, I. (eds), *Complex Hunter-Gatherers.* Salt Lake City UT: Utah University Press, 115–31.

Lepofsky, D., Hallett, D., Lertzman, K., Mathewes, R., McHalsie, A. and Washbrook, K. 2005 Documenting Precontact plant management on the Northwest Coast: An example of prescribed burning in the Central and Upper Fraser Valley, British Columbia. In Deurand, D. and Turner, N.J. (ed.), *Keeping it Living: indigenous plant management on the Northwest Coast.* Seattle WA: University of Washington Press, 218–39.

Leslie, K. and Short, B. 1999 *An Historical Atlas of Sussex.* Chichester: Phillimore.

Levy, P. 2007 *Fellow Travelers: Indians and Europeans contesting the early American Trail.* Gainsville FL: University Press of Florida.

Lewis, H.T. and Ferguson, T.A. 1999 Yards, corridors and mosaics. In Boyd (ed.) 1999, 164–84.

Lewis, J.S.C. and Rackham, J. 2011 *Three Ways Wharf, Uxbridge: a lateglacial and early Holocene hunter-gatherer site in the Colne Valley.* London: Museum of London Archaeology.

Lewis, J.S.C., Brown, L. and Smith, A. 2006 *Landscape Evolution in the Middle Thames Valley: Heathrow Terminal 5 Excavations, Vol 1: Perry Oaks.* Salisbury and Oxford: Framework Archaeology.

Lewis, J.S.C., Leivers, M., Brown, L., Smith, A., Cramp, K., Mepham, L. and Phillpotts, C. 2010 *Landscape Evolution in the Middle Thames Valley: Heathrow Terminal 5 Excavations Volume 2.* Oxford and Salisbury: Framework Archaeology.

Lewis, M.P. 1992 The Prehistory of Coastal South-west Wales 7500–3600 BP: an interdisciplinary palaeoenvironmental and archaeological investigation. University of Wales: Lampeter: Dept of Archaeology, Unpublished PhD thesis.

Liebenberg, G.L. 1990 *The Art of Tracking.* Claremont: David Philips.

Lillie, M. 2015 *Hunters, Fishers and Foragers in Wales.* Oxford: Oxbow Books.

Lillie, M.C. 1997 Women and children in prehistory: resource sharing and social stratification at the Mesolithic–Neolithic transition in Ukraine. In Moore, J. and Scott, E. (eds) 1997, 213–28.

Ling, J. 2004 Beyond transgressive lands and forgotten seas: towards a maritime understanding of rock art in Bohuslän, *Current Swedish Archaeology* 12, 121–40.

Ling, J. 2008 *Elevated Rock Art: towards a maritime understanding of rock art in northern Bohuslän.* Gothenburg Archaeological Thesis: GOTARC Series B.

Ling, J., Stos-Gale, Z., Grandin, L., Billström, K., Hjärthner-Holdar, E. and Persson, P.O. 2014 Moving metals II: provenancing Scandinavian Bronze Age artefacts by lead isotope and elemental analyses, *Journal of Archaeological Science* 41, 106–32.

Lipke, P. 1984 *The Royal Ship of Cheops.* Oxford: British Archaeological Report S225.

Livadie, C.A. 2002 A first Pompeii: the Early Bronze Age village of Nola-Croce del Papa (Palma Campania phase), *Antiquity* 76, 941–2.

Lockley, M.G. and Hunt, A.P. 1995 *Dinosaur Tracks and Other Fossil Footprints in the Western United States.* New York: Columbia University Press.

Lockley, M.G. and Meyer, C. 2000 *Dinosaur Tracks and Other Fossil Footprints of Europe.* New York: Columbia University Press.

Lockley, M.G., Roberts, G. and Kim, J.Y. 2008 In the footprints of our ancestors: an overview of hominid track record, *Ichnos* 15, 106–25.

Lockley, M.G., Vasquez, R.G., Espinoza, E. and Lucas, S.G. 2009 America's most famous human footprints: history, context and first description of Mid-Holocene tracks from the shores of Lake Managua, Nicaragua, *Ichnos* 16, 55–69.

Lødøen, T. and Mandt, G. 2005 *The Rock Art of Norway.* Oxford: Windgather Press.

Lorimer, H. 2006 Herding memories of humans and animals, *Environment and Planning D: Society and Space* 24, 497–518.

Lorimer, H. 2010 Forces of nature, forms of life: calibrating ethology and phenomenology. In Anderson, B. and Harrison, P. (eds), *Taking-Place: Non-Representational Theories and Geography.* London: Ashgate, 55–78.

Louwe Kooijmans, L.P. (ed.) 2001a *Hardinxveld-Giessendam Polderweg. Een Mesolithisch jachtkamp in het rivierengebied (5500–4450 v. Chr).* Amersfoort: RAM 83.

Louwe Kooijmans, L.P. (ed.) 2001b *Hardinxveld-Giessendam De Bruin. Een kampplaats uit het Laat-Mesolithicum en het begin van de Swifterbant-cultur (5500–4450 v. Chr).* Amersfoort: RAM 88.

Louwe Kooijmans, L.P. 2005 The Netherlands in prehistory: retrospect. In Louwe Kooijmans, L.P., van den Broeke, P.W., Fokkens, H. and van Gijn, A.L. (eds), *The Prehistory of the Netherlands* Volume 2. Amsterdam: Amsterdam University Press, 695–720.

Loveday, R. 1998 Double entrance henges: routes to the past? In Gibson, A. and Simpson, D.A. (ed.), *Prehistoric Ritual and Religion.* Stroud: Sutton, 14–31.

Loveday, R. 2006 *Inscribed Across the Landscape.* Stroud: Tempus.

Lovschal, M. 2013 Ways of wandering: the LBA barrow landscape of the Himmerlant area, Denmark. In Fontijn, D., Louwen, A.J., van der Vaart, S. and Wentink, K. (eds), *Beyond Barrows.* Amsterdam: Sidestone Press, 225–50.

Lucas, S.G., Spielmann, J.A. and Lockley, M.G. 2007 *Cenozoic Vertebrate Tracks and Traces* Bulletin 42. Albuquerque NM: New Mexico Museum of Natural History and Science.

Lynch, F. 1993 *Excavations in the Brenig Valley.* Bangor: Cambrian Archaeological Association.

Lynch, F. 2000 The later Neolithic and earlier Bronze Age. In Lynch, F., Aldhouse-Green, S.H.R. and Davies, J.L. (ed.), *Prehistoric Wales.* Stroud: Sutton, 79–138.

Mabey, R. 1972 *Food for Free.* London: Collins.

Mabey, R. 2006 *Gilbert White.* London: Profile Books.

Mabey, R. 2014 *Dreams of the Good Life: the life of Flora Thompson and the creation of Lark Rise to Candleford.* London: Allen Lane.

MacFarlane, R. 2003 *Mountains of the Mind.* London: Granta Books.

MacFarlane, R. 2007 *The Wild Places.* London: Granta Books.

MacFarlane, R. 2012 *The Old Ways: a journey on foot.* London: Penguin.

MacFarlane, R. 2014 Introduction to: Watkins, A., *The Old Straight Track.* London: Head of Zeus.

MacFarlane, R. 2015 *Landmarks.* London: Hamish Hamilton.

MacFarlane, R. 2019 *Underland: A deep time journey.* London: Hamish Hamilton.

MacFarlane, R., Donwood, S. and Richards, D. 2012 *Holloway.* London: Faber and Faber.

Mack, J. 2011 *The Sea: a cultural history.* London: Reaktion Books.

Maddrell, A., Terry, A. and Gale, T. 2015 *Sacred Mobilities: journeys of belief and belonging.* Farnham: Ashgate.

Madgwick, R., Lamb, A.L., Sloane, H., Nederbragt, A.J., Albarella, U., Parker Pearson, M. and Evans, J.A. 2019 Multi-isotope analysis reveals that feasts in the Stonehenge environs and across Wessex drew people and animals from throughout Britain, *Science Advances* 5 (3), DOI: 10.1126/sciadv.aau6078

Madry, S. and Crumley, C. 1990 An Application of Remote Sensing and GIS in a regional archaeological study. In Allen, K., Green, S. and Zubrow, E. (ed.), *Interpreting Space: GIS and Archaeology.* London: Taylor and Francis, 364–81.

Mafart, B. 2006 Anthropological study of Holocene human footprints of the sebkra El Azrag (Islamic Republic of Mauritania), *Anthropologie* 110, 766–87.

Magnani, G., Bartolomei, P., Cavulli, F., Esposito, M., Marino, E.C., Neri, M., Rizzo, A., Scaruffi, S. and Tosi, M. 2007 U-series and radiocarbon dates on mollusc shells from the uppermost layer of the archaeological site of KHB-1, *Journal of Archaeological Science* 34, 749–55.

Maguire, E.A., Gardian, D.S., Johnsrude, I.S., Good, C.D., Ashburner, J., Frackowiac, R.S.J. and Frith, C.D. 2000 Navigation-related structural changes in the hippocampi of taxi drivers, *Proceedings of the National Academy of Sciences of USA* 97 (8), 4398–403.

Malafouris, L. 2004 The cognitive basis of material engagement: where brain, body and culture conflate. In De Marrais *et al.* (eds), 53–62.

Malafouris, L. and Renfrew, C. (ed.) 2010 *The Cognitive Life of Things: recasting boundaries of the mind.* Cambridge: McDonald Institute.

Malim, T. 2015 Roads, routes and ceremonies: the Fenland superhighway. In Vianello, A. (ed.), *Rivers in Prehistory.* Oxford: Archaeopress, 137–50.

Malim, T. and Hayes, L. 2010 An engineered Iron Age road, associated Roman use (Margary Route 64) and Bronze Age activity recorded at Sharpstone Hill, 2009, *Transactions of the Shropshire Archaeology and History Society* 85, 7–80.

Malim, T. and Hayes, L. 2011 The road, *British Archaeology* 120, 14–19.

Malpas, W. 2007 *The Art of Richard Long.* Maidstone: Crescent Moon.

Manker, E. 1953 *The Nomadism of the Swedish Mountain Lapps.* Stockholm: Acta Lapponica 7.

Manley, J. (ed.) 2016 *Secrets of the High Woods: revealing hidden landscapes.* Midhurst: South Downs National Park Authority.

Manninen, M.A. 2009 Evidence of mobility between the coast and the inland region in the Mesolithic of northern Fennoscandia. In McCartan, S., Schulting, R., Warren, G. and Woodman, P.C. (eds), *Mesolithic Horizons.* Oxford: Oxbow Books, 102–8.

Manning, S.W., Pulak, C., Kromer, B., Talamo, S., Bronk Ramsay, C. and Dee, M. 2009 Absolute dating of the Uluburum shipwreck: a very late Bronze Age time-capsule for the East Mediterranean. In Manning, S.W. and Bruce, M.J. (eds), *Tree-rings, Kings and Old World Archaeology and Environment: papers presented in honour of Peter Ian Kuniholm.* Oxford: Oxbow Books, 163–87.

Marcigny, C., Ghesquière, E. and Kinnes, I. 2007 Bronze Age Cross Channel relations: the Lower Normandy, France example. In Burgess, C., Topping, P. and Lynch, F. (eds), *Beyond Stonehenge: essays in the Bronze Age in honour of Colin Burgess.* Oxford: Oxbow Books, 255–67.

Margary, I.D. 1955 *Roman Roads in Britain.* London: Phoenix House.

Margary, I.D. 1965 *Roman Ways in the Weald.* London: Dent.

Margetts, A. 2018a *Wealdbœra: excavations at Wickhurst Green, Broadbridge Heath and the landscape of the west central Weald.* London: Spoil Heap Publications.

Margetts, A. 2018b A world of summer and autumn: the Romano-British to early medieval Weald and signs of continuity, *Archaeology International* 21, 89–94.

Marlon, J.R., Bartlein, P.J., Daniau, A.-L., Harrison, S.P., Maezumi, S.Y., Power, M.J., Tinner, W. and Vanniere, B. 2013 Global biomass burning: a synthesis and review of Holocene paleofire records and their controls, *Quaternary Science Reviews* 65, 5–25.

Marples, N. and Poulton, R. 2019 *Prehistoric and Early Medieval Landscapes at North Park Farm, Bletchingley, Surrey.* London: Spoil Heap Publications.

Martens, J. 1990 Borremose reconstructed, *Journal of Danish Archaeology* 7, 159–81.

Marty, D., Strasser, A. and Meyer, C.A. 2009 Formation and taphonomy of human footprints in microbial mats of present-day tidal-flat environments: implications for the study of human footprints, *Ichnos* 16, 127–42.

Matson, R.G. and Coupland, G. 1995 *The Prehistory of the Northwest Coast.* London: Academic Press.

Matsui, A. 2008 *The Higashimyo wetland site: basket makers of the Jomon period around 7000 BP.* Nara: Meishinsha.

Matsui, A. and Kanehara, M. 2006 The question of prehistoric plant husbandry during the Jomon period in Japan, *World Archaeology* 38 (2), 259–73.

Mayor, A. and Sarjeant, W.A.S. 2001 The folklore of footprints in stone: from classical antiquity to the present, *Ichnos* 8 (2), 143–63.

McAndrews, J.H. 1988 Human disturbance of North American forests and grasslands: the fossil pollen record. In Huntley, B. and Webb, T. (eds), *Vegetation History.* Dordrecht: Kluwer, 673–697.

McComb, A.M.G. 2009 The ecology of hazel (*Corylus avelana*) nuts in Mesolithic Ireland. In McCartan, S., Schulting, R., Warren, G. and Woodman, P.C. (eds), *Mesolithic Horizons: papers presented at the seventh international conference on the Mesolithic in Europe.* Oxford: Oxbow Books, 225–231.

McDermott, C., Moore, C., Murray, C., Plunkett, G. and Stanley, M. 2009 A Colossus of roads: the Iron Age archaeology of Ireland's peatlands. In Cooney, G., Becker, K., Coles, J., Ryan, M. and Sievers, S. (ed.), *Relics of Old Decency: archaeological studies in later prehistory: festschrift for Barry Raftery.* Dublin: Wordwell, 49–65.

McEarlean, T., McConkey, R. and Forsythe, W. 2002 *Strangford Lough.* Belfast: Environment and Heritage Service.

McGlade, J. 1995 Archaeology and the ecodynamics of human-modified landscapes, *Antiquity* 69, 113–32.

McGrail, S. 1978 *Logboats of England and Wales.* Oxford: British Archaeological Report 51.

McGrail, S. 1981 *Brigg 'Raft' and her Prehistoric Environment.* Oxford: British Archaeological Report 89.

McGrail, S. 1983 Cross-Channel seamanship and navigation in the late 1st millennium BC, *Oxford Journal of Archaeology* 2, 299–337.

McGrail, S. 2004 *Boats of the World: from the Stone Age to medieval times.* Oxford: Oxford University Press.

McLaren, D., Fedje, D., Dyck, A., Mackie, Q., Gauvreau, A. and Cohen, J. 2018 Terminal Pleistocene epoch human footprints from the Pacific coast of Canada, PLOS One *https://doi.org/10.1371/journal.pone.0193522.*

McLaren, S.J. and Rowe, P.J. 1996 The reliability of U-series mollusc dating from the western Mediterranean Basin, *Quaternary Science Reviews* 15, 709–17.

McMaster, J.B., Lewis, M.P., Clark, W. and Biddle, N. 1814 *History of the Expedition under the Command of Captains Lewis and Clarke to the Sources of the Missouri, thence across the Rocky Mountains and down the River Columbia to the Pacific Ocean: Performed during the Years 1804–5 by Order of the Government.* New York: Allerton Book Company.

McOmish, D., Field, D. and Brown, G. 2002 *The Field Archaeology of the Salisbury Plain Training Area.* Swindon: English Heritage.

McQuade, M. and O'Donnell, L. 2007 Late Mesolithic fish traps from the Liffey estuary, Dublin, Ireland, *Antiquity* 81, 569–84.

Meadows, J., Barclay, A. and Bayliss, A. 2007 A short passage of time: the dating of the Hazelton long cairn revisited, *Cambridge Archaeological Journal* 17, 45–64.

Mears, R. and Hillman, G. 2007 *Wild Food.* London: Hodder and Stoughton.

Meddens, F.M. 1996 Sites from the Thames Estuary wetlands, England, and their Bronze Age use, *Antiquity* 70, 325–34.

Meddens, F.M. and Beasley, M. 1990 Wetland use in Rainham, Essex, *London Archaeologist* 6, 242–8.

Meddens, F.M. and Sidell, E.J. 1995 Bronze Age trackways in East London, *Current Archaeology* 143, 412–16.

Meehan, B. 1982 *Shell Bed to Shell Midden.* Canberra: Australian Institute of Aboriginal Studies.

Mellars, P. 1976 Fire, ecology, animal populations and man: a study of some ecological relationships in prehistory, *Proceedings of the Prehistoric Society* 42, 15–45.

Mellars, P. 1987 *Excavations on Oronsay: prehistoric human ecology on a small island.* Edinburgh: Edinburgh University Press.

Mellars, P. 2004 Mesolithic Scotland, coastal occupation, and the role of the Oronsay middens. In Saville, A. (ed.), *Mesolithic Scotland and its Neighbours. The Early Holocene Prehistory of Scotland, its British and Irish Context, and some Northern European Perspectives.* Edinburgh: Society of Antiquaries of Scotland, 167–83.

Mellars, P. and Dark, P. 1998 *Star Carr in Context: new archaeological and palaeoecological investigations at the Early Mesolithic site of Star Carr, North Yorkshire.* Cambridge: McDonald Institute.

Meller, H.H. 2004 *Der Geschmiedete Himmel.* Theiss: Landesmuseum für Vorgeschichte.

Melton, N.D., Montgomery, J. and Knüssel, C.J. 2013 *Gristhorpe Man: a life and death in the Bronze Age*. Oxford: Oxbow Books.

Menotti, F. 2012 *Wetland Archaeology and Beyond*. Oxford: Oxford University Press.

Menotti, F. and O'Sullivan, A. 2013 *The Oxford Handbook of Wetland Archaeology*. Oxford: Oxford University Press.

Mietto, P., Avanzini, M. and Rolandi, G. 2003 Human footprints in Pleistocene volcanic ash, *Nature* 422, 133.

Miles, D., Palmer, S., Lock, G., Gosden, C. and Cromarty, A.M. 2003 *Uffington White Horse and its Landscape*. Oxford: Oxford Archaeology.

Millet, M. and McGrail, S. 1987 Archaeology of the Hasholme logboat, *Archaeological Journal* 144, 69–155.

Milner, N. 2002 Oyster, cockles and kitchen middens: changing practices at the Mesolithic/Neolithic transition. In Miracle, P. and Milner, N. (eds), *Consuming Passions and Patterns of Consumption*. Cambridge: McDonald Institute, 89–96.

Milner, N. 2005 Seasonal consumption practices in the Mesolithic: economic, environmental, social or ritual? In Milner, N. and Woodman, P.C. (eds), *Mesolithic Studies at the Beginning of the 21st century*. Oxford: Oxbow Books, 56–68.

Milner, N., Conneller, C. and Taylor, B. 2018 *Star Carr: Volumes 1 and 2*. York: White Rose Press.

Milner, N., Taylor, B., Conneller, C. and Schadla-Hall, T. 2013 *Star Carr: life in Britain after the Ice Age*. York: Council for British Archaeology.

Milner, N., Craig, O.E., Bailey, G.N., Pedersen, K. and Andersen, S.H. 2004 Something fishy in the Neolithic? A re-evaluation of stable isotope analysis of Mesolithic and Neolithic coastal populations, *Antiquity* 78, 9–22.

Mischka, D. 2011 The Neolithic burial sequence at Flintbek LA3, North Germany and its cart tracks: a precise chronology, *Antiquity* 85, 742–58.

Mitchell, F. 1992 Notes on some non-local cobbles at the entrances to the passage graves at Newgrange and Knowth, County Meath, *Journal of the Royal Society of Antiquaries of Ireland* 122, 128–45.

Mitchell, F. 2005 How open were the European primeval forests? Hypothesis testing using palaeoecological data, *Journal of Ecology* 93, 168–77.

Mithen, S. 1991 Ecological interpretations of Palaeolithic art, *Proceedings of the Prehistoric Society* 57, 103–14.

Mithen, S. (ed.) 2000 *Hunter-gatherer Landscape Archaeology: the Southern Hebrides Mesolithic Project 1988–98. Vol. 1: Islay; Vol. 2 Colonsay*. Cambridge: McDonald Institute.

Mithen, S. and Wicks, K. 2018 The interpretation of Mesolithic structures in Britain: new evidence from Criet Dubh, Isle of Mull, and alternative approaches to chronological analysis for inferring occupation tempos and settlement patterns, *Proceedings of the Prehistoric Society* 84, 77–110.

Mobley, C.M. and Eldridge, M. 1992 Culturally modified trees of the Pacific North-West, *Arctic Anthropology* 29, 91–110.

Molloy, K. and O'Connell, M. 1995 Palaeoecological investigations towards the reconstruction of environment and land use changes during prehistory at Céide Fields, West Ireland, *Probleme der Künstenforschung in Südlichen Nordseegebiet* 23, 187–225.

Moloney, A., Bermingham, N., Keane, M., McDermott, C. and O'Carroll, E. 1995 *Blackwater Survey and Excavations: artefact deterioration in peatlands, Lough More, Co Mayo*. Dublin: Transactions of the Irish Archaeological Wetland Unit 4.

Monbiot, G. 2013 *Feral: Searching for Enchantment on the Frontiers of Rewilding*. London: Allen Lane.

Moore, J. and Scott, E. 1997 *Invisible People and Processes: writing gender and childhood into European archaeology*. Leicester: Leicester University Press.

Morgan, R.A. 1988 *Tree-ring Studies of Wood used in Neolithic and Bronze Age Trackways from the Somerset Levels*. Oxford: British Archaeological Report 184.

Morriss, R.K. 2005 *Roads, Archaeology and Architecture*. Stroud: Tempus.

Moss, M.L. 2013 Beyond subsistence: the social and symbolic meanings of shellfish in Northwest Coast societies. In Bailey, G.N., Hardy, K. and Camara, A. (ed.), *Shell Energy*. Oxford: Oxbow Books, 7–20.

Mottershead, D., Pearson, A. and Schaefer, M. 2008 The cart ruts of Malta: an applied geomorphology approach, *Antiquity* 82, 1065–79.

Mowat, R.J.C. 1996 *The Logboats of Scotland*. Oxford: Oxbow Monograph 68.

Muir, R. 2000 *The New Reading the Landscape*. Exeter: Exeter University Press.

Mullan, G. and Wilson, L. 2005 Cave art in Somerset, *Current Archaeology* 197, 227.

Müller, S. 1904 Vei og bygd i sten-og bronzealderen, *Aarbøger for Nordisk Oldkyndighed og Historie* 1904.

Munn, N.D. 1970 The transformation of subjects into objects in Walbiri and Pitjantjatjara myth. In Berndt, R.M. (ed.), *Australian Aboriginal Anthropology: modern studies in the social anthropology of the Australian Aborigines*. Perth: University of Western Australia Press, 141–63.

Munn, N.D. 1973 The spatial presentation of cosmic order in Walbiri iconography. In Forge, J.A.W. (ed.), *Primitive Art and Society.* Oxford: Oxford University Press, 193–220.

Murphy, K., Caseldine, A., Barker, K., Fielding, S., Burrow, S. and Carlsen, S. 2014 Mesolithic human and animal footprints at Lydstep Haven, Pembrokeshire, 2010: the environmental context, *Archaeologia Cambrensis* 163, 23–41.

Murphy, P.L. 2009 *The English Coast.* London: Continuum.

Murphy, P.L. 2014 *England's Coastal Heritage: a review of progress since 1997.* Swindon: English Heritage.

Murray, H.K., Murray, C. and Fraser, S.M. 2009 *A Tale of the Unknown Unknowns.* Oxford: Oxbow Books.

National Maritime Museum Falmouth 2012 *Exhibition: Cornwall and the sea in the Bronze Age.* Falmouth: National Maritime Museum.

Nayling, N. and Caseldine, A. 1997 *Excavations at Caldicot, Gwent: Bronze Age palaeochannels in the Lower Nedern valley.* York: Council for British Archaeology Research Report 108.

Nayling, N. and McGrail, S. 2004 *The Barland's Farm Romano-Celtic Boat.* York: Council for Britih Archaeology Research Report 138.

Needham, S. and Parfitt, K. 2006 *The Ringlemere Cup: precious cups and the beginning of the Channel Bronze Age.* London: British Museum Press Research publication.

Needham, S., Parham, D. and Frieman, C.J. 2013 *Claimed by the Sea: Salcombe, Langdon Bay and other marine finds of the Bronze Age.* York: Council for British Archaeology Research Report 173.

Neelis, J. 2012 Overland shortcuts for the transmission of Buddhism. In Alcock *et al.* (eds) 2012, 12–32.

Nelson, R.K. 1973 *Hunters of the Northern Forest.* Chicage IL: University of Chicago Press.

Newman, P. 2011 *The Field Archaeology of Dartmoor.* Swindon: English Heritage.

Nickel, K. 1948 *The Skolt Lapp Community Suenjelsijd during the year 1938.* Stockholm: Acta Lapponica 5.

Nicosia, C. and Stoops, G. 2017 *Archaeological Soil and Sediment Micromorphology.* Chichester: Wiley Blackwell.

Nielsen, V. 2010 *Oldtidsagre i Danmark: Sjælland, Møn og Lolland-Falster.* Moesgård: Jysk Arkæologisk Selskab.

Nimura, C. 2016 *Prehistoric Rock Art in Scandinavia.* Oxford: Oxbow Books.

Noble, G. 2006 Harnessing the waves: monuments and ceremonial complexes in Orkney and beyond, *Journal of Maritime Archaeology* 1, 100–17.

Noble, G. 2007 Monumental journeys: Neolithic monument complexes and routeways across Scotland. In Cummings and Johnston (eds) 2007, 64–74.

Noble, G. 2017 *Woodland in the Neolithic of Northern Europe.* Cambridge: Cambridge University Press.

Nolan, C. 2019 Prehistoric landscapes as transitional space. In Darvill, T., Barrass, K., Drysdale, L., Heaslip, V. and Staelens, Y. *Historic Landscapes and Mental Well-being.* Oxford: Archaeopress, 163–78.

Norman, C. 2003 Mesolithic to Bronze Age activity at Parchey Sandbatch, Chedzoy, *Proceedings of the Somerset Archaeological and Natural History Society* 145, 9–38.

Noshiro, S. and Sasaki, Y. 2014 Pre-agricultural management of plant resources during the Jomon period in Japan – a sophisticated subsistence system on plant resources, *Journal of Archaeological Science* 42, 93–106.

Nuszdorfer, F. and Kuster, P. 2000 Old and large Douglas Fir and Western Red Cedar in the Squamish Forest District, British Columbia, Canada. Nanaimo, BC: Vancouver Forest Region.

Olalde, I. *et al.* 2018 The Beaker phenomenon and the genomic transformation of northwest Europe, *Nature* 555 (7695) doi: 10.1038/nature25738. Epub 2018 Feb 21., 190–6.

Oldling-Smee, F.J., Laland, K.N. and Feldman, M.W. 2003 Niche construction, *American Naturalist* 147, 641–8.

Oliver, J. 2010 *Landscapes and Social Transformations on the Northwest Coast: Colonial encounters in the Fraser Valley.* Tucson AZ: University of Arizona Press.

Olwig, K.R. 2008 Performing on the landscape. In Ingold and Vergunst (eds) 2008, 81–92.

Onac, B.P., Viehmann, I. and Lundberg, J. 2005 U-Th ages constraining the Neanderthal footprint at Vârtop Cavem Romania, *Quaternary Science Reviews* 10, 1151–7.

Oosthuizen, S. 2011 Anglo-Saxon fields. In Hamerow, H., Hinton, D.A. and Crawford, S. (ed.), *Handbook of Anglo-Saxon Archaeology.* Oxford: Oxford University Press, 377–401.

Oosthuizen, S. 2013 *Tradition and Transformation in Anglo-Saxon England: Archaeology, common rights and landscape.* London: Bloomsbury.

Ordnance Survey 1956 *Map of Roman Britain.* Southampton: Ordnance Survey.

Ordnance Survey 1962 *Map of Southern Britain in The Iron Age.* Southampton: Ordnance Survey.

Ordnance Survey 1973 *Field Archaeology in Britain.* Southampton: Ordnance Survey.

O'Sullivan, A. 2003b The Social and Ideological Role of Crannogs in Early Medieval Ireland. National University of Ireland, Dublin: Unpublished PhD thesis.

O'Sullivan, A. 2007 Exploring past people's interactions with wetland environments in Ireland, *Proceedings of the Royal Irish Academy* 107C, 147–203.

Ottoway, P. 2017 *Winchester: Swithun's 'City of Happiness and Good Fortune': an Archaeological Assessment.* Oxford: Oxbow Books.

Out, W. 2009 *Sowing the Seed? Human Impact and Plant Subsistence in Dutch Wetlands during the Late Mesolithic and Early and Middle Neolithic (5500–3400 cal BC).* Leiden: Leiden University Press.

Ouzman, S., Taçon, M., Mulvaney, K. and Fulagar, R. 2002 Extraordinary engraved bird track from North Australia: extinct fauna, Dreaming being and/or aesthetic masterpiece?, *Cambridge Archaeological Journal* 12, 103–12.

Palmer, R. 1984 *Danebury: an Iron Age Hillfort in Hampshire: an aerial photographic interpretation of its environs.* London: RCHME.

Pare, C.F.E. 1992 *Wagons and Wagon-Graves of the Early Iron Age in Central Europe.* Oxford: Oxford University Committee for Archaeology 35.

Parfitt, K. 2014 Settlement at the Church Whitfield Crossroads. In Bennett, P., Parfitt, K. and Rady, J. (ed.), *Prehistoric and Anglo-Saxon Discoveries on the East Kent Chalklands.* Canterbury: Canterbury Archaeological Trust, 87–180.

Parfitt, S.A., N.M, A., Lewis, S.G., Abel, R.L. and Coope, G.R. 2010 Early Pleistocene human occupation at the edge of the Boreal zone in northwest Europe, *Nature* 466, 229–33.

Parker, A.G., Goudie, A.S., Anderson, D.E., Robinson, M.A. and Bonsall, C. 2002 A review of the mid-Holocene elm decline in the British Isles, *Progress in Physical Geography* 26, 1–45.

Parker Pearson, M. 2012 *Stonehenge: exploring the greatest Stone Age mystery.* London: Simon and Schuster.

Parker Pearson, M. and Godden, K. 2002 *In Search of the Red Slave: shipwreck and captivity in Madagascar.* Stroud: Sutton.

Parker Pearson, M. and Ramilisonina 1998 Stonehenge for the ancestors: the stones pass on the message, *Antiquity* 72, 308–26.

Parkes, D. and Thrift, N.J. 1980 *Times, Spaces and Places: a chronogeographic perspective.* Chichester: Wiley.

Parker Pearson, M., Jay, M., Chamberlain, A., Sheridan, A. and Evans, J. 2019 *The Beaker People: Isotopes, mobility and diet in prehistoric Britain.* Oxford: Oxbow Books.

Pastoors, A., Lenssen-Erz, T., Ciqae, T., Kxunta, U., Thao, T., Bégouën, R., Biesele, M. and Clottes, J. 2015 Tracking in Caves: experience based reading of Pleistocene human footprints in French caves, *Cambridge Archaeological Journal* 25, 551–64.

Pedersen, L., Fischer, A. and Aaby, B. (ed.) 1997 *The Danish Storebælt since the Ice Age.* Copenhagen: Storebælt.

Peglar, S. 1993 The development of the cultural landscape around Diss Mere, *Review of Palaeobotany and Palynology* 76, 1–47.

Peglar, S., Fritz, S.C. and Birks, H.J.B. 1989 Vegetation and land-use history at Diss, Norfolk, UK, *Journal of Ecology* 77, 203–22.

Pemberton, S.G., Sarjeant, W.A.S. and Torrens, H.S. 2008 History of ichnology, *Ichnos* 4, 321–4.

Pétrequin, A.-M. and Pétrequin, P. 1988 *Le Néolithique des Lacs: Préhistoire des lacs de Chalain et de Clairvaux (4000–2000 av J.-C.).* Paris: Editions Errance.

Pétrequin, P., Sheridan, A., Cassen, S., Errera, M., Gauthier, E., Klassen, L., Le Maux, N. and Pailler, Y. 2008 Neolithic Alpine axeheads, from the Continent to Great Britain, the Isle of Man and Ireland, *Analecta Praehistoria Leidensia* 40, 261–79.

Phillips, M., Duncan, H. and Mallows, C. 2009 *Four Millennia of Human Activity along the A505 Baldock Bypass, Hertfordshire.* Bedford: East Anglian Archaeology Report 128.

Philp, R. 2019 Changing tides: the archaeological context of sea-level change in prehistoric South Wales. Cardiff University: Unpublished PhD thesis.

Pickett, S.T.A. and White, P.S. 1985 *The Ecology of Natural Disturbance and Patch Dynamics.* Orlando FL: Academic Press.

Piggott, S. 1983 *The Earliest Wheeled Transport.* London: Thames and Hudson.

Pitts, M. 2015 Blick Mead: discovering the Mesolithic of southern England, *British Archaeology* May/June, 28–35.

Plets, G., Gheyle, W., Verhoeven, G., De Reu, J., Bourgeois, J., Verhegge, J. and Stichebaut, B. 2012 Three-dimensional recording of archaeological remains in the Altai Mountains, *Antiquity* 86, 884–97.

Plunkett, G., McDermott, G., Swindles, G.T. and Brown, D.M. 2013 Environmental indifference? A critique of environmentally deterministic theories of peatland archaeological site construction in Ireland, *Quaternary Science Reviews* 61, 17–31.

Polat, M.A. 2013 Neolithic period. In: Kiziltan, Z. and Baran Çelik, G. (ed.), *Stories from the Hidden Harbour: the shipwrecks of Yenikapi.* Istanbul: Istanbul Archaeological Museums Press, 77–93.

Politis, G.G. 2005 Children's activity in the production of the archaeological record of hunter-gatherers: an ethnoarchaeological approach. In Funari, P.P., Zarankin, A. and Stovel, E. (eds), *Global Archaeological Theory: contextual voices and contemporary thought.* New York: Kluwer, 121–43.

Politis, G.G. 2006 The different dimensions of mobility among the Nukak foragers of the Columbian Amazon. In Sellet *et al.* (eds), 2006, 23–43.

Politis, G.G. and Bayon, C. 1995 Early Holocene human footprints and sea mammals in the tidal zone of the Argentinian sea shore, *Past* 20, 5–6.

Pollard, J. 2000a Ancestral places in the Mesolithic landscape, *Archaeological Review from Cambridge* 17 (1), 123–38.

Pollard, J. 2000b Neolithic occupation practices and social ecologies from Rinyo to Clacton. In Ritchie, A. (ed.), *Neolithic Orkney in its European Context.* Cambridge: McDonald Institute, 363–9.

Pollard, J. 2017 The Uffington White Horse geoglyph as sun-horse, *Antiquity* 91, 406–21.

Pollard, T. 1996 Time and tide: coastal environments, cosmology and ritual practice in early prehistoric Scotland. In Pollard, T. and Morrison, A. (eds), *The Early Prehistory of Scotland.* Edinburgh: Edinburgh University Press, 198–212.

Polzer, M.E. 2011 Early shipbuilding in the Eastern Mediterranean. In Catsambis, A., Ford, B. and Hamilton, D.L. (eds), *Maritime Archaeology.* Oxford: Oxford University Press, 349–78.

Powesland, D. 2001 The Heslerton Parish Project: an integrated multi-sensor approach to the archaeological study of Eastern Yorkshire, England, *Remote Sensing in Archaeology XI: Ciclo di Lezioni sulla ricerca applicata in archeologia,* 233–56.

Preston, P.R. 2013 Bones, stones or ethnography? Challenging the Mesolithic mobility models for northern England. In Preston, P.R. and Schorle, K. (eds), *Mobility and Change in Prehistory and Classical Antiquity.* Oxford: British Archaeological Report S2534, 29–48.

Price, N. 1995 Houses and horses in the Swedish Bronze Age, *Past* 20, 10–12.

Pringle, H. 2015 Could these be the oldest human footprints in North America?, *Hakai Magazine* June 22, hakaimagazine.com.

Pryor, F. 1991 *Flag Fen Prehistoric Fenland Centre.* London: Batsford.

Pryor, F. 1998a *Farmers in Prehistoric Britain.* Stroud: Tempus.

Pryor, F. 1998b Welland Bank Quarry, South Lincolnshire – a tale of a sausage sandwich, *Current Archaeology* 160, 139–45.

Pryor, F. 2001 *The Flag Fen Basin: Archaeology and environment of a Fenland landscape.* Swindon: English Heritage.

Pryor, F. and Bamforth, M. 2010 *Flag Fen, Peterborough: excavations and research 1995–2007.* Oxford: Oxbow Books.

Pulak, C. 1988 Uluburun shipwreck: an overview, *International Journal of Nautical Archaeology and Underwater Exploration* 27, 188–224.

Pyne, S.J. 2012 *Fire: Nature and Culture.* London: Reaktion Books.

Rackham, O. 1980 *Ancient Woodland.* London: Arnold.

Rackham, O. 1986 *The History of the Countryside.* London: Dent

Rackham, O. 1988 Wildwood. In Jones, M. (ed.), *Archaeology and the Flora of the British Isles.* Oxford: Oxford Committe for Archaeology Monograph 14, 3–6.

Raftery, B. 1990 *Trackways through Time.* Dublin: Headline Publishing.

Raftery, B. (ed.) 1996 *Trackway Excavations in the Mountdillon Bogs, Co Longford 1985–1991.* Dublin: Irish Archaeological Wetland Unit.

Raftery, B. 1999 Paths, tracks and roads in Early Ireland: viewing the people rather than the trees. In Harding, A. (ed.), *Experiment and Design: archaeological studies in honour of John Coles.* Oxford: Oxbow Books, 170–82.

Rahtz, P. and ApSimon, A.M. 1962 Excavations at Shearplace Hill, Sydling St Nicholas, Dorset, England, *Proceedings of the Prehistoric Society* 28, 329–61.

Rasmussen, P. 2010 Geelskov: Vegetationshistorisk undersøgelse i forbindelse med oldtidagersystemet. In Nielsen, V. and Clemmenson, N. (eds), *Oldtidsagre i Danmark: Sjælland, Møn og Lolland-Falster.* Moesgård: Jysk Arkæologisk Selskab, 223–33, 391–3.

Rasmussen, P. 1995 Mid-Holocene vegetation development at the inland Ertebølle settlement of Ringkløster, Eastern Jutland, *Journal of Danish Archaeology* 12, 65–85.

Rector, C. 1979 5,000-year-old footprints on the Mojave River, California, USA, *Antiquity* 53, 149–50.

Rector, C. 1999 *Human and Animal Trackway at Oro Grande.* San Bernardino CA: San Bernardino County Museum Association.

Rector, C., Swenson, J.D. and Wilke, P.J. 1983 *Archaeological Studies at Oro Grande, Mojave Desert, California.* Redlands CA: San Bernardino County Museum Association.

Rednap, M. 2011 *Discovered in Time: treasures from early Wales.* Cardiff: National Museum of Wales.

Regnell, J., Gaillard, M.-J., Bartholin, T.S. and Karsten, P. 1995 Reconstruction of environment and history of plant use during the late Mesolithic/Ertebolle culture at the inland settlement of Bikeberg III, south Sweden, *Vegetation History and Archaeobotany* 4, 67–91.

Reilly, S., Nolan, C. and Monkton, L. 2018 *Wellbeing and the Historic Environment.* Swindon: Historic England: https://historicengland.org.uk/images-books/publications/wellbeing-and-the-historic-environment/wellbeing-and-historic-environment/.

Remondino, F., Rizzi, A., Girardi, S., Petti, F.M. and Avanzini, M. 2010 3D Ichnology – recovering digital 3D models of dinosaur footprints, *Photogrammetric Record* 25, 266–82.

Renfrew, C. 2003 *Fighting It Out: the parallel visions of artists and archaeologists.* London: Thames and Hudson.

Renfrew, C. 2004 Towards a theory of material engagement. In De Marrais *et al.* (eds) 2004, 23–31.

Rentzel, P., Nicosia, C., Gebhardt, A., Brönnimann, D., Pümpin, C. and Ismail-Meyer, K. 2017 Trampling, poaching and the effect of traffic. In Nicosia and Stoops (eds) 2017, 281–98.

Reynolds, A. and Langlands, A. 2011 Travel as communication: a consideration of overland journeys in Anglo-Saxon England, *World Archaeology* 43, 410–27.

Rhodes, E.J. 2011 Optically Stimulated Luminescence dating of sediments over the past 200,000 years, *Annual Review of Earth and Planetary Sciences* 39, 461–88.

Richards, J. 1990 *The Stonehenge Environs Project.* London: HBMC.

Richards, J. 2007 *Stonehenge: the story so far.* Swindon: English Heritage.

Richards, M.P. and Mellars, P.A. 1998 Stable isotopes and the seasonality of the Oronsay Middens, *Antiquity* 72, 178–84.

Richards, M.P. and Sheridan, J.A. 2000 New AMS dates on human bone from Mesolithic Orkney, *Antiquity* 74, 313–15.

Richards, M.P., Schulting, R.J. and Hedges, R. 2003 Sharp shift in diet at onset of Neolithic, *Nature* 425, 38–66.

Riedman, M. 1997 *Sea Otters.* Monterey CA: Monterey Bay Aquarium.

Rippon, S. 1996 *Gwent Levels: the evolution of a wetland landscape.* York: Council for British Archaeology Research Report 105.

Rippon, S. 2012 *Historic Landscape Analysis.* York: Council for British Archaeology.

Rippon, S., Smart, C. and Pears, B. 2015 *The Fields of Britannia.* Oxford: Oxford University Press.

Robb, J. 2004 The extended artefact and the monumental economy: methodology for material agency. In De Marrais *et al.* (eds) 2004, 131–40.

Robbins, L.M. 1985 *Footprints: Collection, Analysis and Interpretation.* Springfield, Il.: Thomas.

Robbins, W.G. 1999 Landscape and environment: ecological change in the intermontane Northwest. In Boyd (ed.) 1999, 219–54.

Roberts, A. 1996 Late Upper Palaeolithic and Mesolithic hunting-gathering communities. In Kain, R. and Ravenhill, W. (eds), *Historical Atlas of South-West England.* Exeter: University of Exeter Press, 47–50.

Roberts, D.L. 2008 Last Interglacial hominid and associated vertebrate fossil trackways in coastal eolianites, South Africa, *Ichnos* 15, 190–207.

Roberts, G. 2014 *The Lost World of Formby Point.* Liverpool: Alt Press.

Roberts, G. and Worsley, A. 2008 Evidence of human activity in mid-Holocene coastal palaeoenvironments of Formby, North West England. In Lewis, J.M. and Stanistreet, J.E. (eds), *Sand and Sea: Sefton's coastal heritage.* Bootle: Sefton Council, 28–43.

Roberts, G., Gonzalez, S. and Huddart, D. 1996 Intertidal Holocene footprints and their archaeological significance, *Antiquity* 70, 647–51.

Roberts, M.R. 2010 'Footprints in the concrete': a study of the Chemin des Juifs (Jews' Road), Jewish slave labour camps, and related sites, in the Nord-de-Calais, France, *The Historic Environment* 1, 70–102.

Roberts, O.T.P. 1992 Brigg 'raft' reassessed as a round bilge Bronze Age boat, *International Journal of Nautical Archaeology and Underwater Exploration* 21, 245–58.

Robertson, D. 2016 A second timber circle, trackways, and coppicing at Holme-next-the-Sea Beach, Norfolk: use of salt- and freshwater marshes in the Bronze Age, *Proceedings of the Prehistoric Society* 82, 227–58.

Rowley-Conwy, P. 2004 How the West was lost: a reconsideration of agricultural origins in Britain, Ireland, and Southern Scandinavia, *Current Anthropology* 45, S83–113.

Rowley-Conwy, P. and Layton, R.H. 2011 Foraging and farming as niche construction: stable and unstable adaptations, *Philosophical Transactions of the Royal Society* B 366, 849–62.

Rudling, D. 2002 *Downland Settlement and Land-use: the archaeology of the Brighton Bypass*. London: English Heritage.

Rudling, D. 2016 Impact of Rome. In Moore, D., Allen, M.J. and Rudling, D. (eds), *Archaeology of the Ouse Valley, Sussex, to AD 1500*. Oxford: Archaeopress, 73–94.

Russell, J. 2009 *Ravilious in Pictures: Sussex and the Downs*. Sparham: Mainstone Press.

Ryan, P. and Blackford, J.J. 2010 Late Mesolithic environmental change at Black Heath, south Pennines, UK: a test of Mesolithic woodland management models using pollen, charcoal and non-pollen palynomorph data, *Vegetation History and Archaeobotany* 19, 545–58.

Salisbury, E.J. 1964 *Weeds & Aliens*. London: Collins New Naturalist 43.

Samson, A., Cooper, J. and Caamonso-Dones, J. 2016 European visitors in native spaces: using palaeography to investigate religious dynamics in the New World, *Latin American Antiquity* 27, 443–61.

Samuels, S.R. 1994 *Ozette Archaeological Project Research Reports Vol 1: house structure and floor midden*. Seattle WA: Washington State University Report 63.

Sandom, C.J., Ejrnaes, R., Hansen, M.D.D. and Svenning, J.-C. 2013 High herbivore density associated with vegetation diversity in interglacial ecosystems, *Proceedings of the National Academy of Sciences of the United States of America* 111 (1), 4162–7.

Sands, R. 2013 Portable wooden objects from wetlands. In Menotti, F. and O'Sullivan, A. (eds), *The Oxford Handbook of Wetland Archaeology*. Oxford: Oxford University Press, 305–22.

Saville, A. 1990 *Hazelton North, Gloucestershire1979–82: the excavation of a Neolithic long cairn of the Cotswold-Severn group*. London: English Heritage Archaeological Report 13.

Saville, A. 2000 Orkney and Scotland before the Neolithic. In Ritchie, A. (ed.), *Neolithic Orkney in its European Context*. Glasgow: McDonald Institute, 91–100.

Scales, R. 2006 Prehistoric Coastal Wetland Exploitation: the evidence of footprint-tracks and animal bones, with reference to the Severn Estuary. University of Reading: Dept of Archaeology, Unpublished PhD.

Scales, R. 2007 Footprint-tracks of people and animals. In Bell (ed.) 2007, 139–59.

Scarre, C. 2002 A Pattern of Islands: the Neolithic monuments of north-west Brittany, *European Journal of Archaeology* 5, 24–41.

Schaepe, D.M. 2001 Stó:ló communication and Transportation routes c 1850. In Carlson, K.T. (ed.), *A Stó:ló Coast Salish Historical Atlas*. Vancouver: Douglas & McIntyre, 60–1.

Schama, S. 1995 *Landscape and Memory*. London: Harper Collins.

Schelvis, J. 1992 The identification of archaeological dung deposits on the basis of remains of predatory mites Acari:Gamasida, *Journal of Archaeological Science* 19, 677–82.

Schlichtherle, H. and Strobel, M. 1999 *Archaeology and Protection of Nature in the Federsee Bog*. Stuttgart: Landesdenkmalamt Baden-Württemberg.

Schou Jørgensen, A.M. 1988a Færdsel over stenalderfjorden: om den ældste vej i Tibirke, *Fortidsminder og Kulturhistorie Antikvariske Studies* 9, 157–67.

Schou Jørgensen, A.M. 1988b Vej, Vejstrøg og vejspærring. In Morlensen, P. and Rassmussen, B.M. (ed.), *Jernalderens Stammesamfund*. Aarhus: Jutland Archaeological Society, 101–16.

Schulting, R. 2003 The marrying kind: evidence for a patrilocal postmarital residence pattern in the Mesolithic of Southern Brittany? In Larsson, L. (ed.), *Mesolithic on the Move*. Oxford: Oxbow Books, 431–41.

Schulting, R. 2005 'Pursuing a rabbit in Burrington Combe': new research on the Early Mesolithic burial cave of Aveline's Hole, *Proceedings of the University of Bristol Spelaeological Society* 23, 171–265.

Schulting, R. and Richards, M.P. 2001 Dating women and becoming farmers: palaeodieting and AMS dating evidence from the Breton Mesolithic cemeteries of Téviec and Hoëdic, *Journal of Anthropological Archaeology* 20, 314–44.

Schulting, R. and Richards, M.P. 2002 Finding the coastal Mesolithic in southwest Britain: AMS dates and stable isotope results on human remains from Caldey Island, Pembrokeshire, South Wales, *Antiquity* 76, 1011–25.

Schulting, R. and Bradley, R. 2013 'Of human remains and weapons in the neighbourhood of London': New AMS 14C dates on Thames 'River Skulls' and their European context, *Archaeological Journal* 170, 30–77.

Schulting, R.J., le Roux, P., Gan, Y.M., Pouncett, J., Hamilton, J., Snoeck, C., Ditchfield, P., Henderson, R., Lange, P., Lee-Thorpe, J.A., Gosden, C. and Lock, G. 2019 The ups and downs of Iron Age animal management on the Oxfordshire Ridgeway, south-central England: A multi-isotope approach, *Journal of Archaeological Science* 101, 199–212.

Sealy, J. 2001 Body tissue chemistry and palaeodiet. In Brothwell, D.R. and Pollard, T. (eds), *Handbook of Archaeological Sciences*. Chichester: Wiley, 268–79.

Sebire, H. 2005 *The Archaeology and Early History of the Channel Islands*. Stroud: Tempus.

Seitsonen, O., Houle, J.-L. and Broderick, L.G. 2014 GIS approaches to past mobility and accessibility: an example from the Bronze Age Khanuy Valley, Mongolia. In Leary (ed.) 2014, 79–112.

Sellet, F., Greaves, R. and Yu, P.-L. 2006 *Archaeology and Ethnoarchaeology of Mobility*. Gainesville FL: University Press of Florida.

Servais, G.R. 2007 *La Tène, un Site, un Mythe 1: Chronique en images (1857–1923)*. Neuchâtel: Office et musée cantonal d'archéologie de Neuchâtel.

Sheets, P. 2009 When the construction of meaning preceded the meaning of construction: from footpaths to monumental entrances in ancient Costa Rica. In Snead *et al.* (eds) 2009, 158–79.

Sheldon, I. and Hartson, T. 1999 *Animal Tracks of British Columbia*. Edmonton: Lone Pine.

Shennan, S., Lambeck, K., Flather, R., Horton, B., McArthur, J., Innes, J.B., Lloyd, J., Rutherford, M. and Wingfield, R. 2000 Modelling western North Sea palaeogeographies and tidal changes during the Holocene. In Shennan, S. and Andrews, J. (ed.), *Holocene Land-Ocean Interaction and Environmental Change around the North Sea*. London: Geological Society Special Publication 166, 299–319.

Sheridan, A. 2003 French conections I: spreading the *marmites* thinly. In Armit, I., Murphy, E.M., Nelis, E. and Simpson, D.A. (ed.), *Neolithic Settlement in Ireland and Western Britain*. Oxford: Oxbow Books, 3–17.

Sheridan, A. 2004 Neolithic connections along and across the Irish Sea. In Cummings, V. and Fowler, C.S. (ed.), *The Neolithic of the Irish Sea: materiality and traditions of practice*. Oxford: Oxbow Books, 9–21.

Sheridan, A. 2008 Upper Largie and Dutch-Scottish connections during the Beaker period, *Analecta Praehistoria Leidensia* 40, 246–60.

Sherratt, A. 1996 Why Wessex? The Avon route and river transport in later British prehistory, *Oxford Journal of Archaeology* 15, 211–34.

Shields, G. 2005 The Roman roads of the Portslade/Aldrington area in relation to a possible Roman port at Copperas Gap, *Sussex Archaeological Collections* 143, 135–50.

Shillito, L.-M. 2017 Gas Chromatography – Mass Spectrometry (GC/MS). In Nicosia and Stoops (eds) 2017, 399–402.

Simmons, A.H. 2014 *Stone Age Sailors: Paleolithic Seafaring in the Mediterranean*. Walnut Creek CA: Left Coast Press.

Simmons, I.G. 1993 *Interpreting Nature: cultural constructions of the environment*. London: Routledge.

Simmons, I.G. 1996 *The Environmental Impact of Later Mesolithic Cultures: the creation of moorland landscape in England and Wales*. Edinburgh: Edinburgh University Press.

Simmons, I.G. and Tooley, M.J. 1981 *The Environment in British Prehistory*. London: Duckworth.

Simpson, S.J. and Pankova, S. 2017 *Scythians: warriors of ancient Siberia*. London: British Museum.

Sjögren, K. 2006 *Ecology and Economy in Stone Age and Bronze Age Scania*. Stockholm: Riksantikvarieämbetet.

Skoglund, P., Nimura, C. and Bradley, R. 2017 Interpretations of footprints in the Bronze Age rock art of south Scandinavia, *Proceedings of the Prehistoric Society* 83, 289–303.

Smedstad, I. 2001 Research on wooden trackways in Norway. In Cross, S., Murray, C., Ó Néill, J. and Stevens, P. (eds), *Recent Developments in Wetland Research*. Dublin: Seandálaíocht, 191–200.

Smith, A.G. and Cloutman, E.W. 1988 Reconstruction of Holocene vegetation history in three dimensions at Waun-Fignen-Felen, an upland site in South Wales, *Philosophical Transactions of the Royal Society B* 322, 159–219.

Smith, B.D. 1995 Seed plant domestication in eastern North America. In Price, T.D. and Gebauer, A.B. (eds), *Last Hunters – First Farmers*. Santa Fe NM: School of American Research Press, 193–213.

Smith, B.D. 2011 General patterns of niche construction and the management of 'wild' plant and animal resources by small-scale pre-industrial societies, *Philosophical Transactions of the Royal Society B* 366, 836–48.

Smith, D.N., Whitehouse, N., Bunting, M.J. and Chapman, H. 2010 Can we characterise 'openness' in the Holocene palaeoenvironmental record? Analogue studies from Dunham Massey deer park and Epping Forest, England, *The Holocene* 20, 215–29.

Smith, K., Coppen, J., Wainwright, G.J. and Beckett, S.C. 1981 The Shaugh Moor Project: third report – settlement and environmental investigations, *Proceedings of the Prehistoric Society* 47, 205–73.

Smith, O., Momber, G., Bates, R., Garwood, P., Fitch, S., Pallen, M., Gaffney, V. and Allaby, R.G. 2015 Sedimentary DNA from a submerged site reveals wheat in the British Isles 8000 years ago, *Science* 347, 998–1001.

Smith, R.J.C., Healy, F., Allen, M.J., Morris, E.L., Barnes, I. and Woodward, P.J. 1997 *Excavations along the Route of the Dorchester By-pass, Dorset, 1986–8.* Salisbury: Wessex Archaeology Report 11.

Smithson, M. 1993 *Olympic: Ecosystems of the Peninsula.* Helena: American and World Geographical Publishing.

Snead, J.E. 2011 The 'secret and bloody war path'; movement, place and conflict in the archaeological landscape of North America, *World Archaeology* 43 (3), 478–92.

Snead, J.E., Erickson, C.L. and Darling, J.A. 2009 *Landscapes of Movement: trails, paths and roads in anthropological perspective.* Philadelphia PA: University of Pennsylvania.

Snoeck, J., Pouncett, J., Claeys, P., Goderis, S., Mattielli, N., Parker Pearson, M., Willis, C., Zazzo, A., Lee-Thorp, J.A. and Schulting, R.J. 2018 Strontium isotope analysis on cremated human remains from Stonehenge support links with west Wales, *Science Report* 8, 10790.

Soberl, L., Pollard, J. and Evershed, R. 2009 Pots for the afterlife: organic residue analysis of British Bronze Age pottery from funerary contexts, *Past* 63, 6–8.

Sobolik, K.D. 1996 Nutritional constraints and mobility patterns of hunter-gatherers in the northern Chihuahuan Desert. In Reitz, E.J., Newsom, L.E. and Scudder, S.J. (ed.), *Case Studies in Environmental Archaeology.* London: Plenum Press, 195–214.

Sohn, Y.K., Yoon, W.S., Ahn, U.S., Kim, G.B., Lee, J.-H., Ryu, C.K., Jeon, Y.M. and Kang, C.H. 2015 Stratigraphy and age of the human footprint-bearing strata in JeJo Island, Korea: controversies and new findings, *Journal of Archaeological Science Reports* 4, 264–75.

Solnit, R. 2011 *A Book of Migrations.* London: Verso.

Solnit, R. 2014 *Wanderlust: a history of walking.* London: Granta.

Spindler, K. 1993 *The Man in the Ice.* London: Weidenfeld and Nicolson.

Sprackland, J. 2012 *Strands: a year of discoveries on the beach.* London: Jonathan Cape.

Spratling, M. 1979 The debris of metalworking. In Wainwright, G.J. (ed.), *Gussage All Saints.* London: HMSO, 125–49.

Spratt, D. 1982 *Prehistoric and Roman Archaeology of North-east Yorkshire.* Oxford: British Archaeological Report S104.

Stafford, E., Goodburn, D. and Bates, M. 2012 *Landscape and Prehistory of the East London Wetlands: investigations along the A13 DBFO Roadscheme, Tower Hamlets, Newham, Barking and Dagenham, 2000–2003.* Oxford: Oxford Archaeology Monograph 17.

Standish, C.D., Dhuime, B., Hawkesworth, C.J. and Pike, A.W.G. 2015 A non-local source of Irish Chalcolithic and Early Bronze Age gold, *Proceedings of the Prehistoric Society* 81, 149–77.

Starr, H. and Harris, S. 2009 Are we there yet? Using GIS to investigate paths and landmarks in the Mesolithic of south-west Germany. In McCartan, S., Schulting, R., Warren, G. and Woodman, P.C. (eds), *Mesolithic Horizons.* Oxford: Oxbow Books, 89–95.

Stead, I. 1979 *The Arras Culture.* York: The Yorkshire Philosophical Society.

Stein, J.K. 1992 *Deciphering a Shell Midden.* London: Academic Press.

Stein, J.K. 1993 Scale in archaeology, geosciences, and geoarchaeology. In Stein, J.K. and Linse, A.R. (eds), *Effects of Scale on Archaeological and Geoscientific Perspectives.* Boulder CO: Geological Society of America Special Paper 283, 1–10.

Stein, J.K. 2000 *Exploring Coast Salish Prehistory: the archaeology of San Juan Island.* Seattle WA: University of Washington Press.

Stevens, C.J. and Fuller, D.Q. 2015 Alternative strategies to agriculture: the evidence for climatic shocks and cereal declines during the British Neolithic and Bronze Age (a reply to Bishop), *World Archaeology* 47 (5), 856–75.

Stewart, H. 1977 *Indian Fishing: early methods on the Northwest Coast.* Seattle WA: University of Washington Press.

Stewart, H. 1984 *Cedar.* Seattle WA: University of Washington Press.

Stilgoe, J.R. 2015 *What is Landscape?* Cambridge MA: MIT Press.

Stoertz, C. 1997 *Ancient Landscapes of the Yorkshire Wolds.* Swindon: RCHME.

Strachan, D. 2010 *Carpow in Context.* Edinburgh: Society of Antiquaries of Scotland.

Stringer, C. 2006 *Homo Britannicus: the incredible story of human life in Britain.* London: Allen Lane.

Stroud, C. 2017 *The Wild Other.* London: Hodder and Stoughton.

Stuart, C. and Stuart, T. 1998 *A Field Guide to the Tracks and Signs of Southern and East African Wildlife.* Cape Town: Struik.

Stukeley, W. 1740 *Stonehenge: A Temple Restor'd to the British Druids.* London: W. Innnys and R. Maney.

Sutikna, T., Tocheri, M.W., Morwood, M.J., Wahyu Saptomo, E., Jatmiko, RokusDueAwe, SriWasisto, Westaway, K.E., Aubert, M., BoLi, Jian-XinZhao, Storey, M., Alloway, B.V., Morley, M.W., Meijer, H.J.M., van den Bergh, G.D., Grün, R., Dosseto, A., Brumm, A., Jungers, W.L. and Roberts, R.G. 2016 Revised stratigraphy and chronology for *Homo floresiensis* at Liang Bua in Indonesia, *Nature* 532, 366–9.

Svenning, J.-C. 2002 A review of natural vegetation openness in north-western Europe, *Biological Conservation* 104, 133–48.

Sylvester, M. 2009 The Haugvik Boat – a pre-Roman Iron Age boat find from northern Norway. In Bockius, R. (ed.), *Between the Seas: transfer and exchange in nautical technology. Symposium at Mainz.* Mainz: Römisch-Germanisches Zentralmuseum, 52–8.

Synge, J.M. 1979 *The Aran Islands.* Oxford: Oxford University Press.

Taylor, C. 1975 *Fields in the English Landscape.* London: Dent.

Taylor, C. 1979 *Roads and Tracks of Britain.* London: Dent.

Taylor, C. 1983 *Village and Farmstead.* London: George Phillip and Son.

Terberger, T. 2006 The Mesolithic hunter-fisher-gatherers on the Northern German Plain. In Møller Hansen, K. and Buck Pedersen, K. (eds), *Across the West Baltic.* Vordingborg: Sydsjællands Museum, 111–84.

Thomas, E. 1909 (2009 edition) *The South Country.* Toller Fratrum: Little Toller Books.

Thomas, E. 1916 *The Icknield Way.* London: Constable.

Thomas, G. 2013 Life before the Minster: the social dynamics of monastic foundation at Anglo-Saxon Lyminge, *Antiquaries Journal* 93, 109–45.

Thomas, J. 1991 *Rethinking the Neolithic.* Cambridge: Cambridge University Press.

Thomas, J. 1996 *Time, Culture and Identity.* London: Routledge.

Thomas, J. 2006 On the origins and development of cursus monuments in Britain, *Proceedings of the Prehistoric Society* 72, 229–41.

Thomas, J. 2013 *The Birth of Neolithic Britain.* Oxford: Oxford University Press.

Thomas, K.D. 1982 Neolithic enclosures and woodland habitats on the South Downs in Sussex, England. In: Bell, M. and Limbrey, S. (ed.), *Archaeological Aspects of Woodland Ecology.* Oxford: British Archaeological Report S146, 147–70.

Thorne, A. 2016a Secrets of the High Woods: landscapes and lasers, *Past* 83, 1–3.

Thorne, A. 2016b Landscapes and lasers, *Current Archaeology* 317, 42–3.

Tilley, C. 1994 *A Phenomenology of Landscape.* Oxford: Berg.

Tilley, C. 2010 *Interpreting Landscapes, Geologies, Topographies, Identities: explorations in landscape phenomenology.* Walnut Creek CA: Left Coast Press.

Tipping, R. 1996 Microscopic charcoal records, inferred human activity and climate change in the Mesolithic of northernmost Scotland. In Pollard, T. and Morrison, A. (ed.), *The Early Prehistory of Scotland.* Edinburgh: Edinburgh University Press, 39–61.

Tomalin, D.J., Loader, R.D. and Scaife, R.G. 2012 *Coastal Archaeology in a Dynamic Environment: a Solent case study.* Oxford: British Archaeological Report 568.

Tooley, M.J. 1985 *Sea-level Changes in Northern England during the Flandrian Stage.* Oxford: Clarendon Press.

Tree, I. 2018 *Wilding.* London: Picador.

Tuck-Po, L. 2008 Before a step too far: walking with Batek hunter-gatherers in the forests of Pahang, Malaysia. In Ingold, T and Vergunst (eds) 2008, 21–34.

Turner, D.J. 1980 The North Downs trackway, *Surrey Archaeological Collections* 72, 1–13.

Turner, N. 1999 'Time to burn'. In Boyd (ed.) 1999, 185–218.

Turner, N.J. 1995 *Foodplants of Coastal First Peoples.* Vancouver: UBC Press.

Turner, N.J. 1998 *Plant Technology of First Peoples in British Columbia.* Vancouver: UBC Press.

Turner, N.J. 2005 *The Earth's Blanket.* Vancouver: Douglas and McIntyre.

Turner, N.J. 2014 *Ancient Pathways, Ancestral Knowledge. Vol 1: The History and Practice of Indigenous Plant Knowledge; Vol 2: The Place and Meaning of Plants in Indigenous Cultures and Worldviews.* Montreal: McGill-Queen's University Press.

Turner, N.J., Davidson-Hunt, I.J. and O'Flaherty, M. 2003 Living on the edge: ecological and cultural edges as sources of diversity for social-ecological resilience, *Human Ecology* 31 (3), 439–61.

Turner, N.J., Ari, Y., Berkes, F., Davidson-Hunt, I.J., Ertug, Z.F. and Miller, A.M. 2009 Cultural management of living trees: an international perspective. In Lepofsky, D. (ed.) Indigenous Resource Management: Past, present and future. Special issue, *Journal of Ethnobiology* 29, 237–70.

Turner, N.J., Deur, D. and Lepofsky, D. 2013 Plant management systems of British Columbia's First Peoples, *BC Studies* 179, 107–11.

Tuttle, R.H. 1987 Kinesiological inferences and evolutionary implications from Laetoli bipedal trails. In Leakey and Harris (eds) 1987, 503–23.

Tuttle, R.H. 2008 Footprint clues in hominid evolution and forensics: lessons and limitations, *Ichnos* 15 (3–4), 158–65.

Tuttle, R.H., Webb, D., Weidl, E. and Baksh, M. 1990 Further progress on the Laetoli trails, *Journal of Archaeological Science* 17, 347–62.

van Asperen, E.N. 2017 Fungal diversity on dung of tropical animals in temperate environments: implications for reconstructing post mega faunal populations, *Fungal Ecology* 28, 25–32.

Van de Noort, R. 2004 *The Humber Wetlands: the archaeology of a dynamic landscape.* Macclesfield: Windgather Press.

Van de Noort, R. 2011 *North Sea Archaeologies: a maritime biography, 10,000 BC–AD 1500.* Oxford: Oxford University Press.

Van de Noort, R. and Ellis, S. 1999 *Wetland Heritage of the Vale of York, an archaeological survey.* Hull: Humber Wetlands Project.

Van de Noort, R. and O'Sullivan, A. 2006 *Rethinking Wetland Archaeology.* London: Duckworth.

Van de Noort, R., Cumby, B., Blue, L., Harding, A., Hurcombe, L., Hansen, T.M., Wetherelt, A., Wittamore, J. and Wyke, A. 2014 Morgawr: an experimental Bronze Age type sewn-plank craft based on the Ferriby boats, *International Journal of Nautical Archaeology and Underwater Exploration* 43, 292–313.

van den Broeke, P.W. 2005 Gifts to the gods: rites and cult sites in the Bronze Age and Iron Age. In Louwe Kooijmans, L.P., van den Broeke, P.W., Fokkens, H. and van Gijn, A.L. (eds), *The Prehistory of the Netherlands* Volume 2. Amsterdam: Amsterdam University Press, 659–77.

van der Sanden, W.A.B. 2002 Veenwagen in Drenthe: enkele nieuwe dateringen, *Nieuwe Drentse Volksalmanak* 2002, 101–12.

van Geel, B., Buurman, J. and Waterbolk, H.T. 1996 Archaeological and palaeoecological indications of an abrupt climate change in The Netherlands, and evidence for climatological teleconnections around 2650 BP, *Journal of Quaternary Science* 11, 451–60.

Vancouver, G. 1801 *A Voyage of Discovery to the North Pacific Ocean and round the World.* London: J.S. Stockdale.

Vanwalleghem, T., van den Eeckhaut, M., Poesen, J., Deckers, J., Nachtergaele, J., van Oost, K. and Slenters, C. 2003 Characteristics and controlling factors of old gullies under forest in a temperate humid climate: a case study from the Meerdaal Forest (Central Belgium), *Geomorphology* 56, 15–29.

Vanwalleghem, T., Bork, H.R., Poesen, J., Dotterweich, M., Schmidten, G., Deckers, J., Scheers, S. and Martens, M. 2006 Prehistoric and Roman gullying in the European loess belt: a case study from central Belgium, *The Holocene* 16, 393–401.

Vera, F.W.M. 2000 *Grazing, Ecology and Forest History.* Wallingford: CABI.

Victoria State Government 2008a *Aboriginal Scarred Trees.* Melbourne: Victoria State Government leaflet.

Victoria State Government 2008b *Aboriginal Mounds.* Melbourne: Victoria State Government leaflet.

Viner, S., Evans, J., Albarella, U. and Parker Pearson, M. 2010 Cattle mobility in prehistoric Britain: strontium isotope analysis of cattle teeth from Durrington Walls (Wiltshire, Britain), *Journal of Archaeological Science* 37, 2812–20.

von Thunen, J.H. 1826 (1966 edition) *The Isolated State: with respect to agriculture and political economy.* Oxford: Pergamon.

Wachsmann, S. 2009 *Seagoing Ships and Seamanship in the Bronze Age Levant.* College Station TX: Texas A & M University.

Waddington, C. 1998 A Landscape Archaeological Study of the Mesolithic/Neolithic in the Millfield Basin, Northumberland. Unpublished D. Phil Thesis, Durham University.

Waddington, C. 2005 Yeavering in its Stone Age landscape. In Frodsham, P. and O'Brien, C. (ed.), *Yeavering: people, power and place.* Stroud: History Press, 84–97.

Waddington, C. 2007 *Mesolithic Settlement in the North Sea Basin: a case-study from Howick, North-east England.* Oxford: Oxbow Books.

Waddington, C. 2014 Battling the waves, *British Archaeology* Nov/Dec, 34–9.

Wainwright, G.J. 1989 *The Henge Monuments.* London: Thames and Hudson.

Wainwright, G.J. and Longworth, I.H. 1971 *Durrington Walls.* London: Society of Antiquaries.

Walker, M.J.C. 2005 *Quaternary Dating Methods.* Chichester: Wiley.

Walker, T. 2018 *The Gwithian Landscape: molluscs and archaeology in Cornish sand dunes.* Oxford: Archaeopress.

Wallace, P.A.W. 1965 *Indian Paths of Pennslyvania.* Harrisburg PA: Pensylvania Historical and Museum Commission.

Walsh, K. 2013 *The Archaeology of Mediterranean Landscapes.* Cambridge: Cambridge University Press.

Walsham, A. 2010 Footprints and faith: religion and the landscape in early modern Britain and Ireland, *Studies in Church History* 46, 169–83.

Warren, G., Davis, S., McClatchie, M. and Sands, R. 2014 The potential role of humans in structuring the wooded landscapes of Mesolithic Ireland: a review of data and discussion of approaches, *Vegetation History and Archaeobotany* 23, 629–46.

Watkins, A. 1925 (2014 edition) *The Old Straight Track.* London: Head of Zeus.

Watkins, T. 1980 A prehistoric coracle in Fife, *International Journal of Nautical Archaeology and Underwater Exploration* 9, 277–86.

Watson, A. 2001 Round barrows in a circular world: monumentalising landscape in Early Bronze Age Wessex. In Brück, J. (ed.), *Bronze Age Landscapes: tradition and transformation.* Oxford: Oxbow Books, 207–16.

Watson, P.J. 1969 *The Prehistory of Salts Cave, Kentucky.* Chicago: Illinois State Museum Reports of Investigations 16.

Watson, P.J. 1974 *Archaeology of Mammoth Cave area.* New York: Academic Press.

Watson, P.J., Kennedy, M.C., Willey, P., Robbins, L.M. and Wilson, R.C. 2005 Prehistoric footprints in Jaguar Cave, Tennessee, *Journal of Field Archaeology* 30 (1), 25–43.

Webb, D. 2014 Oldest human footprints, *Current World Archaeology* 67, 10.

Webb, D., Robu, M., Moldovan, O., Constantin, S., Tomus, B. and Naeg, I. 2014 Ancient human footprints in Ciur-Izbuc Cave, Romania, *American Journal of Physical Anthropology* 155, 128–35.

Webb, S. 2007 Further research of the Willandra Lakes fossil footprint site, southeastern Australia, *Journal of Human Evolution* 52, 711–15.

Webb, S., Cupper, M.L. and Robins, R. 2006 Pleistocene human footprints from the Willandra Lakes, southeastern Australia, *Journal of Human Evolution* 50, 405–13.

Webb, T., Cushing, E.J. and Wright, H.E. 1984 Holocene changes in the vegetation of the Midwest. In Wright, H.E. (ed.), *Late Quaternary Environments of the United States.* Volume 2. London: Longman, 142–65.

Weiser, A.L. and Lepofsky, D. 2009 Ancient land use and management of Ebey's Prairie, Whidbey Island, Washington, *Journal of Ethnobiology* 29, 184–212.

Welinder, S. 1990 Mesolithic forest clearance in Scandinavia. In Bonsall, C. (ed.), *The Mesolithic in Europe.* Edinburgh: John Donald, 362–6.

Wells, O.N. 1987 *The Chilliwacks and their Neighbours.* Vancouver: Tulon Books.

Westerdahl, C. 1992 The maritime cultural landscape, *International Journal of Nautical Archaeology and Underwater Exploration* 21, 5–44.

Westerdahl, C. 2006 The relationship between land roads and sea routes in the past: some reflections, *Deutches Schiffartsarchiv* 5, 59–114.

White, G. 1788 (1890 edition) *The Natural History of Selbourne.* London: George Bell.

White, R. 1999 Indian land use and environmental change. Island County, Washington: a case study. In Boyd (ed.) 1999, 36–49.

Whitehouse, N. and Smith, D. 2010 What is 'Natural'? Forest composition, open-ness and the British 'Wildwood': implications from palaeoentomology for Holocene development and landscape structure, *Quaternary Science Reviews* 29, 539–53.

Whitehouse, N. and Smith, D.N. 2004 'Islands' in Holocene forests: implications for forest openness, landscape clearance and 'culture steppe' species, *Environmental Archaeology* 9, 203–12.

Whittle, A. 1997 Moving on and moving around: Neolithic settlement mobility. In Topping, P. (ed.), *Neolithic Landscape.* Oxford: Oxbow Monograph 86, 15–22.

Whittle, A. and Wysocki, M. 1998 Parc le Breos Cwm transepted long cairn, Gower, West Glamorgan: date, content and contexts, *Proceedings of the Prehistoric Society* 64, 139–82.

Whittle, A., Healy, F. and Bayliss, A. 2011 *Gathering Time: dating the Early Neolithic enclosures of south Britain and Ireland.* Oxford: Oxbow Books.

Whittle, A., Pollard, J. and Grigson, C. 1999 *The Harmony of Symbols: the Windmill Hill causewayed enclosure, Wiltshire.* Oxford: Oxbow Books.

Whittle, A., Atkinson, R.J.C., Chambers, R. and Thomas, N. 1992 Excavations in the Neolithic and Bronze Age complex at Dorchester-on-Thames, Oxfordshire, 1947–52 and 1981, *Proceedings of the Prehistoric Society* 58, 143–201.

Wickham-Jones, C.R. 2005 Summer walkers? Mobility and the Mesolithic. In Milner, N. and Woodman, P.C. (eds), *Mesolithic Studies at the Beginning of the 21st century.* Oxford: Oxbow Books, 30–41.

Wicks, K. and Mithen, S. 2014 The impact of the abrupt 8.2 ka cold event on the Mesolithic population of western Scotland: A Bayesian chronological analysis using 'activity events' as a population proxy, *Journal of Archaeological Science* 45, 240–69.

Wicks, K., Pirie, A. and Mithen, S. 2014 Settlement patterns in the late Mesolithic of western Scotland: the implications of Bayesian analysis of radiocarbon dates and inter-site technological comparisons, *Journal of Archaeological Science* 41, 406–22.

Wiesbauer, H. and Zettel, H. 2014 *Hohlwege und Lössterrassen in Niederösterreich.* Vienna: Wissenschaft Forschung Niederösterreich.

Wilkes, E. 2007 Prehistoric sea jouneys and port approaches: the south coast and Poole Harbour. In Cummings and Johnston (eds) 2007, 121–30.

Wilkinson, T.J. 1992 Off-site archaeology: land-use diversity in Bronze Age Upper Mesopotamia, *National Geographic Research and Exploration* 8, 196–207.

Wilkinson, T.J. 2003 *Archaeological Landscapes of the Near East.* Tucson AZ: University of Arizona Press.

Wilkinson, T.J. and Murphy, P.L. 1995 *The Archaeology of the Essex Coast, Vol 1: the Hullbridge survey.* Chelmsford: East Anglian Archaeology 71.

Willey, P., Watson, P.J., Crothers, G. and Stolen, J. 2009 Holocene human footprints in North America, *Ichnos* 16, 70–5.

Williams, G. 1988 *The Standing Stones of Wales and South West England.* Oxford: British Archaeological Report 197.

Williams, J. 2006 *Clam Gardens: aboriginal mariculture on Canada's west coast.* Vancouver: New Star Books.

Williams, M. 1990 Clearing of the forests. In Conzen, M.P. (ed.), *The Making of the American Landscape.* Boston MA: Unwin Hyman, 146–68.

Williams, M. 2003 *Deforesting the Earth.* Chicago IL: University of Chicago Press.

Williams-Freeman, J.P. 1915 *An Introduction to Field Archaeology as Illustrated by Hampshire.* London: MacMillan.

Williamson, T. 1987 Early co-axial field systems on the East Anglian Boulder Clays, *Proceedings of the Prehistoric Society* 53, 419–31.

Williamson, T. 2008 Co-axial landscapes: time and typography. In Rainbird, P. (ed.), *Monuments in the Landscape.* Stroud: Tempus, 123–35.

Williamson, T. and Bellamy, L. 1983 *Ley Lines in Question.* Tadworth: World's Work.

Williams-Thorpe, O., Jones, M.C., Potts, P.J. and Webb, P.C. 2006 Preseli dolerite bluestones: axe-heads, Stonehenge monoliths and outcrop sources, *Oxford Journal of Archaeology* 25, 29–46.

Wills, W.H. 1995 Archaic Foraging and the beginning of food production in the American southwest. In Price, T.D. and Gebauer, A.B. (ed.), *Last Hunters – First Farmers.* Santa Fe NM: School of American Research Press, 215–43.

Winn, R. 2018 *The Salt Path.* London: Michael Joseph.

Witcher, R. 1998 Roman roads: phenomenological perspectives on roads in the landscape. In Forcey, C., Witcher, R. and Hawthorne, J. (ed.), *TRAC: Proceedings of the 7th Theoretical Roman Archaeology Conference 1997*. Oxford: Oxbow Books, 60–70.

Wohl, E., Gerlak, A.K., Poff, N.L. and Chin, A. 2014 Common core themes in geomorphic, ecological and social systems, *Environmental Management* 53, 14–27.

Wood, W. 1635 *New England's Prospect.* (1865 edition) Boston MA: Prince Society.

Woodman, P.C. 1985 *Excavations at Mount Sandel, 1973–77.* Belfast: HMSO, Northern Ireland Archaeological Monograph 2.

Woodman, P.C., Andersen, E. and Finlay, N. 1999 *Excavations at Ferriter's Cove, 1983–95: last foragers, first farmers in the Dingle Peninsula.* Bray: Wordwell.

Woodman, P.E. 2000 A predictive model for Mesolithic site location on Islay using logistic regression and GIS. In Mithen, (ed.) 2000, Vol 2, 445–64.

Woodward, A. 2000 *British Barrows: a matter of life and death.* Stroud: Tempus.

Woodward, A. and Woodward, P. 1996 The topography of some barrow cemeteries in Bronze Age Wessex, *Proceedings of the Prehistoric Society* 62, 275–91.

Woodward, P. 1991 *The South Dorset Ridgeway: survey and excavation 1977–84.* Dorchester: Dorset Natural History and Archaological Society Monograph 2

Wooldridge, S.W. and Goldring, F. 1953 *The Weald.* London: Collins.

Worster, D. 1990 The ecology of order and chaos, *Environmental History Review* 14, 1–18.

Wright, E. 1990 *The Ferriby Boats: seacraft of the Bronze Age.* London: Routledge.

Wylie, A. 2002 *Thinking from Things.* Berkeley CA: University of California Press.

Wymer, J. 1962 Excavations at the Maglemosian sites at Thatcham, Berkshire, England, *Proceedings of the Prehistoric Society* 28, 329–62.

Wysocki, M. and Whittle, M. 2000 Diversity, lifestyle and rites: new biological and archaeological evidence from British earlier Neolithic mortuary assemblages, *Antiquity* 74, 591–601.

Yates, D.T. 2007 *Land, Power and Prestige: Bronze Age field systems in southern England.* Oxford: Oxbow Books.

Yen, D.E. 1989 The domestication of environment. In Harris, D.R. and Hillman, G.C. (ed.), *Foraging and Farming.* London: Unwin Hyman, 55–75.

Yi-Fu, T. 1977 *Space and Place: the prospective of experience.* London: Arnold.

Zeder, M.A. 2012 The Broad Spectrum Revolution at 40: resource diversity, intensification and an alternative to optimal foraging explanations, *Journal of Anthropological Archaeology* 31, 241–64.

Zeller, A.C. 1987 A role for children in hominid evolution, *Man* 22, 528–557.

Zhang, D.D. and Li, S.H. 2002 Optical dating of Tibetan human hand and footprints: an implication for the palaeoenvironment of the last glaciation of the Tibetan Plateau, *Geophysical Research Letters* 29 (5), 16.1–3.

Zvelebil, M. 1994 Plant use in the Mesolithic and its role in the transition to farming, *Proceedings of the Prehistoric Society* 60, 35–74.

Index

by Jennifer Foster

Abbot's Way, trackway 142
Aboriginal Australian 51, burning 72, modified trees 57, myths 10
acorn leaching 42
acrobat, rock art 214, 215
Adam's Grave, Neolithic tomb 195
aDNA (see Ancient DNA)
adult footprint-tracks 95, 101, cave 103, Happisburgh 87, Redwick 99, size of footprint 83, **83**, 84, Uskmouth 91, 95, Willandra Lakes 88
Adur river 221, 226, **228**, 239
aeolian footprint contexts 88
age of people, footprint-tracks 82, **83**, 84, 94, 95, **105**, 106, 107
age-group of population 75
Agent Based Modelling 72–3
Ageröd, Sweden, fishtraps 59
air photography 2, 7, 175, 188, **189**, 192, **194**, 195, 245
Akeman Street, Roman road 199
Alaska, colonisation 102, 108
alignments, of barrows 19, 163–73, **164**, **166**, **169**, **171**
alignments, of cooking pits 165
alignments, of monuments, Neolithic 114
All Cannings, multiple hollow ways 185, **185**
Allen, JRL 91
Allerød interstadial 67, 210
Alpine Lakes, wheels 136, 138
Alps, French, grazing animals in 6
Alps, jadeite axe from 141, **141**
Alvastra, Sweden, Neolithic causeway 140
amber 157, 165, 170, 202, 211, **212**, 217
America, mobility 5, 27–47, footprint-tracks 102–3, South West 242, waggon trails 14
Amesbury Archer, mobility 113
Amino Acid Racemisation, dating 81
amphorae 213
Anbarra, Australia 51, 107
ancestor paths 45, 121
ancestors, Aboriginal Australian 12, 51, 53
Ancholme valley 209
anchors, Iron Age 214
Ancient DNA, evidence for mobility 114
Anglo-Saxon 223, **225**, 247, burials 234, 236, routes 232, royal centre, Lyminge 237
animal footprint-tracks 70, **76**, **81**, **82**, 85–6, 87, **89**, **93**, 95, 99, **100**, **105**, 108, associated with human 85, **86**, **93**, 95, **98**, 99, **100**, 102, 103, **105**, 108
animal movement 6, 23, 151, animal-created paths 25, **25**, construction of routeways 253, paths, use of by people 43, 249, track guides 85
animal, or human, footprints 82–3
animals, grazing 5, 6
anthropological perspective 42, **52**, 251–3
anthropology 24, 216, 217, First Nations 44, Ken Basso 45, North-west Coast 29, routes 9, Tim Ingold 11–13
antler cheek pieces, horses 183
antler mattock, Mesolithic 60, 91
Antonine Itinerary 197
Aoccamonfina, Italy, footprint trails 87
Apache 45, 46
archaeogeomorphology 248
Argon/Argon dating of footprint trails 87
Arran Islands 205, 216
Arras culture, Yorkshire 184, cart burial 188
art, America 103, Australia 104, Caergwle boat 205, Creswell Crags 53, Gundestrup 143, Kung Bushmen 104, Palaeolithic, Chauvet Cave 90, Stonehenge 128
art, rock 214–16, **215**, **250**, Bronze Age rock art 104, cosmological interpretation 250, footprint-tracks **105**, 106, horses 193, Italy 112, **112**, landscape 21–2, **22**, on trackways 143, Scandinavian 206, 215–6, wheeled vehicles in 136
artefact distribution 2, 223, Gwernvale **118**, movement 15, 170, Scandinavia 210, **210**
artefacts 69–70, at Flag Fen 155, cluster, Mesolithic 69, connections 211, 213, dating 248, dating of footprints 82, evidence of mobility 113, for dating 248, from tree-throw pits 119, in bog 146, in wetland 154, ritual deposition 201, Star Carr 63, use in dating 182, vegetation disturbance 62, **61**, with Irish trackways 152
artiodactyl footprint-tracks 103
Arun river 221, 239
Ascott-under-Wychwood, chambered cairn 117

Aston Clinton, Icknield Way 197, Iron Age track 190
Atlantic connectivity 211, 213, 214, 216
aurochs 54, 95, 98, 250
Australopithecus afarensis, Laetoli 86
Avebury 11, 64, 113, 123–5, **124**, 128, 195, 244, 246, 248, Case study 123–5, **124**
Aveline's Hole 53, 71
Avellino footprint-tracks 102
Avenues, stone, Neolithic 122, 126, Avebury 123, **124**, Stanton Drew 128, Stonehenge **124**, 126, 128, 202
Avon river, Wiltshire 123, 125, 126, 202, 209
axe factory, Graig Lwyd 170, **171**
axes 111–22, 116, 119, 128, 131–2, 140, 141, 152, 157, 158, 170, 210, 215
axes of movement 2, 46, 83, 108, 154, 171, 243, Goldcliff **92**, henges 122, Melton 150, Mojave river 103, Ridgeway 195, Severn Estuary 150, Uffington hillfort 193
axial alignment 232
axial route, Jutland 245
axis of routeways, Veluwe 169, **169**
axis of trackways, Heathrow Terminal 5 190, **190**
axles, vehicle, Grosse Moor 139–40, pivoting 143, ritual deposition at Flag Fen 155

Bad Buchau, trackway 143
Badbury Rings 199
Bagendon 199
Balbridie, Neolithic house 70
ballast 213
Baltic 204, 206, 215
barefoot footprint-tracks 12, **14**, 75, 84, **86**, 90, 95, 103, 104
Barger-Oostervold cult site 146, **147**
bark boats 203, 204
Barr, Kirsten, footprint-tracks 81, **82**
barrow 226
barrow alignment 3, 163–73, **164**, **166**, **169**, **171**, 192, 251, Bishopstone 170, Bronze Age 3, Flintbek 167, Himmerland, Denmark 19, Neolithic 3, on routeways **164**, 163–7, Stonehenge landscape 125, 172
barrows 223, **225**, 226, 236, 246, 251, along Icknield Way 195, and routes, Netherlands 11, around Avebury **124**, construction on grassland 192, Dorchester 129, **130**, linear routes 168, on cursus 120
Barway trackway 155
baskets, North-west Coast 42
Båsmyr trackway 143
Bayesian statistical modelling 62, 114, 140, **198**, 199
Beachy Head 221, 223, **225**
Beacon Hill Park, burial cairns 40
Beaker period 15–16, 113, 114, 206, 211, 222, 223, **225**, 252
bear 35, **36**, 90, 101
beavers 17, 20, **25**, 36, 43, 46, 56, 63–5, 202
beaver-gnawed wood 56, **56**
Beckhampton Avenue 123, **124**, 125
beetles 55, 149
Belle Tout 223, **225**
Belloc, Hilaire 19, 192, 232
Belmarsh Track, London 140, 142
berries 32, 33, 40
Bexhill, Sussex 69
biodiversity 253
biological evidence, for routeways 5
biomechanics 77
biomolecular studies 6
bipedal footprint-tracks 83, 84, 87, 106–7, hominin footprints 86, hominin trail, Lake Turkana 86, Laetoli 108
bird footprint-tracks 86, 95, 98, 103, 104
Bishopstone Sussex 119, 170, 223, **224**, 225, 246, Case study 223, **224**
bison 45, 87, 90
Bjerre, Denmark, amber 170
Blair Drummond, Scotland, wheel 136
blazes 47
Bletchingly, Surrey 69, 232
Blick Mead, Mesolithic 125
blue stones, Stonehenge 117, 125, 126, 132, 202
Bluestone Henge, West Amesbury 126
boat burials 251
boat cleat, Test valley 158
boat models 205
boats 250, colonisation of Norway 210, in rock art 206, 214, **250**, log 201, **203**, **204**, 203–5
boat-shaped burials 205
Bodmin Moor, field system 186
bog bodies 143, 146, **147**, 160
Bohuslan rock art **250**
Bonfield Gill, Yorkshire **50**, 62
Borremose, causeway 143, **144**
Borrow, George, *Wild Wales* 19
Borth footprint-tracks 101
Bos bones, earliest 211
Boscombe Bowmen, mobility 113
Bourtranger Moor 144–7, **147**, 161–2
bovid footprint-tracks, Swanscombe 87
Boyne valley 211
boys, size of footprint 83, **83**, 84
bracken fern rhizomes, eaten 42
brain, pathways 18
Bramber Castle 239
Brenig Valley, seasonal grazing 172
bridges 154, 157–8, 159–60, 209, 247, 251
Brigg boat 206, 208, 209
Brittany 15, 205–8, 209, 217

Broad Spectrum Revolution (BSR) 59, 73, 108
Broighter boat 205, 217
Bronze Age, boats 206–8, **207**, 209, drove roads, foot-print-tracks 102, fields 236, fields and trackways 175, isotopic evidence 15, long house, Ullunda 101, ritual deposition at Flag Fen 155, 162, rock art 104, settlement, Redwick 99, **100**, technology 206
Bronze Age barrows 128, alignments 163–73, **164**, **166**, **169**, **171**, 188
Bronze Age trackways 142–62, **137**, 148–9, **148**, 161–2, Bourtranger Moor 146
Brookland log boats 204
Broomend of Crickie, henge and avenue 128
Broskov 144, **145**, 184
brushwood trackway 136, 142, 150–1, **150**, 152, 159–60, **198**, 199
Buddhist, route to enlightenment 7
buffalo 86
Bulbury, anchor 214
Bulgaria, stone axes 210
Bullock Down 223, **225**, 226, 248, Case study 223, **225**
Bunyan, John *Pilgrim's Progress* 19
burials 40, 51, 53, 113, 205
burning 5, **6**, 49, 72, 251, at Waun-Fignen-Felen 67, Bronze Age Alps 112, by First Nations group 28, 45, Mesolithic 98, of vegetation 44, on continent 67, **67**, Thatcham 64, use of 38–9, 42, 43
Bushmen trackers, footprint interpretation 84, 90, 107

Cabot, John 43
caching 70, 119
Caergwle boat 205, 213, 217, 250
Caesar 213, 214, 252
cairns 33, **34**, 201
Caldicot 158, 206, 208, 209
California 42, 43
Callanish, henge and avenue 128
Calvert Island 102, 108
camas lily bulbs 33, 40, **41**
Campemour trackway 139
Canada, North-west Coast 27–47
Candidula intersecta 237, **238**
canoes 202
carbon isotopes 71
cargo, in boats 205, 206–7
Carpathians 249
Carron, river, dugout 204
carts 136, 146, 165, 169, 183–4
case studies, archaeological 4, 5, 11
Case study: Avebury 123–5, **124**
Case study: Bishopstone 223, **224**
Case study: Bullock Down 223, **225**
Case study: Kennet valley 63–5, **64**
Case study: Kilen 165–7, **166**
Case study: Lyminge, Kent 237–8
Case study: Mesolithic paths in the Severn Estuary 90–7, **92**, **93**, **96**, **97**
Case study: Seasonal pastoralists in Severn Estuary 99–101, **100**
Case study: Severn Estuary trackways 149–51
Case study: Somerset Levels 148–9, **148**
Case study: Stonehenge 125–8, **127**
Case study: The Ice Man 111–13
Case Study: Veulwe barrow roads 168–9, **169**
Case study: Weald 219–39, **220**
Case study: Wiltshire and Oxfordshire Ridgeway 193–5, **194**
Casparie, Will, trackways 133
Cassivellaunus, chariots 184
cattle 86, 98, 99, **100**, 101, 102, 112, **112**, 113, 115–16, 139, 146, 155, 170, 199, 211, 245
cauldron, Gundestrup 143–4
causewayed enclosure 115, 119, 141, 195
causeways 200, Danish 143, **144**, Flag Fen 199–200, for Tibirke track 140, Molbjerg 165, **166**, 167, Neolithic, Alvastra 140, Roman, Fiskerton 157, stone, Ireland 152, Swedish 143, Thrupp 160, Yarnton 160, **160**
cave art 53, 103
caves, footprints in 89–90, 103, 107
cedar, red, North-west Coast 29
Celtic coins 214
Celtic fields 144, **147**, 222, 223, 226, 227, cut by Ridgeway 195, Netherlands and Germany 170, Veluwe 168
Celtic name 202
cereal-type pollen 115
cervid footprint-tracks, Swanscombe 87
Châines operatoires 72
chalkland landscapes 21, **22**, 57
chambered tombs 117–19, **118**, 250
Channel Islands 214
Channel, English 206, 209, 211, 213, 214
Channel Tunnel 234, 244
charcoal 5, **6**, 45, 60, **61**, 62–3, 67, 68, 111, 125, 182
Chariot of the Sun, Trundholm 215, 217, 250
chariots 119, 184, 251
Charles II, Monarch's Way 253
Chaucer, G *Canterbury Tales* 19
Chauvet Cave, footprints 90
Cheddar Gorge, Somerset 53
Chedzoy, raw materials 72
Cherbourg peninsula 213
Cherbury hillfort, wheel ruts 184
Cheshunt Park, Hertfordshire 180, **181**
Chew River, Somerset 128
Chichester 230
Chideock, hollow way 284

children footprint-tracks 84, 87, 88, 89–90, 91, 94, 95, 98, 99, 101, 102, 103, 107
children in prehistory 107, 108, 215, 243, **243**
Christian footprints in rocks 104
circumnavigation, Britain 213
Cistercian abbeys 183
Ciur-Izbuc, Romania, footprint-tracks 90
Clairvaux lake village, human footprint-tracks 98
clam gardens 34
Clare Bay, Australia 80, 95
clearing enclosures **181**, 232
Cleaven Dyke, Scotland, cursus 120
Cloonbony, Neolithic trackway 142
clothing, Ice Man 111, Kwaday Dan Ts'inchi 33
Cnatro Ciengas footprint-track 102
Coasts, Nicholas Crane 21
coaxial field system 102, 149, 151, 155, 180, **181**, 182, 186, **187**, 188–90, **190**, 200, 213, 232, 236, 245, 247, 251
Cobham hollow ways 234
cognitive studies 18–19
Colchester Roman road 199
Coles, Bryony 133
Coles, John 133, 148, 242
colluvium 177, **179**, **180**, 182, 199, 222, 225, 237
Colne river 64
colonisation 216, America 83, 102, 103, Britain 210, human 108, Ireland 210, Norway 210
communication, by trackway 143, by water, Neolithic 167, in prehistory, by river 202, long distance in Bronze Age 208
connectivity 143, 219, 232, 239, 249, 254, between sites 49, by sea 210, by water 215, 216, cross-Channel 213, 214, island 216, landscape 243, ridgetops and valley 246, rivers 49, sites 243, TRB across North Europe 170, upland/ lowland 172
continuity 73, 236–7, **236**, boundaries 180, **181**, in landscape 190, 191–2, in routeways 223, 244, 245, 246, 250, Iron Age roads 197–200, of coaxial fields 190–2, of road patterns 7, of routes 167, 168, 169, 170, 173, 175
convergence of routes 244, of trackways 148–9, Goldcliff 150, **150**, 151
Cook, James 31, 32, 42, 44
cooking paces *fulachtafiadh* 152
Cooleeny trackway 152, 162
coppicing 42, 59, 68, 142
coprolites 102, 103
coracles 206
corduroy track 134, **135**, 139, 142
Corlea 134, **135**, 138, 152–4, **153**, 161, 162, 243
Cornu aspersum 237, **238**
cosmology 7, 12, **13**, 163, 171, 173, 215–6, 250
Cottenham, wheel 136
Cowlitz trail 46
Cranborne Chase, Dorset 57
Crathes, Mesolithic post pits 70
Crawford, OGS 3
Creswell Crags, Derbyshire 53, **53**
Creswellian hunters 53
cross dykes, Buckinghamshire 197
crossroads burial 234
cross-Channel trade 209, 213, 214
crossings, rivers 43, 56, 155, **156**, 157, 158, 160, **160**, 165, 167, 170, 201–2, 209, 239
cross-ridge dykes 221, 246
cross-topographic routes 239, 244, 246
Cuckmere river 221
cultivation 167, First Nations 42, New England 43, on middens 70, under barrows 167
Culturally Managed Trees (CMT) 29, 35–8, **37**, 57
Cunobelin 214
cup-marked stone 144, **145**
cups, Bronze Age 211, **212**
currachs, hide boats 205, 216
cursus monuments 116, 119–121, **120**, **122**, 126, 128, 129, **131**, 172

daggers, Stonehenge 128
Damalina bovis 101
Danebury hillfort 183, 188, 245
Danube 202
Danubian farmers 210
Dartmoor landscape 7, 186, **187**
date, of the wheel 136
dating 2, footprints 81, 87, 88, 90, 102, 103, hollow ways 175, 242, 248, 249, Irish trackways 151–2, landscape features 8, molluscs 237, Neolithic 114, of dugouts 203, of ridgeways 192, 193, 195, 253, of trackways and footprints 22–3, rock art 215, Roman roads 197, routeways 175, 176, 247, 248, trackways 4, 14, 134, 136, **137**, 138, 178–83, **179**, **181**, vegetation disturbance 3
daughter settlements 221
deer 62, 64, 87, 95, 103
defecation 6, 47
deforestation 14, 146
demography 108
dendrochronology, dating 23, 71, 197, 203
dendrochronology, trackways 136, 138, 139, 140, 142, 143–4, 148, 149, 151, 152, 154, 155, 157, 158, **158**, 161, 184
Denmark, barrow alignments on routeways 163–7, **164**, hollow ways 184, marine resources 71, Mesolithic fishtraps 57–9, Neolithic trackway 140, shell middens 66, 68–9, trackways 143–4, wheels 136, 138
deposition, ritual 63, 146, **147**, 154, 155, 157, 160, 162
Derryaghan trackway 152
Derryoghil trackway 152

Derryville trackways 152, 161
Devil's Arrows 122
Devil's trails, Palaeolithic footprints 87, 104
diet, isotopes 71
dinosaur footprint-tracks 77, 104, 106
disc wheels, single piece 136, 140, 152
Diss Mere clearance 190
distributions, artefacts 211, **212**, 239
disturbance, vegetation 73
disturbed ground plants 115
DNA analysis 6, 15-16, 248
dog 29, 33, 90, 98
Doggerland 20, 54, **67**, 68, 73, 210
Dogrib group, North-west Coast 32, 45
dolerite 132
Domesday Book 221
domestic animal footprints 98, 99, **100**, 101
domestication 46, 59, 114, 211
Dorchester cursus 119, 120
Dorchester, Dorset, Neolithic monuments 129, **130**
double ditch trackway 236
double entrance henges 129, 131
double lynchet trackway 177, 178, **180**, 182, 222, 223, 227, **228**, 230, **231**, 248
double post alignments 158, **158**, 209
Douglas, James 28, 40
Douglas map 28, **28**, 38, 46, 57, 244, 248
Dover 206–7, **207**, 208, 232, 239
Downholland Moss 98
downland landscape 21, **22**
draft animals 136, 139
Drake, F 43
Drayton cursus 120
Dreamtime myths 10, 12
drove roads 102, 131, 219, 221, 226, 227, **228**
droveways 2, 23, 101, 151, 155, **156**, 175, 177, 178, **179**, 188–90, 199, 209, 232, 234, **235**, 239, 245, 246–7, 247, 251
Druid's Circle, Penmaenmawr 170, **171**
dryland routes 163, 167
dryland/wetland connections **147**, 149, 152, 155, **156**, 168, 202, 206, 243
dry valley 221, 225
dugout boats 202, 203, **203**, 204–5, **204**, 209, 217
dung 6, **6**, 33, 43, 60, 115–16, 141, 149, 190, 199
Durrington Walls 202, 209, cattle and pig mobility 113, henge, on River Avon 126, River Avon 125
Duvensee, Germany, hazelnuts 60
dykes, cross-ridge 221, Icknield Way 197, Silchester 197–9

East Dean Woods 227, **231**
Ebbsfleet Valley 142, 159–60, **159**
ecology 49, 54, 77, 225
Edercloon, Ireland 136, 152, 162
eel 66, 204
Elbe river, crossing points 170
elephant footprint-tracks 87
elites 183, 197, 217, 250, 251
elm decline 62, 115, 140
Emmen, Netherlands **147**, 168
empathy with landscape 20, 241, 242
emu footprint-tracks 95
Engare Sero, footprint-tracks 87
environmental archaeology 2, 5–7, 14, 17, 44–5, 49, 215, 248
ergonomics 77
Ermine Street, Roman road 180, **181**, 190
erosion 94, 101, 222, along paths 14–15, **15**, of hollow ways 165, 175, 177, **179**, **180**, 184, 249
Ertebolle burials 217
escarpments 197, 226, 234, 239, 246
estuarine silts **243**
ethnoecology, First Nations 29
ethnographic evidence 51–3, **52**, 107, 242, 244
ethnohistory 24, 247, perspectives 216, sources 7
Eton Rowing Lake 160
evolution, human footprint-tracks 108
excavation 227, 230, 247, 248
exchange, of goods 113, 119, 170, 209, 215
exogamy, Mesolithic Brittany 15
exotic objects 208, 209, 214, 217
experimental archaeology 24
experimental boats **207**, 208, 214
extinct animals, in art 104
extinction of molluscs, use in dating 182
eye motif, on boats 205

faecal spherulites 5, 6, **6**
faeces 47, 90, 103, 251
Falmouth Harbour 208
farming, spread of 114
Federsee, Germany 73, 139, 140
feet, in rock art **250**
female figures, absence of 215
female footprint-tracks 84, 86, 90, 95, 98, 103, 107
female gene line in Europe 114
Fen causeway 155, **156**, 199–200
Fengate 101, 129, **131**, 154–5, **156**, 190, 247
Fenland, route of Icknield Way 195
Ferriby boats 206, **207**, 208, 209
Ferriter's Cove, Ireland 211
fertilisation of patches 40
field banks, prehistoric 7
fields 7, 144, **145**, 177, **180**, 223, **225**, 226, **229**, **230**, Avebury 123, Bronze Age 175, Bronze Age, under sand 165, coaxial 186, **187**, Flag Fen 154–5, **156**, Heathrow Terminal 5 190, **190**, later prehistory 175, LiDAR 227, **231**, 245, Neolithic 115, Overton and

Fyfield Down 195, prehistoric 222, **222**, with routes, Broskov 184
Fife, Scotland, coracle burial 206
figurines, carved wooden 143
Finistère, pots 213
fire 55, 56, 143, Africa 62, destruction of Must Farm 155, disturbance, Mesolithic 60, 62, effect on wildwood 55, 56, reduction in the Neolithic 62, use of, East Coast 43, use of, First Nations 38–9, use of, in California 42
First Nations people 28–47, **28**, 107
fishtraps 34, **36**, 57–9, 66, 68, 69, 95, **97**, 98, 133, 150, 151, 204, 205, 216
Fiskerton, Lincolnshire 157
Flag Fen 134, 136, 146, 154–5, **156**, 157, 162, 199–200, 202
Flagstones, Neolithic pits 129, **130**
flea, human 101
flint axe, Sweet Track 141
Flintbek 167, 173
floods 55, 56, 72, 230, **233**
flora distinct on routes 36
Fontanet Cave, France 90
foot length, and height 84
foot size and age 83, **83**, 84
footpaths 242
footprint trails 82, 86, **86**, 94–5, **96**, **97**
footprints 1, 12, **13**, 24, 70, 77, 242, 243, **243**, animal, Shaugh Moor 186, Australia 51, cattle 146, 155, 199, dating 22–3, domestic animals 98, evidence of mobility 75, Formby 21, in Bronze Age rock art 104, in caves 89–90, Redwick 150
footprint-tracks **2**, 23, 101, 106, 107, 243, America 102–3, dates **105**, 106, formation process 77, 80, French caves 90, human 75–109, **76**, **78–9**, **81**, **82**, **86**, **88**, **89**, **92**, **93**, 95, **96**, **97**, **100**, **105**, Laetoli 85–6, **85**, male or female 103–4, methodology 77, **81**, Mojave River 103, 108, mythology 104, recording 81, **82**, Severn Estuary 90–7, terminology 80–81, **81**
foraging 46, 47, 51
forensics 24, 77, 84, 107
Formby, human footprint-tracks 98, 106, 107
Fort Vancouver, Washington 40
forts, Roman, Usk valley 199
French caves, footprints in 89–90
Friarton dugout 204
fungal spores 5, **6**, 55
Funnel Beaker tombs, Germany and Netherlands 167, 169
funnel-shaped entrances 14, 188, **189**, 227, **231**
fur trade, North-west Coast 31, 32
Fure 232
Fyfield Down sarsens 123, 195

gabbroic clay 213
Garton Slack cemetery 188, **189**
gas chromatography 6
Gaul 213, 214
genetic origin, farmers in Europe 114
geoarchaeology 24, 106, 202, 239, 248
geocaches 168, **169**, 252
geochemistry 2, 211, 213, 248
Geographic Information Systems (see GIS)
geology 5, 46, 57, 246, sourcing of lithics 71–2, sourcing of Neolithic axes 131–2, stone movement 131–2, Weald 219–39, **220**
geomorphic context, footprint-tracks 106
geophysics 175
Germany, barrow alignments 163, 167, Bronze and Iron Age trackways 142–3, Neolithic trackways 139–40, trackways 162, wheels 136, 138
ghost routeways 223, **224**
Giraldus Cambrensis, walking 19
girls, footprint 83, **83**, 84
GIS 11, 72, 117, 167, 183
glacier melting, the Ice Man 111
globular urn 221
goat, dispersal of weed seeds 115–16
gold 205, 208, 211, **212**, 213
Goldcliff, Wales 69, 91, axes of movement 108, boat planks 206, 208, footprint-tracks **2**, 76, **76**, **82**, **92**, **93**, **96**, **97**, 108, **243**, Iron Age site 101, Island 91, **92**, 94, submerged forest 54, **54**, trackways **135**, 149, 150–1, **150**
Gormley, A, intertidal images 22
Goughs Cave 53
Graeco-Roman anchor 214
Graig Lwyd axe factory 132, 170, **171**
grassland 120, 125, 138, 192
grazing 6, 54, 63, 64, 149, 193, 221, 225, 246, and plantain 115, barrows 192, effect on woodland 54–5, 56
grease trails 38
Green Way 232
Greylake post alignments 158
Gristhorpe burial, Yorkshire 205
Grosse Moor trackways 139, 143
Guernsey, middens and tombs 70
Gundestrup cauldron 143–4
Gussage All Saints, chariot fittings 184
Gussage/Pentridge cursus 119
Gwernvale megalithic tomb 117, **118**
Gypsey Race river, cursus **120**, 121

Hadrians Wall 253
hafod 247
Haida Gwaii, Canada 29, 32, **31**
Hambledon Hill causewayed enclosure 119
Hampnas boat 214
Hangleton deserted Medieval village 226
Happisburgh, Norfolk, footprint-tracks 87, **88**

Harbour, Poole, Iron Age 209
Hardinxveld, Netherlands 72, 203, **203**
Harlyn, Cornwall lunula 211, **212**
Harter's Hill post alignment 158, **158**
Harvejen, Denmark 165
Hasholme log boat 205
Haslov, Sweden, routes and barrows 167
Hatfield Moors, corduroy track 142
Haugvik, boats 214
Hayen, H, trackways 133, 138
Hayes Storage Depot track 151
hazel 54
hazel coppice 59, 68, 142
hazelnuts, Mesolithic 60
Heathery Burn cart fittings 184
heathland 11, 169, **169**
Heathrow Airport Terminal 5, trackways 190, **190**, 247
Hebrides 73, 210
height from footprint-tracks 107
hendref 247
henges 113, 116, 120, 121–31, **122**, 123–5, **124**, 126, **196**, 197
Hengistbury Head 213
herd animals, and hollow ways 178
herders 18, 21, 83, 112, 250
Herepath, Avebury 195
hererotopia 217
heritage conservation 252
hide boats 203, 204, 205–6, 214
High Speed 1 (HS1) 234, 244
High Street Ridgeway, Lincolnshire 192, 209
hillforts 183, 185, 192, 193, **194**, 199
Himmerland, Denmark, barrow alignments 19
hippocampus 18
hippopotamus 86
Historic Environment Records 252
historic landscape study 2, 7, 221, 243
Hjortspring boat 214
hoards, Bronze Age 213
Hoddersrug route, Netherlands 245
Hoedic, Brittany 71, 107
Holland, wheels 136, 138
hollow ways 1, 14–15, **15**, 23, 60, 163, 175, 177, 178, **179**, 182, 184–5, **185**, 200, 209, 222, 226, **229**, 234, 236, **236**, 239, 244, 245, 246, 248, Dorchester 129, Hondsrug 168, Lyminge 237–8, **238**, Neolithic 131, Orams Arbour 199, Penmaenmawr 170, **171**, conservation 253–4, dating 242, Denmark 144, **145**, erosion by water 25, 249, Iraq 15, **16**, Kilen 165, **166**, 167, Ravilious 21, **22**, Rother valley 230–2, **233**, Swaledale 247, theoretical perspectives 9, Uffington 193, Vale of Pewsey 195
Holloway **15**, 21
Holme-next-the-Sea, henges 197, **196**
Holne Moor, coaxial fields 186, **187**
Holywood North, cursus 120
Homer, *Odyssey* 21, 217
hominins 85, **85**, 87, 243
Homo antecessor 87
Homo erectus 86
Hondsrug 144, 146, 168, 170
Horley, Surrey 232
horse head prows 214, 215
horses 8, 20, 43, 45, 64, **82**, 98, 101, 102, 104, 115–16, 183–4, 193, **194**, 200, 215, 250, 251
Hoskins, WG, *The Making of the English Landscape* 7, **8**, 9
Household, Geoffrey 15, 184
Howick, Northumberland 69
Hudson, H 43
human agency in landscape change 14, 62
human bone, ritual deposition 157, 158
human figurines, Wittemoor 162
human footprints 24, 75–109, **76**, **78–9**, **81**, **82**, 83, **83**, 84, **86**, 87, **88**, 88, **89**, **92**, **93**, **96**, **97**, 98, 99, **100**, 101, 102–3, **105**, 106–7, 108, **243**, 250
human remains, evidence for mobility 113–14, isotope evidence 71
human skulls, ritual deposition 140, 201
human/animal interaction 28, 47, 51, 56, 73, 108, 109, 243
Humber river, boats 206
hundreds, Anglo-Saxon 234, 236
Hunebed tomb alignments 167, 169
hunter-gatherers 13, 244, 251, movements 51–3, **52**, path, created by animals 25, **25**, plant use 59, 60, routes 49–73, vegetation disturbance, Britain 60–63, **61**
hunting routes 33, **34**, 35, 53
Hunybedder Centre, Netherlands, walking 12, **14**
hurdle trackways 134, **135**, 142
husbandry, animal 59, 63, 188, 190, **190**, 245
hyena footprint-tracks 88

Iberia 205–8, 209, 211, 213
ibex, footprint-tracks, Chauvet Cave 90
Ice Man 111–13
ice melting 111, 112
Iceland 213
ichnology 24, 77
Ichnos 77
Icknield Way 4, 19, 21, **176**, 192, 195–7, **196**
Ileret, human footprints 86
Ingold, Tim 12–13, 241
inscribed stones 8, 138, 188
insects 136
intersecting routes 116, 244
intertidal archaeology 106, 142, 216
Inuit movements 51
Ipwegermoor 142

Ireland 171, 211, Bronze Age and Iron Age trackways 151–4, **153**, Mesolithic vegetation disturbance 63, Neolithic trackways 142
Irish diaspora 20
iron 170, 208, 214, bog ore 143, 146, industry, Weald 221
Iron Age 101, 172, 184, 188, **189**, 197–200, **198**, 206, 223, **225**, 226, **230**, 232
Iron Age trackways 142–62, **150**, Bishopstone 170, Bourtranger Moor 146, Germany 143, lack of Irish trackways 151
Isla de Mona, Puerto Rica, footprint-tracks and art 103
Isle of Ely trackway 155–7
Isle of Thanet, Kent 213
Isle of Wight 213
isostatic land rise, Sweden 101, 215
isotope studies 15, 23, 49, 71, 99, 113, 136, 193, 200, 215, 248
Istanbul Metro, Neolithic human footprint-tracks 98
Italy, the Ice Man 111–13
Itford Bottom/ Hill 222, **222**, 248

jadeite axes, Neolithic 131, 141, **141**
Jaguar Cave, Tennessee 103
Järrestad, Sweden, footprints in art 104
Jeju Island, Korea, volcanic ash with footprints 102
Jethauser Moor trackway, dating 142
Jewish slave workers, footprints 104
Jurassic Way **176**, 192, 245

kangaroo footprint-tracks 95
Kenfig footprint-tracks 101
Kennet Valley, Wiltshire 57, **58**, 63–5, **64**, 123, 195, Case study 63–5, **64**
Kerivoa lunula 211
Khanty people, Siberia 13, 51, 57, 73
Kilen, Denmark 165–7, **166**, 184–5, 245, 252, Case study 165–7, **166**
Killoran trackway 152
Kilnsea boat 206
Kimmeridge shale, Caergwle boat 205
kingship boundaries 160
Kit's Coty House 234, **235**
Klinchon people, Canada 45
Knap Hill causewayed enclosure 119, 195
Knowledge, The, taxi drivers 18
Knowth, Boyne, megalithic tombs 132
Kolhorn, Netherlands, cattle hoof prints 101
Koobi Fora Formation, human footprints 86
Kung Bushmen 84, 90, 104, 107
Kutchin, Alaska 51, **52**
Kwaday Dan Ts'inchi 33

La Olla, Argentina, footprint-tracks 103
La Tène 134, 154
Laetoli 77, 84, **85**, 86, 108
Lake Managua, Mexico footprint-tracks 103, 108
Lake Neuchatel, La Tène 154
Lake Turkana 86
lake-edge platforms 139
laminated sediments 80, 82, 87, 88, **88**, 90, 94, 98, 102, 106
landscape archaeology 4, 5, 7–8, **8**, 9, 239
landscape art 21–2, **22**
landscape 3, 7, 9, 11, 13, 14–16, 123, 125, **127**, 175, 192, 253
Langdale axes 132
Langdon Bay wreck 208, 239
Langebaan footprint trails 88
Langstone Harbour Survey 216
Lapps, movement of reindeer **52**, 132
laser scanning 87
Lea Valley 190
Leakey, Mary 86
least-effort paths 193
Lechlade cursus 120
lesser celandine, Mesolithic 60
lesser transhumance 23, 191–2, 247
ley lines 3
lice, cattle 101
LiDAR 165, 184–5, 227, **231**, 245
Lifeworld 10, 12, **13**
Liffey River, Mesolithic fish traps 59
Limfjord 170
liminal places 201, 216–7
lineage, Bronze Age barrows 171
linear ditches 24, 172, 188, **189**, 193
linear heathland strips 248
linear manors 232
linear Neolithic monument complex 129, **131**
linear patterns of movement 248
linear trackway, Yorkshire 188, **189**
Linearbandkeramik 114, 139
Lisheen, mine and trackways 152
lithic artefacts 33, 70, 71–2, 170
living landscape 250
Ljubljana marshes, multiple hollow ways 185
loess, hollow ways 185
log boats 201, 203–5, **203**, **204**, 251
London Basin 219
London, East, trackways 151
long barrows, Neolithic 115, 125, 126, 129, 165
Long-Blade sites 64
long-distance routeways 23, 197, 221, 252, 253
longevity of boundaries 232
longevity of routes 26, 116, 143, 144, 146, 149, 167, 168, 169, 170, 173, 175, 223, 225, 234, 236–7, 244, 245, 246, 250
Lough Neagh log boats 204
Lunulae 211, **212**

lurs, in rock art 215
Lydstep Mesolithic footprint-tracks 95
Lyminge 22, 182, 230, 237–8, **238**, 239, 247, 248, Case study 237–8
lynchets 144, **145**, 177, **179**, 226, 245, 248

MacFarlane, Robert 3, 20–1, 45, 89, 184, 241
macrofossils, charred 5, **6**
Madagascar 126, 132
Magdalenian footprints 89–90
Magrethes Naes, Denmark 59, 204
male emphasis, rock art 214–5
male footprint-tracks 84, 86, 90, 95, 98, 103, 107
Malta, wheel ruts 184
manuring, evidence for 115, 168, 182
maps 2, 3, 7, 19, 43–4, 245
Marden, cattle and pig mobility 113
marine food in diet 71, 113
Maritime Beakers 211
maritime communication 202, 206, 210, 211, **212**, 213–16, 214, 242, 251–2
maritime transport 17, 201, 205–8, 248
marking routes 33, 34, **34**, **35**, 242
marks on trees 251
Marlpost 232
mass spectroscopy 6
mattocks, Mesolithic 60
Maumbury Rings 129, **130**
Maya art 103
Meare Heath 148, **148**, 149, 161
Medieval horse 183
Medieval landscape reconstruction 7
Medieval moorland routes 187–8
Medieval ships 98
Medieval transhumance, Dartmoor 186–8
Mediterranean 112, 214, 216, 252
Medway river 234, **235**
Meerdaal Forest, hollow ways 185
Meerhauser Moor trackway 139
megalithic tombs 113, 117–19, **118**, 132, 169, 211, 217, 234, **235**, 244
Melton, axis of movement 150
Mendip, source of lead 113
menhirs 211
mental benefits of walking 253–4
Mesolithic 4, 69, 117, **118**, 123, 250, boats 204, 210, burials 53, burning, continent 67, **67**, diet, plants 60, evidence under barrows 116–7, **118**, footprint-tracks **2**, 76, **76**, **82**, 90–97, **92**, **93**, **96**, **97**, 108, Goldcliff **92**, 94 pits, Stonehenge 70, 125, 244, Rodbyhavn 98, role of children 107, routeways 56, shell middens 66, 68–9, Star Carr 63, 69, territories 210, trackways, inconclusive 138–9, vegetation disturbance 49–73, **50**, 115
metals, transport of 206, 209, 211, **212**, 213–16
metalwork, Bronze Age, from wrecks 208
metalwork deposition 155, 154, 157, 160, 234
micromorphology 5–6
micro-residues, Beaker pots 114
middens 68, 69, 70, 71, 117
midwinter sunset, Dorset cursus 120
migration 12–13, 20, 211
Milfield Basin, Northumberland 128
Milheeze, Netherlands 67
milk products in Beaker pots 114
mirror burial, Iron Age 199
mobility 5, 9, 15, 23, 161, 210, 219, 241, 248, and isotopes 71, DNA analysis 15, early farmers 111–13, effect on skeleton 83, First Nations 32–3, isotope evidence 15, 113, Mesolithic 66, **66**, Native Americans 45, Neolithic 114, 116, 132, prehistoric 246
model carts 136
models, of boats 251
modified trees and plants 28, 201
Mojave 43, 103, 108
Mokhali Cave, Lesotho, footprint-tracks 104
Molbjerg causewayed enclosure 165, **166**
mollusc evidence, lynchet 125, 182, 237, **238**, 248
Monarch's Way 253
Monarcha cantiana 237, **238**
Monks Trod, linking abbeys 183
Monkton-up-Wimborne ritual site 113
Monte Bego rock art 112, **112**, 136
Monte Hermoso, footprint-tracks 102–3, 106, 107
Monte Verde, Chile footprint trails 88, 102,
moorlands 60, 186
mosaic environment 56, 57
Mount Batten 213
Mount Pleasant Neolithic henge 113, 129, **130**
Mount Sandel, Ireland 69
mountain walk, Nicholas Crane 21, 22
mountains, transhumance 23
Mountdillon Bog trackways 133, 152, 161
movement, patterns 1, 10–11, **10**, 172, anthropological evidence 51–3, **52**, effect on skeleton 83, linear 168, microscopic evidence 5–6, **6**, of artefacts 71–2, 136, 200, 210, Stonehenge stones 132
Mud Bay fish trap, Washington state 34, **36**
multi-disciplinary approach 2
multiple routes 185, **185**, 226, **229**, 246, 247
multiproxy approach 24, 84
muscles, gender differentiation 83
Must Farm 101, 136, 155, **157**, 205
Mynydd Rhiw axes 132
myth and landscape 20
mythology 9, 10, 12, 46–7, 70, 104

Nahoon footprint trails 88
Naj Tunich, Guatemala, ritual in caves 103

Nanaimo, Vancouver Island, petroglyphs 34, **35**
National Maritime Museum, Falmouth 208
natural ecology 249
natural places 25
"natural" routeways 14, 53, **53**, 64, 68, 72, 123, 131, 165, 168, 193, 195, 199, 221, **228**, 232, 242, 245–6
nature conservation and routes 253–4
nature reserves, connectivity 253
navigation 213
Neanderthals 53
Near East farming 59
Near East, hollow ways 14–15, **16**
Nebra sky disc 213, 217, 250
Neckera pollen 112
Nedern Brook, Caldicot boat 208
negative lynchets 177, 178, **180**
Neolithic 11, 16, 57, 117, **118**, 211, 234, axe factory 170, **171**, barrow alignment, Yorkshire 188, cultivation 70, Durrington Walls 113, early 234, enclosures 165, **166**, 167, henges 121–31, **122**, hollow way 131, human footprint-tracks 98, Irish log boats 204, isotopic evidence of mobility 15, landscapes 114–16, long barrow 172, pits 119, 223, **225**, **227**, routes 16, 199, 200, 234–6, skeletal evidence for mobility 113–4, Stonehenge 18, 113, tombs 70, 195, 234, **235**, trackways 139–42, **137**, 143, 146
Neolithisation 114
Netherlands, barrow alignments 168–9, barrows and routes 11, Bronze and Iron Age trackways 144–7, **147**, Neolithic trackway 140, prehistoric routeways 242, trackways 134, trackways for wheeled vehicles 138
nettles, cultivation 42
New England, cultivation 43
New Grange, Boyne, megalithic tombs 132
Newbridge chariot burial, Edinburgh 184
Niaux Cave, footprint-tracks 90
niche construction 16–18, 25, 29, 34, 42, 56, 59, 64, 67, 73, 108, 112, 242, 248, 249–51
Nisqually prairies 44, trail 46
nitrogen isotopes 71
Non-pollen palynomorphs (NPP) 5, 55, 63, 248
Nootka Sound, Vancouver Island **28**
Normanton Down barrow alignments 172
Nors, Denmark boat models 205
North Downs 219–39, **220**, 232–7, **235**
North Downs Way **176**, 192, **220**, 232, 234, **235**, 236
North Gill, Yorkshire 62, **50**
North York Moors **50**, 62, 172, 252
Northton, Harris, midden 70
North-west Coast, America 1, 19, 28–47, **28**, **30**, **31**, **34**, **35**, **36**, **37**, **39**, **41**, 244, pollen analysis 44
Norway, Iron Age trackways 143
NPP (see Non-pollen palynomorphs)
Nunamint, Alaska, seasonal movements 51, **52**
Nuu-chah-nulth 42

oars 206
oculus, on boats 205
Odyssey, Homer 21, 217
Offa's Dyke 253
Old Sarum 199
Oldtidsvejen road, Denmark **164**, 165
Olduvai Gorge, footprint-tracks 87
Olympic Mountains, USA 44
Olympic Peninsula, USA 28
open woodland, Mesolithic 54, 57, 125
opossum footprint-tracks 103
oppida 197, 199, 248, 251
optical laser scanning 81
Optically Stimulated Luminescence dating (see OSL dating)
oral history 46
Orams Arbour 199
Ordnance Survey maps **176**, 192, 197, 221, 232
Oregon trail 38
orientation of footprint-trails **93**, 94, 95, **96**
orientation of roads 197, 199
origins of Roman roads 197
Orkney 122, 210, 216
Oronsay shell middens 69, 71
OSL dating 23, 88, 98, 102, **179**, 182, 193, 230, 237, **238**, 248
Ostrya carpenifolia pollen 112
otter paths, Norway 25
Otzi 111–13
Ouse river 221, 225, 239
Oversley Farm, hollow way 131
Overton 246
Overton Down, Ridgeway 193–5
overtraces, of footprints 80, 82, 83
ownership, land/ resources, Mesolithic 59, 139
Ozette, Washington State 29

paddlers, boats 206, 214
paddles, Mesolithic **204**
palaeochannels 150, **157**
palaeoenvironmental evidence 11, 44–5, 54, 57, 60, **61**, 84, 98, 99, 106, 117, 119, 120, 123, 125, 149, 244, 248
Palaeolithic art 53, 90
Palaeolithic burning 67
Palaeolithic caves 53
Palaeolithic footprint-tracks 83, 84, 85–90, **85**, 86, 87, **88**, **89**, 104, 107
Palingas Schittas 232
parallel droveways 190, 191, 221, 245
parallel routes 186, 188, **189**, 246
parasites, human 5–6, **6**
Parc le Breos, Wales 113
parent/daughter settlements 219–21, 230, 232, 246
parish boundary 234

pastoralists 7, 12, **13**, 18, 23, 25, **25**, 51, **52**, 73, 99–101, **100**, 114, 116, 132, 186, 250
patterns of movement 243, 248
paved roads, Denmark 143–4, **144**
Peacehaven 225, **227**
peat bogs **92**, 94, 99, 101, 134, 136, 138, 143, 146, 149, 150, 151, 161
Peddars Way, Roman road 195, **196**
Pembrokeshire chariot burial 184
Penmaenmawr, Bronze Age cairns 170, **171**
Pennines, burning 62
Pech Merle cave, footprint-tracks 90
Pesse dugout 203
petroglyphs, marking routes 34, **35**
petrological analysis, axes 170
phenomenology 11, 46, 72, 116–17, 242
photogrammetry 81, **82**, 87
physical anthropology 77
pig 99, **100**, 115–16, 211
Pilgrim's Way 200
pilgrimage 7, 21
Pilgrims Way 19, 234, **235**, 236, 246
pine nut collection 42–3
Pintubi, Aboriginal footprint interpretation 84
pit alignments, Catholme 129, **131**
pit line 122, Mesolithic Stonehenge 125
pits, cooking, alignments 165
place names 21, 45, 221, 232, 247
plaggen soils 168
plank boats 206, **207**, 209, 214
plant communities, effect of humans 242, 244, Happisburgh 87, shell middens 70
plant macrofossils 136, Tybrind Vig 66
plant manipulation 44
plant use, First Nations 28, hunter-gatherer 59, 60, North-west Coast 29–31, 41–2
Plantago lanceolata along routeways 43, 115
plants along routeways 46, 60, 244
plants, clues on tracks 34, **34**
platforms, wood 157, 139, 205
Pleistocene site, Jeju, Korea 102
plough, rock art, Italy 112, **112**
ploughmarks under barrows 167
Plumpton Plain 222, **222**
Plym valley, reaves 186
Poling 232
pollen evidence 5, 11, 98, 115, 136, 149, 167, 186, barrows, Brenig Valley 172, Bourtranger Moor 146, impact of First Nations 44, Irish trackways 152, Kwaday Dan Ts'inchi's clothes 33, North America 45, Stonehenge landscape 125, Sweet Track 140–1, the Ice Man 111, turves, Veluwe 168, Tybrind Vig 66, vegetation disturbance 60, 62, 63, **61**, woodland 54–55, 57
Pomatias elegans 237, **238**
Poole Harbour 202, 209, 213
population replacement, based on aDNA 114
population structure 243
Port's Road 226, **228**, 246
portages 209
Porth Felen, anchor 214
Ports Way 239
positive lynchets 177, 178, **180**
post alignments 134, 158, **158**
post pits, Mesolithic 70
post-processualists 9
Potassium-Argon dating 81, 86
potlatch ceremonies 32
pottery 114, 141, 146, 182, 211, 222, 226, **229**, 237
prairies 28, **28**, 33, 38, 39, 44, 46
prehistoric origins of Roman roads 197
prehistoric routeways 7, 167, 221, 232, 239
Preseli, Wales, source of bluestones 117, 132
preservation, of footprints 80, 106, 107, 243
preservation of routeways 252
Prestatyn, Wales, 70
prestige goods 217
processions 120–1, 122, 125, 129, 146
processualists 9
Puget Sound, pollen 44
Pulex irritans 101
Pytheas of Massalia 213

quartz, Knowth and New Grange 132
querns, ritual deposition 146

Rackham, Oliver 7
radial network of paths 108
radiocarbon dating 22, **61**, 62, 81, 102, 103, 106, 114, **131**, 136, 138, 151, 161, 182, **198**, 199, 203, **207**, 248
Raftery, Barry 133
ranch boundaries for husbandry 245
Raunds, Neolithic long mound 117
Ravenna Cosmology 197
Ravvejen, Denmark 165
raw material movement 170, 208, 209
razors 193, 215, 217
reaves, on prehistoric field systems 186, **187**
recording of footprint-tracks 81, **82**
red deer 54, 90, **91**, 94, 95, 98
Redwick, Bronze Age settlement 99, **100**, 101, 107, 149
reindeer 18, 53, 64
retrogressive landscape analysis 2, 176, 178, **179**, 180, **181**, 245
rewilding 20, 254
Rhine 202, 214
rhinoceros 86, 87
Rhum, raw materials 72

ridgeways 4, 19, 20, 21, **22**, 57, 123, **124**, 175, **176**, 183, 192–7, **194**, 200, 202, 219–39, **220**, 221, 222, 244, 245–6, 252, 253
Ring of Brodgar, Orkney, henge 122
Ringkloster, Denmark 66
rites of passage 107
ritual deposition 144, 158, 217, crossing places, Denmark 165, Flag Fen 155, Hjortspring boat 214, in watery places 146, 154, 201, metalwork 155, 157, of artefacts 98, Sweet Track 142, Vauxhall bridge 159
ritual, in caves 103
ritual post pits, Stonehenge 125
ritual site, Monkton-up-Wimborne 113
ritual use of trackways 143
ritualised paths, cursus monuments 121
river connectivity 49, 239, 251–2
river crossings 120, 122, 123, 131, 199
River Kennet, Mesolithic occupation 63–5, **64**
river transport 167, 201, 202–5, 248
river valley mobility 64, 66–7, **66**, 68
riverine movement 126, 242
roads 4, 7, 23, 183, 197–8, 199, 200, 242, 251
roc bird legend, Algeria 104
Roccamonfina, Italy, Devil's trails 104
rock art 14, 112, **112**, 136, 193, 206, 213, 214–16, **215**, 217, 250, **250**
Rodbyhavn, Denmark 98
roe deer footprint-tracks 95, 98
Rogue Male 21, **15**, 184
Roman 149, 170, 213, 214, 219, 232, causeway, Fiskerton 157, iron industry 221, river ports 239, ships 98
Roman roads 4, 117, **118**, 129–31, 155, **156**, 170, **171**, **176**, 190, 195–7, **196**, 197–200, **198**, 209, 226, 227, 230, 232, 234, 239, 244, 251
Romano-British lynchets 226
Romano-Celtic boat 214, 252
Ronas Skov, Denmark, fishtraps 57–9
Rother river valley 184, 230–2, **233**, 239, 246, 249
Round Hill, Sussex 226
routes 7, 11, 26, 239, 248, ancestral 12, around Stonehenge 172, cursuses aligned along 120–1, henges along 122, long distance 8, 122, 183, social significance of 45–6, theoretical perspectives 9, distinct flora 36
routeways 2, 3, 6, 23, 42, 56, 146, 165, 175, 176, **176**, 209, 219, 225, 226, 227, **230**, 232, 241, 242, 245, 249, 252, along ridges 192–7, ceremonial sites on 116, continuity 72, Corlea 152–4, **153**, environmental evidence 5, for wheeled vehicles 138, henges along 128, Neolithic 131, pilgrimage 7, plants along 60, 115, 244, Stonehenge 125, symbolism 7, Veluwe 168, **169**, Yorkshire Wolds 188, **189**
routeways marked by barrows 11, 128, 163, **164**, 165, 167, **171**
Rowfold, routeway dating 180, **181**
Rowlands trackway 142
royal centres, Corlea 243
Rudston, Yorkshire, cursus **120**, 121

sacred landscape, Khanty 51
Saddlescombe, Sussex 226, **228**, **229**
sail 205, 206, 214, 252, leather 214
saints' footprints in rocks 104
Salcombe, wreck 208
salmon 33, 66
salt production, Netherlands and Germany 170
saltmarsh grazing 24, 86, 99, 108, 150, 243, 247
Salts Cave, Kentucky 103
Saltwood Tunnel, Kent 236–7, **236**, 239, 244, 245
Sami people, mobility **52**, 53, 71
Sanctuary, Avebury 123, **124**, 195
sarsen stones 123, 125, 126, 132, 195, 202, 234
Saxon estates 234
Saxon settlement, Lyminge 239
Saxony trackways 134, **135**, 138, 142–3, 162, 170
Scandinavian rock art 214–16, **215**
Scandinavian, Bronze and Iron Age trackways 143–4, **145**
Scilly Isles 210, 216
Scole-Dickleburgh, coaxial fields 190
Scotland, Neolithic routes 116
sea level rise 54, 73, 94
sea mammals, eating 103
sea otter, hunted for fur 32
sea-going boats 205–8, 209
sea-henges **196**, 197
Seamer Carr, Yorkshire 69
seasonal footprints 85–6, 95
seasonal grazing 99–101, **100**, 149–50, 151, 168, 172, 186–8, 188–90, 226, 247
seasonal mobility 32–3, 246
seasonal movements 51, **52**, 239, of cattle, Uffington 193, Neolithic 116, North-west Coast 29, 32
seasonality 66, 69, 82, 107, 108, 113, 243
Secrets of the High Woods 226, 245
sedentary, Mesolithic 68–9
sediments 2, 23, 77, 80, 81, **105**, 106, 249
sedimentary contexts 5, 36, 56, 70, 106, 243, 248
Seekirch Stockwiesen disc wheel 140
Selborne, Hampshire 19
Severn Estuary 64, 81, **82**, 86, 107, 108, 123, 158, 208, 213, 214, 243, 247,
human footprints **76**, **82**, 90–97, **92**, **93**, **96**, **97**, 99–101, **100**, 106, 107, seasonal pastoralists 99–101, **100**, Case studies: Mesolithic paths 90–7, **92**, **93**, **96**, **97**, seasonal pastoralists 99–101, **100**, trackways 149–51
Severn river 202
sewn plank boats 8, 206, **207**, 209, 214, 251
sex, of footprint-tracks 107
shale 211, **212**, 213

Shapwick **56**, 149, 190
Shardlow log boat 205
Sharpstone, road **198**, 199, 244, 248
Shaugh Moor 102, 186
sheep 86, 98, 193, 211, 245
sheep/goat 99, **100**, 101, 102, 211
shell middens 66, 68–9, 70, 71
shellfish collected, Anbarra 107
Shetland 210
ships 14, 98
ships, Bronze Age rock art 104
shod footprints 84, 98, 103, 104
shoe last axes 210
Silbury Hill 125, 172
significance, of routes 45–6
Silchester oppidum 197–8, 199
Silk Route, movements of artefacts 15
silver cups 211, **212**
skin boat 209, 251
Skolt Lapps, Finland 51, **52**
Skye, raw materials 72
Slovakia, stone axes 10
smallpox epidemics, North-west Coast 32
Smilde trackways 146
social interaction, from footprint-tracks 108
social role, ships 217
social theory 2, 9, 133
solar disc 215
solar observance, rock art 104
Soletorp, Sweden, rock art 214, **215**
Solomon Islands, unshod feet 84
solstice alignment 120, 122
Somerset Levels 140–2, 148–9, **148**, 158, 244, Case study 148–9, **148**
soot marks in caves 90, 103
South Africa, tracking 35
South Dorset Ridgeway, territories 172
South Downs, 57, 219–39, **220**, National Park 221, Way **220**, 221
South Hornchurch droveway 190
South Street long barrow 125
speed of walking, from footprint-tracks 107
Speghhoje, track 143
spherulite particles, Sharpstone road 199
Squaxin Indians, trail blazing 36
St Kilda 211, 216
standing stones 117, **118**, **120**, 121, 122, 167, 170, **171**, 201
Stane Street 227, 230
Stanton Drew, Somerset, stone circles 128
Stanwell cursus 121
Star Carr, Yorkshire 62, 63, 68, 69, 138–9, 204
Starçevo gene pool 114
stature, footprint-tracks and skeletal remains 107
status 108, 152, 250
Stedham bridge **233**
Stoasnaig, Colonsay, hazelnuts 60
stone avenues 116, 252
stone causeways, wetland contexts 136
stone circles, Penmaenmawr 170, **171**
stone roads, Broskov 144, **145**
Stonehenge 11, 70, 113, 120, 121, 126, **127**, 128, 132, 172, 192, 195, 202, 209, 211, **212**, 244, 248, Case study 125–8, **127**
Storebaelt dugouts 204
storms 55, **55**, 56, 72
storytelling 46–7
straight paths in prehistory 128
Strangford Lough Survey 216
stride length 80
structuration of landscape 73
Stukeley, cursus monuments 119
Stuntney trackway 155
submerged forests 54, **54**, 66, 94
sun motifs 104, 205, 215
sunken ways (see hollow ways) 175
Supplementary Appendix 3.1 **50**, 60
Supplementary Appendix 4.1 77, 90, **105**
Supplementary Appendix 6.1 134, 138
Sussex, prehistoric routes 3
Swaledale, hollow ways 247
Swanscombe, animal footprint-tracks 87
Sweden, causeways 143
Sweden, hollow ways 184
Sweden, Lapps 51
Sweet Track, Somerset Levels 134, **135**, 136, 140–2, **141**, 162
swine cott 232
swine pasture 219, 221
Switzerland 214
symbolic routes 144
symbolism, in art 215–6
symbolism, routeways 7

taboo areas 7
Tabula Peutingeriana 197
Tal-y-fan, Roman road 199
Tamu-Mai, Nepal, migration 12–13
Tarrent, Celtic name 202
taxi drivers, The Knowledge 18
Taylor, Chris 4, 7, 33
Tell el Hawa, Iraq, hollow ways 15, **16**
temperate woodland, difficulty of igniting 62
Tenterden 219
terminology 23–4, of footprint-tracks 80–81, **81**
terminus ante quem 182
terminus post quem 182
terp sites 170
Terra Amata, France 87

terrace way 21, **22**, 177, 178, 226, **229**
territorial boundaries, barrows on 172
Test Valley post alignments 157–8
testate amoebae 136, 161
Testwood Lakes boat 206, 208
Tevic, Brittany 71, 107
Thames Estuary 142, 151, 234, 236
Thames river 201, 214, 232
Thames valley 64, 159, 188–90, **190**, 192, 209
Thanet 219
Thatcham 62, 63–5, **64**, 68, 69
The Old Ways, Macfarlane 20–1
Theopetra Cave, footprints in 89
theoretical archaeology 9–11, 241
Thomas, Edward 192
Thompson, Flora 19
Thornborough henges, Yorkshire 122, **122**, 128
Throwleigh Common, Dartmoor 186, **187**
Thrupp, Iron Age causeway 160
Thundersbarrow Hill 226, **228**, **230**, 239, 246
Tibet, human colonisation 108
Tibirke trackway 140, 143
tidal range, Severn Estuary 90
time, movement in 10–11, **10**
time, perception 23
tin 205, 208, 213
Tinney's Trackway 149
Tlicho people, Canada 45
tombs, Neolithic, location 116
topography, role of 7, 10, 46, 53, 11, 71, 77, 184, 232, 242, 243, 246, barrow alignments 170, Neolithic monuments 116, the Weald 219–39, **220**
torcs, Bronze Age 211
totem poles, North-west Coast 29, 30
trace fossils 77
tracking 24, 84, 103, 107
tracks, animal 35, **36**
tracks, origins 7
tracks, wetland 243
trackways 1, 4, 7, 17, 23, 154–5, 160–2, 134, **135**, 170, 197, 222, **222**, 225, 226, 227, **227**, **230**, 239, Heathrow Terminal 5 190, **190**, and horses 183, Bronze and Iron Age 142–62, **142**, **145**, **147**, **148**, **150**, **156**, 175, dating 22–3, 178–83, **179**, **181**, edge colluvium 177–8, Ferriby 209, Flag Fen 155, **156**, for vehicles 138, 143,
Ireland 151–4, Neolithic 98, Netherlands 144–6, **147**, *Old Straight Track* 2–4, Somerset Levels 148–9, **148**, the Weald 219–39, **220**, typology 177–8, **179**, under barrows, Hondsrug 168, wetland 133–62, **131**
trade 213, Bronze Age 170, cross-Channel 213, 214, networks, First Nations 33, 38, tin 213
trading 33, 209
trail blazing 35–6, **37**, 251
trails 23, 28, 44, 80, 83, 101
trails, footprint, Willandra Lakes 88, Formby 98, Goldcliff **92**, **93**, 94, 95, Lake Managua 103
transhumance 6, 20, 23, 112, 113, 172, 186–8, 191–2, 247
transport of materials 15, bog iron 143, by boat 32–3, 205, 209, wheeled, early evidence 167
travertine, footprint trails in 88
TRB 167, 170
tree modification 14, 29, 35–8, **37**, 57–9
trees, rewilding 20
tree-throw **55**, 57, **58**, 70, 119, 120, 138
Trent, Celtic name 202
Trevisker style pots 213
Trichuris 6, **6**
Trods 188, 252
Trundholm sun waggon 136, 215, 217
Tuc D'Audoubert, footprint-tracks 90
tufa, preserving footprints 102
turves, used for barrows 168, 172
two-wheeled cart, wheel ruts 184
Tybrind Vig 59, 66, 204, **204**, 210

Uffington hillfort 193, **194**, 246
Uffington White Horse 20, 192, 193, **194**
Ullunda, Sweden footprint-tracks 101
Underhill Road 221
undertraces, of footprints 80, 82, 83
unmetalled roads, erosion 249
Upton, Derek 91
Uranium Series dating 23, 81, 84, 87, 91, 102, 237, **238**, 248
urnfields, linear routes 168
Usk valley Roman roads 199
Uskmouth, footprint-tracks 91, 95

Val Camonica 136
Vale of Pickering, Mesolithic sites 69
valleys, natural routeways 53
Valseqvillio, Mexico 82–3
Valtherbrug trackway 146
Vancouver, George 31, 32, 39, 44
Vancouver Island, Canada 28, 44
Vanguard Way, Weald 253
Vartop Cave, Romania footprints in 89
Vauxhall bridge, double post alignment 159, 209
vegetation disturbance 44, 49–73, **50**, 66, 68, 115
vehicle trackways 143, 162
vehicles found with trackways 139–40
Veluwe 168–9, **169**, 172, 248, 252, Case study 168–9, **169**
Veneti, Brittany 214
Venford Terminal Reave, Dartmoor 186, **187**
Vera, H 54–5
Verulamium, Roman road 199

volcanic ash, preserving footprint-tracks 44, **85**, 86, 87, 102, 103, 106
Vroue Hede, Jutland, routes and barrows 167

waggon graves, Urnfield and Hallstatt 136
Walbiri, Aborigines, Australia 12, **13**, 51
Wales 66–7, **66**, 247
walking the landscape 7, 12, 19, 20, 192, as knowing landscape 241, 243, role in marking the landscape 9–10, **10**, therapeutic 253
wallaby footprint-tracks 95
Walvis Bay, Namibia 102
Wansdyke, Overton 195
war canoe, Hjortspring 214
War Coppice 232
Warren Field, Neolithic house 70
warriors, mounted, late Bronze Age 183
water erosion, on hollow ways 184, 249
water, transport by 201
watering holes, Heathrow Terminal 5 190, **190**
Watkins, Alfred 2–4, 117, 128, 226, 252
Waun-Fignen-Felen 67
Wawcott 69
Wayland's Smithy 193, **194**
Weaber Range, Australia 104
Weald 239, 246, 247, Case study 219–39, **220**, 245
weapons, in rock art 112, **112**, 214–5
weapons, ritual deposition 146, 154, 155, 157, 201, 214
weeds 43, 60, 115–16
Weimar-Niederweimar, Germany 67
well, Belle Tout 223, **225**
Welland Bank, wheel ruts 184
Welsh Way droveway 23
Wessex 211, 232, 245
West Amesbury, Bluestone Henge 126
West Kennet Avenue, Avebury 123, **124**, 195
West Kennet chambered tomb 125
West Penwith, field system 186, 188
Wetwang Slack cemetery 188, **189**
wheatsheaf motif, Denmark 210, **210**
wheel ruts 134, **135**, 160, 162, 165, 167, 168, 178, 184, 199
wheeled vehicles 23, 136, 138, 155, 167, 183–4, 200, 215, 250, 251, 252
White Horse, Uffington 192, 193, **194**
White, Gilbert 19
Whitehorse Stone Roman road 200, **220**, 234, **235**, 239, 244, 246
wildfires, natural 62
wildlife on routeways 253–4
wildlife sustainability 254
wildwood 54–7
Willamette valley 28, **28**, **30**, 38, **39**, 46, 57, 64
Willandra Lakes, Australia footprint trails 80, 83, 84–5, 88, **89**
willow coppice, Mesolithic 59
Wiltshire Avon river 202
Wiltshire/Oxfordshire Ridgeway 176, 192, 193–5, **194**, 200, 221, 232, 245–6, Case study 193–5, **194**
Winchester, Orams Arbour 199
Windmill Hill causewayed enclosure 119
Winkhurst Green, 232
winter base camp, Mesolithic 67, **66**
winter camp, Lapps 51, **52**
winter solstice alignment 120, 122
Winterbourne Stoke crossroads, barrow alignment 172
Wittemoor 143, 162
wolf print, Chauvet Cave 90
wooden platforms 139
Woodland Road lynchet 248
woods, rewilding 20
Wool Street, Roman road 199
Wooton-Quarr, Neolithic trackways 142
World Heritage Site, Stonehenge 125

X-ray Fluorescence 33

Yarnton, Iron Age causeway 160, **160**, 162, 184
Yorkshire Iron Age barrows 172
Yorkshire Wolds, prehistoric landscape 188, **189**, 245
Yupik Inuit cosmology 12

Zutphen, Netherlands, burning 67